Landscape
Sensitivity

British Geomorphological Research Group Symposia Series

Geomorphology in Environmental Planning
Edited by **J. M. Hooke**

Floods
Hydrological, Sedimentological and
Geomorphological Implications
Edited by **Keith Beven** and **Paul Carling**

Soil Erosion on Agricultural Land
Edited by **K. Boardman, J. A. Dearing** and **I. D. L. Foster**

Vegetation and Erosion
Processes and Environments
Edited by **J. B. Thornes**

Lowland Floodplain Rivers
Geomorphological Perspectives
Edited by **P. A. Carling** and **G. E. Petts**

Geomorphology and Sedimentology of Lakes and Reservoirs
Edited by **J. McManus** and **R. W. Duck**

Landscape Sensitivity
Edited by **D. S. G. Thomas** and **R. J. Allison**

Landscape Sensitivity

Edited by
D. S. G. Thomas
Department of Geography
University of Sheffield, UK

and

R. J. Allison
Department of Geography
University of Durham, UK

JOHN WILEY & SONS
Chichester · New York · Brisbane · Toronto · Singapore

Copoyright © 1993 by John Wiley & Sons Ltd,
Baffins Lane, Chichester,
West Sussex PO19 1UD, England

Other Wiley Editorial Offices

John Wiley & Sons, Inc., 605 Third Avenue,
New York, NY 10158-0012, USA

Jacaranda Wiley Ltd, G.P.O. Box 859, Brisbane,
Queensland 4001, Australia

John Wiley & Sons (Canada) Ltd, 22 Worcester Road,
Rexdale, Ontario M9W 1L1, Canada

John Wiley & Sons (SEA) Pte Ltd, 37 Jalan Pemimpin #05-04,
Block B, Union Industrial Building, Singapore 2057

Library of Congress Cataloging-in-Publication Data

Landscape sensitivity / edited by D.S.G. Thomas and R.J. Allison.
    p. cm. — (British Geomorphological Research Group symposia
series)
   Includes bibliographical references and index.
   ISBN 0-471-93636-7
   1. Geomorphology. 2. Man—Influence on nature. I. Thomas, David
S.G. II. Allison, R.J. (Robert J.) III. Series.
GB406.L35 1993
551.4′1—dc20   92—26901
                                                    CIP

British Library Cataloguing in Publication Data

A catalogue record for this book is available from the British Library

ISBN 0-471-93636-7

Typeset in 10/12pt Times by MHL Typesetting Ltd, Coventry.
Printed and Bound in Great Britain by Bookcraft (Bath) Ltd

Contents

Part 2   LAND USE AND LANDSCAPE SENSITIVITY

Part 3 SENSITIVITY AND BUILT ENVIRONMENTS

List of Contributors

Robert J. Allison Department of Geography, Science Laboratories, University of Durham, Soult Road, Durham DH1 3LE, UK

Malcolm G. Anderson Department of Geography, University of Bristol, University Road, Bristol BS8 1SS, UK

Philip Barker Department of Geography, Loughborough University of Technology, Loughborough LE11 3TU, UK

Armelle Billard Laboratoire de Géographie Physique, URA 141 CNRS, Meudon, France

Mark A. Blumler Department of Geography, University of California and Berkeley, Berkeley, California 4720, USA

John Boardman School of Geography, University of Oxford, Mansfield Road, Oxford OX1 3TB, UK

Sue M. Brooks Department of Geography, University of Bristol, University Road, Bristol BS8 1SS, UK

Denys Brunsden Department of Geography, King's College, University of London, The Strand, London WC2R 2LS, UK

Tim P. Burt School of Geography, University of Oxford, Mansfield Road, Oxford OX1 3TB, UK

Edward Derbyshire Department of Geography, Royal Holloway College, University of London, Egham, Surrey TW20 0EX, UK

Tom A. Dijkstra Faculty of Geographical Sciences, University of Utrecht, The Netherlands

Peter W. Downs Department of Geography, University of Southampton, Southampton SO9 5NH, UK

Robert Evans Department of Geography, University of Cambridge, Downing Place, Cambridge CB2 3EN, UK

A. John W. Gerrard Department of Geography, University of Birmingham, Edgbaston, Birmingham B15 2TT, UK

Ken J. Gregory Department of Geography, University of Southampton, Southampton SO9 5NH, UK

Jack Hardisty Beach Research Group, School of Geography and Earth Resources, University of Hull, Hull HU6 7RX, UK

Nick Harvey Mawson Graduate Centre for Environmental Studies, The University of Adelaide, GPO box 498, Adelaide, South Australia 5001, Australia

Nick E. Haycock   Silsoe College, Silsoe, Bedford MK45 4DT, UK

A. Louise Heathwaite   Department of Geography, University of Sheffield, Sheffield S10 2TN, UK

Young-Jin Li   Geological Hazards Research Institute, Giansu Academy of Sciences, Lanzhou, China

Tatiana Muxart   Laboratoire de Géographie Physique, URA 141 CNRS, Meudon, France

Jeremy S. Perkins   Department of Environmental Science, University of Botswana, Private Bag 0022, Gaborone, Botswana

Tom A. Quine   Department of Geography, University of Exeter, Amory Building, Rennes Drive, Exeter EX4 4RJ, UK

Keith S. Richards   Department of Geography, University of Cambridge, Downing Place, Cambridge CB2 3EN, UK

Neil Roberts   Department of Geography, Loughborough University of Technology, Loughborough LE11 3TU, UK

Paul A. Shaw   School of Geology, Luton College of Higher Education, Luton LU1 3JU, UK

Ian J. Smalley   Department of Civil Engineering, Loughborough University of Technology, Loughborough LE11 3TU, UK

Iain S. Stewart   Department of Geography and Geology, West London Institute, Lancaster House, Borough Road, Isleworth, Middlesex TW7 5DU, UK

David S. G. Thomas   Department of Geography, University of Sheffield, Sheffield S10 2TN, UK

Stephen T. Trudgill   Department of Geography, University of Sheffield, Sheffield S10 2TN, UK

Heather A. Viles   Jesus College, Oxford OX1 3DW, UK

Des E. Walling   Department of Geography, University of Exeter, Amory Building, Rennes Drive, Exeter EX4 4RJ, UK

Series Preface

The British Geomorphological Research Group (BGRG) is a national multidisciplinary Society whose object is 'the advancement of research and education in geomorphology'. Today, the BGRG enjoys an international reputation and has a strong membership from both Britain and overseas. Indeed, the Group has been actively involved in stimulating the development of geomorphology and geomorphological societies in several countries. The BGRG was constituted in 1961 but its beginnings lie in a meeting held in Sheffield under the chairmanship of Professor D. L. Linton in 1958. Throughout its development the Group has sustained important links with both the Institute of British Geographers and the Geological Society of London.

Over the past three decades the BGRG has been highly successful and productive. This is reflected not least by BGRG publications. Following its launch in 1976, the Group's journal, *Earth Surface Processes* (since 1989 *Earth Surface Processes and Landforms*), has become acclaimed internationally as a leader in its field, and to a large extent the journal has been responsible for advancing the reputation of the BGRG. In addition to an impressive list of other publications on technical and educational issues, including 30 *Technical Bulletins* and the influential *Geomorphological Techniques*, edited by A. Goudie, BGRG symposia have led to the production of a number of important works. These have included *Nearshore Sediment Dynamics and Sedimentation* by J. R. Hails and A. P. Carr, *Geomorphology and Climate* edited by E. Derbyshire, *Geomorphology, Present Problems and Future Prospects*, edited by C. Embleton, D. Brunsden and D. K. C. Jones, *Mega-geomorphology* edited by R. Gardner and H. Scoging, *River Channel Changes* edited by K. J. Gregory, and *Timescales in Geomorphology* edited by R. Cullingford, D. Davidson and J. Lewin. This sequence of books culminated in 1987 with a publication in two volumes, of the *Proceedings of the First International Geomorphology Conference* edited by Vince Gardiner. This international meeting, arguably the most important in the history of geomorphology, provided the foundation for the development of geomorphology into the next century.

This current BGRG Symposia Series has been founded and is now being fostered to help maintain the research momentum generated during the past three decades, as well as to further the widening of knowledge in component fields of geomorphological endeavour. The series consists of authoritative volumes based on the themes of BGRG meetings, incorporating, where appropriate, invited contributions to complement chapters selected from presentations at these meetings under the guidance and editorship of one or more suitable specialists. While maintaining a strong emphasis on pure geomorphological research, BGRG meetings are diversifying, in a very positive way, to consider links between geomorphology *per se* and other disciplines such as ecology, agriculture, engineering and planning.

The first volume in the series was published in 1988. *Geomorphology in Environmental Planning*, edited by Janet Hooke, reflected the trend towards applied studies. The second volume, edited by Keith Beven and Paul Carling, *Floods — Hydrological, Sedimentological and Geomorphological Implications*, focused on a traditional research theme. *Soil Erosion on Agricultural Land* reflected the international importance of the topic for researchers during the 1980s. This volume, edited by John Boardman, Ian Foster and John Dearing, formed the third in the series. The role of vegetation in geomorphology is a traditional research theme, recently revitalized with the move towards interdisciplinary studies. The fourth in the series, *Vegetation and Erosion — Processes and Environments*, edited by John Thornes, reflected this development in geomorphological endeavour, and raised several research issues for the next decade. The fifth volume, *Lowland Floodplain Rivers — Geomorphological Perspectives*, edited by Paul Carling and Geoff Petts reflects recent research into river channel adjustments, especially those consequent to engineering works and land use change. The sixth volume, *Geomorphology and Sedimentology of Lakes and Reservoirs*, edited by John McManus and Robert Duck, continues the interdisciplinary theme, so much a feature of this BGRG Symposia Series. It reflects the need to know how quickly such bodies of water are being infilled and provides a stimulating mixture of pure and applied research that should appeal to a wide audience.

The present volume *Landscape Sensitivity*, edited by David Thomas and Robert Allison, the seventh in the series, addresses a vital geomorphological topic. This concerns the way in which landscape and landforms respond to external changes. Not only are these themes important for understanding landform development; they are crucial to an appreciation of human-induced response in the landscape. This volume presents a carefully selected blend of pure and applied studies and should appeal to a wide audience.

John Gerrard
BGRG Publications

Preface

This volume is the outcome of the British Geomorphological Research Group (BGRG) sessions at the Annual Conference of the Institute of British Geographers (IBG) held at The University of Sheffield in January 1991. The theme of the BGRG sessions, 'Landscape Sensitivity', was selected for two main reasons: first, in an attempt to address an important issue in studies of geomorphology and environmental change; and second, to provide a conference topic that crossed the thematic boundaries in geomorphology and therefore appealed to a wide audience.

In the event there were 24 oral presentations in the Landscape Sensitivity conference modules at Sheffield, presented over two days. These were divided amongst four groups that reflected the ways in which presenters had interpreted the sensitivity issue or the manner in which sensitivity could be best addressed or understood in a geomorphological context: 'modelling landscape sensitivity', 'catchment scale sensitivity', 'sensitivity to human disturbance' and 'geomorphological processes, landscape sensitivity and climatic change'.

Sixteen of the chapters in this book began life as presentations at Sheffield. The remaining chapters were not submitted for presentation at Sheffield but were subsequently supplied in manuscript form. This has contributed to the organization of the book differing from that of the conference, with chapters divided over three sensitivity sections. The first relates landscape sensitivity to geomorphic processes and climatic change, including overviews, case studies and modelling; the second concerns sensitivity in relation to agricultural and pastoral land use issues; and the third contains two chapters that deal will sensitivity in relation to built environments.

All the contributions have been refereed. In this respect we would like to thank the following for the help that they provided in this capacity and the way in which they completed their tasks efficiently, making our job of editing the volume relatively simple:

Yvan Biot, John Boardman, Bill Carter, John Cooke, Ron Cooke, Nick Cox, John Dearing, Ian Douglas, R. Duck, Bob Evans, Paul Farres, Ian Foster, Louise Heathwaite, Trevor Hoey, Nick Middleton, Frank Oldfield, Julian Orford, Rob Parkinson, Tony Parsons, John Rice, Keith Richards, Neil Roberts, Paul Shaw, Pete Smart, Bernie Smith, Mike Summerfield, Colin Thorne, Steve Trudgill, Heather Viles, Claudio Vita-Finzi, Des Walling.

Additionally, we would like to thank Helen Bailey at John Wiley and Sons, the local organizing committee of the IBG conference at Sheffield and the committee of the BGRG for their support. The index was prepared by Phil Porter.

David Thomas
Robert Allison

1 The Sensitivity of Landscapes

ROBERT J. ALLISON
University of Durham, UK

and

DAVID S. G. THOMAS
University of Sheffield, UK

INTRODUCTION

Environmental change has been a topic of geomorphological, earth science and environmental science research for a number of years but its importance is more widely recognised today than in the past. Interest in environmental change is as much a consequence of political debate and public perception as it is of scientifically driven discovery. As a consequence, much remains to be understood in the environmental change context and there are still many misunderstandings and misrepresentations. Despite much current concern being focused on the present-day impact of human activities on the environment, there is much of importance to be learned from the study of evidence of past changes. In some cases evidence from the past provides inputs for modelling and predicting future changes.

Understanding change within the natural environment frequently requires a fragmentation of natural systems for the sake of simplification. Individual components are studied, rather than adopting a holistic approach. Thus, in the past, for example, vegetation changes have been examined without recourse to soil process changes and soil changes in isolation of geomorphological variations (see Morgan & Rickson, 1988, for example). The science of environmental change none the less has certain common elements, regardless of the scale of investigation or the system being examined. Climate is frequently the driving force in many systems and human activity acts as a modifying influence (see Nisbet, 1991, for example). Understanding and predicting change is not, however, merely a matter of understanding the mechanics of the change process. It requires the recognition and comprehension of the nature of the links between individual system components. Concepts such as thresholds of change (Schumm, 1977), response time and magnitude, rates and paths of change and recovery, are all important if the full nature of environmental changes are to be appreciated (see Brunsden, 1990, for example).

THE NATURE OF SENSITIVITY

There is little doubt that the resistance of natural systems to change, be they environmental, geomorphological or climatological, is a complex and poorly understood subject. Research

Landscape Sensitivity. Edited by D. S. G. Thomas and R. J. Allison
© 1993 John Wiley and Sons Ltd

is constantly being undertaken in an attempt to elucidate how processes interact to induce modifications to earth systems. One area still requiring much attention is the sensitivity of landscapes to change. Sensitivity can be defined in various ways. Brunsden (1990), for example, notes that there is in a landscape a wide spatial variation in the ability of landforms to change and that this is known as the sensitivity to change. On the other hand, landscapes may be able to resist or absorb impulses of change and this is another form of sensitivity (Brunsden & Thornes, 1977; Huggett, 1988). Ecologists frequently use the term to define the susceptibility of a system to disturbance. In other words, the use implies fragility. Sensitivity can be employed more widely, however, depending on the system to which it is being applied and its components.

Landscape sensitivity can manifest itself in a variety of ways but in each case the crux of the problem is understanding the response to external influences. The external factors are varied and include natural and human induced phenomena. An example of the former might be the discharge level at which a river undergoes major modifications to channel morphology. The potential for a coastline to undergo irrecoverable alteration in morphology due to the interference by man in the natural near-shore sediment cycle is an example of the latter.

The question of sensitivity thus focuses on the potential and likely magnitude of change within a physical system and the ability of a system to resist change. A cause/effect relationship can be identified where external processes control, influence and dictate change.

The nature of landscape sensitivity can vary through space and time (Thornes & Brunsden, 1977). Change through time can be best summarised as the difference between slow, gradual modification on the one hand and sudden, catastrophic modification on the other. Geomorphological systems can display both of these phenomena but in the past the sensitivity question has tended to be interpreted in a sudden and catastrophic way rather than focusing on the gradual event (Selby, 1974; Smith, 1992). Nevertheless, sensitivity is a topic for consideration at both levels. A system might be highly sensitive to small changes in its controlling parameters, for example, but with the modifications manifesting themselves as no more than small, minor perturbations to the system itself. On the other hand, thresholds might exist where for only a small change in key variables, a major change takes place.

At another temporal level, the nature of landscape sensitivity can be determined by the time span under investigation. Behavioural aspects of a geomorphological system may be regarded as sensitive on one time base but not on another. De Boer (1992) implies this in his propositions for process geomorphology. Because of the slow response to change in the environment, the morphology, functioning and evolution of many large-scale geomorphic systems are best and frequently can only be analysed over long time spans. Small-scale geomorphic systems, on the other hand, equilibrate much more rapidly with the environment, enabling analysis over shorter time spans. In other words, sensitivity at one scale is different to sensitivity at another scale.

The whole question of landscape sensitivity can be addressed from a variety of angles and, as the chapters in this book indicate, the term can be interpreted in many broad ways. Perhaps one of the most relevant at present is the sensitivity of geomorphological systems to climatic change. Global warming scenarios tend to focus on issues such as rises in sea level, changes in the general circulation of the atmosphere and shifts in world climatic belts. This is partly because such subjects are presently on the political agenda, partly because of media interest and also because they are topics of concern recognised by a large sector of the world community. However, the difference between political perception on the one hand and scientific unknowns on the other is considerable, with the complexity of natural systems frequently precluding the establishment of politically desirable, simple, straightforward solutions.

Virtually no direct records exist where the implications of global warming on geomorphological processes and the sensitivity of systems to changes are discussed. Scenarios are usually founded on data from palaeo-studies. Take fluvial environments, for example. A change in sea-level, for which future estimates are the subject of constant revision, will result in a new base level for many rivers throughout the world. As a consequence the regrading of many river long profiles will commence. The sensitivity question will have fundamental implications for many aspects of river management and control. The need for further research is essential.

In another context, one of the most important sensitivity issues revolves around the human impact on the landscape, the likely effects of human activity and the extent to which natural systems can be irrecoverably modified (Goudie, 1986). Key issues in this context include changes to tropical ecosystems, water use and abuse in arid and semi-arid lands, and land surface degradation. It can be argued, for example, that many years will be necessary for the recently disrupted desert surfaces of the Arabian Peninsula to return to an equilibrium form. Indeed, it may be virtually impossible for such a scenario to be reached. In many cases the effects of any human modification on the environment are likely to be felt not just in the immediate vicinity but also in the surrounding landscape and the more sensitive the environment, the greater the likely sphere of influence over which the effects will be seen.

It should be remembered that the sensitivity issue is not just as simple as understanding, interpreting and predicting change for one process or system in isolation. To understand fully the complexities surrounding the sensitivity question links between systems and between specific components of systems must be recognised and understood. Here lies one reason for oversimplification in some recent studies and in some key areas, such as desertification, for example (Goudie, 1983; Binns, 1990; Hellden, 1991). It is impossible to understand processes such as sheet wash, soil erosion and land surface degradation without considering the influence of vegetation and soil material properties (Duck & McManus, 1990). Of course the precise controls on change and the linkages between components in such an interaction of variables is complex but this alone is not good enough reason to ignore the associated problems.

Many palaeo-studies can throw light on the associations between past events, current processes and the implications both past and present have for future events (Mulder & Hagerman, 1989; Dawson, 1992). The stratigraphic record in many recent Quaternary and Holocene deposits can be used to understand landscape response to changing climatic environments. Using the palaeo-record is particularly useful in attempting to overcome the temporal problems associated with working with short time series data sets. More accurate scenarios can be developed for predictive modelling at all scales, from plot and catchment studies to regional and global levels.

Any attempt to unravel the sensitivity question will require the measurement, collection and analysis of rigorous, high-resolution data. To understand changes in rates of landscape development and in key parameters for a range of processes, such as soil moisture content and rock surface temperature, the appropriate variables must be carefully measured, both in terms of a clearly defined temporal sampling framework and carefully located sampling points. Explaining the sensitivity of any process will only be as good as the data available for analysis and interpretation. The spacing of records, either regular or irregular, the frequency of data collection and the duration of collection will all be contributory factors in reaching an understanding of the sensitivity question (Anderson & Burt, 1990a). A temporal sampling framework must be able to collect data at critical points when results are most likely to shed light on characteristics such as behaviour thresholds. Of equal importance are the techniques chosen for data analysis (Anderson & Burt, 1990b). If sensitivity depends on discrete, serially independent events, there is no point in analysing data as serially dependent information. In addition to this, the importance of time functions such as periodicity and persistence are important in understanding sensitivity and the rates of operation of geomorphological processes.

THIS VOLUME

As the following chapters in this book illustrate, the theme of landscape sensitivity can be interpreted in many ways. Many of the questions posed above cross-cut individual examples where sensitivity has to be taken into account in an attempt to understand natural events, be they the weathering of building stone on the one hand, nitrate leaching through soil or the dynamics of slope stability on the other.

Despite the diversity embraced within the theme for this text, and within the context of the foregoing discussion, three main areas have been identified for considering the sensitivity question. The first centres on sensitivity to climatic change past, present and future. Second, there is landscape sensitivity to human activities and human influences. Finally, there is sensitivity and built environments.

REFERENCES

Anderson, M. G. and Burt, T. P. (1990a). Process determination in space and time. In Goudie, A. S. (ed.), *Geomorphological Techniques*, Unwyn Hyman, London, pp. 17−29.

Anderson, M. G. and Burt, T. P. (1990b). Methods of geomorphological investigation. In Goudie, A. S. (ed.), *Geomorphological Techniques*, Unwyn Hyman, London, pp. 3−16.

Binns, A. (1990). Is desertification a myth? *Geography*, **75**, 106−113.

Brunsden, D. (1990). Tablets of stone: towards the Ten Commandments of geomorphology. *Zeitschrift für Geomorphologie, Suppl.*, **79**, 1−37.

Brunsden, D. and Thornes, J. B. (1977). Landscape sensitivity and change. *Transactions of the Institute of British Geographers*, **NS 4**, 463−484.

Dawson, A. G. (1992). *Ice Age Earth*, Routledge, London.

de Boer, D. H. (1992). Hierarchies and spatial scale in process geomorphology: a review. *Geomorphology*, **4**, 303−318.

Duck, R. W. and McManus, J. (1990). Relations between catchment characteristics, land use and sediment yield in the Midland valley of Scotland. In Boardman, J., Foster, I. D. L. and Dearing, J. A. (eds), *Soil Erosion on Agricultural Land*, Wiley, Chichester, pp. 285−299.

Goudie, A. S. (1983). *Environmental Change*, 2nd ed., Clarendon Press, Oxford.

Goudie, A. S. (1986). *The Human Impact on the Natural Environment*, Blackwell, Oxford.

Hellden, U. (1991). Desertification − time for reassessment. *Ambio*, **20**, 372−383.

Huggett, R. J. (1988). Dissipative systems: implications for geomorphology. *Earth Surface Processes and Landforms*, **13**, 45−49.

Morgan, R. P. C. and Rickson, R. J. (1988). Soil erosion − importance of geomorphological information. In Hooke, J. M. (ed.), *Geomorphology in Environmental Planning*, Wiley, Chichester, pp. 51−62.

Mulder, E. F. J. and Hageman, B. P. (eds) (1989). *Applied Quaternary Research*, Balkema, Rotterdam.

Nisbet, E. G. (1991). *Leaving Eden. To Protect and Manage the Earth*, Cambridge University Press, Cambridge.

Schumm, S. A. (1977). Geomorphic thresholds: the concept and its applications. *Transactions of the Institute of British Geographers*, **NS 4**, 485−515.

Selby, M. J. (1974). Dominant geomorphic events in landform evolution. *Bulletin of the International Association of Engineering Geologists*, **9**, 85−98.

Smith, K. (1992). *Environmental Hazards: Assessing Risk and Reducing Disaster*, Routledge, London.

Thornes, J. B. and Brunsden, D. (1977). *Geomorphology and Time*. Methuen, London.

2 Barriers to Geomorphological Change

DENYS BRUNSDEN
King's College, London, UK

INTRODUCTION

A central problem for the prediction of geomorphological change is that the data obtained from empirical field experiments are variable. Even though two landform components or two systems appear to be very similar, there is uncertainty about whether they will behave in the same way. This may be regarded as randomness, indeterminance or the noise in a data set produced from unknown error, observation or circumstances beyond the control of the experiment. In assessing the stability of a slope, civil engineers prudently cover this uncertainty by applying an addition to the factor of safety. This is really an allowance for possible error in the understanding, measurement, calculation and validity of the equations employed.

The experimental problem is compounded by the fact that no two natural systems are exactly the same. This means that morphology and causal processes cannot be exactly specified and that the histories of the systems are not similar. Thus two hillsides in the same river valley, on the same rock type, in the same climate and ecosystem, may be very similar even to the trained eye or measurement system. Yet they may respond to the same impulse of change in different ways and at different rates. One reason for this singularity (Schumm, 1985) or uniqueness (Brunsden, 1990; Baker & Twidale, 1991) in behaviour is the sensitivity of the landform or system to impulses of change.

SENSITIVITY TO CHANGE

Brunsden & Thornes (1979) expressed the sensitivity of a landscape to change as 'the likelihood that a given change in the controls of a system will produce a sensible, recognisable and persistent response'. A fundamental proposition of geomorphology was proposed in which landscape stability was described as a function of the temporal and spatial distributions of the resisting and disturbing forces. This in turn was described as a landscape change safety factor which, like the civil engineering safety term, is the ratio of the mean magnitude of barriers to change to the mean magnitudes of the disturbing forces. A stable landscape is therefore one in which the controlling resistances of the system are such as either to prevent an impulse from having any effect (resistance) or to be so arranged as to restore the system to its original state (feedback). This chapter examines the idea of landscape resistance. A discussion of the disturbing forces, formative events and aggressive stress can be found in Brunsden & Thornes (1979) and Brunsden (1987, 1990).

Landscape Sensitivity. Edited by D. S. G. Thomas and R. J. Allison
© 1993 John Wiley and Sons Ltd

BARRIERS TO CHANGE

Strength Resistance

All landscapes have a variable framework of rocks from which landscape are produced. In a passive sense these are merely the mass from which the landforms are made. In an active sense, they possess properties such as strength and erodibility and they respond to stress in a liquid, plastic or brittle way depending on the circumstances. At a coarse or broad landscape scale the rock properties, particularly the structural fabric produced by the stress history of the area, provide a limited range of geometrical outcomes such as joint spacing, orientation and continuity. They thereby impose a certain order and regularity to the gross form of the landscape. This is the stress fabric relief produced by the adjustment of the processes to the structure (Sanders, 1970). In any region it is possible to recognise and map an overall family of forms with not too many members and rare abberations. At the same time and at all scales, there is some variation within the family of outcrops, rock properties and stress fabrics and different potential rates of change over space and time. This means that there is an underlying set of probabilities of occurrence of component combinations which will allow specific, even anomalous, response patterns to emerge even if the processes are uniformly applied. Within these constraints divergence of form becomes more likely. In a mathematical sense this may be modelled by bifurcation and catastrophe theory methods. The fractal and chaos ideas of order and pattern at all scales is also apparent.

Morphological Resistance

The propensity of a system to change is usually determined by its morphological configuration. In the simplest sense that is the recognition that landscapes have a variable slope, relief and elevation. They have variable potential energy, which not only varies in space but also changes dynamically in either a positive or a negative sense over time. Location, the energy of position, is a paramount determinant of sensitivity. In crude terms it is very easy to roll downhill but more difficult to climb out of a hole! Neighbouring particles, river banks, canyon walls or mountain interfluves all possess relative relief and therefore are barriers that have to be overcome by any moving fragment of landscape if change is to occur. Again it should be noted that these barriers can all change through time and so can the landscape change safety factor.

The way in which the available potential energy is mobilised is important. This is, in part, determined by the process−form relationships. Slope shape, in plan and profile, provides combinations that control the process of change. Obvious examples are the concentrations of water in hollows or the limiting angles of mass movement. Schumm (1973) has shown how morphological controls act as variable thresholds in gullies, and Iida & Okunishi (1983) have determined critical values for debris sliding based on regolith thickness. All of these controls are dynamic in time and space.

Structural Resistance

The actual design of a geomorphological system can facilitate or hold up change. Some systems are rigid and others flexible. Some are simple with few components, links, controls and thresholds and others are very complex. All allow feedback to restore existing system states.

The first aspect of structural resistance is location, in terms of the closeness of sensitive elements to the processes initiating change. The closeness to base level is an obvious example. As Lester King (1953) said, change to a landscape takes place when 'news of the uplift' is brought to it. A useful general term is *locational resistance*. This idea cannot be separated from the ability of a landscape to transmit the impulse of change. A rock that is well jointed, a dense drainage pattern, a steep slope, will all promote the rapid transmission of energy. This has been termed *transmission resistance* (Brunsden & Thornes, 1979; Brunsden, 1990).

The structure of the system is described by the direction and strength of the relationships between the components and process domains. This is known as coupling and takes place at important junctions, joints and boundaries. The idea directs attention at the functions of each domain, the nature of the links between them and any changes in behaviour that take place across the boundary. A good analogy is to describe the 'couples' as the gear boxes, transmission links and shock absorbers of the system. In practical geomorphological terms there are three types of linkage.

First, a linkage is said to be not coupled when a discontinuity exists between two systems or process domains even though they meet at a boundary. The boundary may also mark a sudden change between landsurfaces of different origin and age. At the boundary one system feeds upon and destroys the other. Secondly, a coupled linkage occurs at a boundary or link across which there is a free transmission of energy, material and messages of an impulse of change. There are gradients or gearings defined by the degree of difference between the frequencies of the systems. Thirdly, a decoupled linkage is one that is temporally inactive because a spatial transmission barrier has been imposed between two potentially linkable subsystems. This is usually achieved by the storage of material within the system. A floodplain between a channel and a hillslope or a beach between the sea and a cliff are examples. Through time all types of linkage can evolve, change, strengthen or weaken. Interception coupling is common, as in the obvious case of river capture.

Filter Resistance

Each system has the ability to manage the way in which kinetic energy is transmitted through the landscape. This idea has five aspects. First, all systems can absorb a proportion of the energy supplied to it without obvious change taking place — this is the shock absorber principle. For example, void spaces fill with water to be followed by drainage. On a beach, much wave energy is dissipated by drainage of the water through the sediments with no obvious beach form change. Energy is also utilised in turbulence merely in overcoming the roughness of the bed. If a transport threshold is not overcome the energy is utilised as heat, again with no gross landform change. Of course, just as a shock absorber weakens with stress fatigue, so subtle changes take place in landscape which may, in time, have a cumulative or chaotic effect and promote punctuated change. This is a research frontier about which we have little information.

Secondly, mobile components, such as a gravel bed river, use up energy by adjusting the position of individual particles or storage. While some abrasion takes place, the main effect is to adjust the mean form over time to accommodate the energy specification of the system. Winter and summer storm beach profiles may be regarded as adjustments about an equilibrium state that are designed to filter energy and to avoid the need for progressive or lasting adjustments to form.

Thirdly, the adjustment of individual particles usually increases the armouring of the system.

On beaches, the imbrication packing of particles serves to make the surface resistant to wave energy and erosion. On semi-arid slopes, the washing away of fines leads to a progressive covering by concentrations of coarse clasts. Desert pavements that erode by sand winnowing and salt heave, polygonal structuring in periglacial areas or gravel bed evolution in rivers are other examples. In rivers, recent research (Wolcott, 1990) has shown that bed surface coarsening, or textural structure (*textural resistance*) is followed by a movement of the clasts into more stable configurations or geometric structures (*geometric resistance*). This is a nice illustration of the geomorphological proposition that nature seeks out stable forms (Scheidegger, 1987; Brunsden, 1990).

Fourthly, energy filtering takes place as subsystem boundaries are crossed. Every component or domain has a different storage capacity, heat exchange mechanism, reaction response, relaxation time, resilience and tolerance. This means that, at a boundary, some impulses cross easily and remain unchanged. Others are modified, diffused, stored or are quite ineffective. Perhaps the most exciting idea is the way in which process events can change in scale across filter boundaries. Brunsden & Jones (1980) showed how the sensitivity changed in a coastal landslide system from immediate response at a fully coupled cliff foot, to a > 1 year response on the vertical sea cliff, to a seasonal change programme in a middle zone of mudslides, to ~ 1:40 year landslides on an undercliff and to a ~ 1:100+ year failure at the head of the system. Finally, across a millimetre of space, a major decoupling takes place. The top of the landslide scar forms a discontinuity to a plateau surface that is still capped by the remnant of an early Tertiary tropical soil (Isaac, 1983). The differing response mechanisms, within the same landslide system, have acted to accumulate the continuous small impulses received from the sea into episodic large-scale responses in the system above. Another example of this is the storage of base level change at a waterfall. In this case energy is only transmitted if it is sufficient to drive the system or if a threshold boundary magnitude is exceeded.

De Boer (1992) believes that this may be typical of any hierarchy of interlocking systems and proposes a fundamental proposition in a theoretical framework for spatial scale considerations in geomorphology:

> Low-magnitude, high-frequency processes operating on a small scale in a geomorphic system may, given sufficient time, provide the conditions for the occurrence of a high-magnitude, low-frequency event, operating on a much larger scale at the same hierarchial level.

When the boundary filter energy crosses into a new system or hierarchical level, however, attention focuses on the discontinuities and interfaces in the landscape. Morphological change is usually abrupt and associated with a change in process domains. In the case of the landslide sea cliff discussed above, the upper limit passes from one of the most sensitive, active systems in Great Britain, to a surface that has taken 50 million years to create and where soil erosion processes are extremely slow. At such disjunctions between sensitivity—insensitive, labile—sluggish (Trudgill, 1976) units, modification, filtering or diffusion are not suitable words. All that happens is the destruction of a historical record as the old surface is removed to be replaced by something entirely new. It is, in effect, a catastrophe, a thorough overturning of the existing order.

A final aspect of filter resistance occurs with the introduction of space scales as an impulse of change diffuses across a system. This is the essence of Schumm's (1973, 1979) complex response statement. If it were possible for the base level of a stream to drop abruptly at its mouth, instead of for the sea just to go out and for the stream to get longer, the result would be a waterfall at the mouth, or point source. In time this impulse, of say 20 m of relief, would be transmitted upstream to form a gorge. Thus the 20 m point would become a linear impulse of change ranging from 0 m to 20 m deep over a given distance. If the gorge is coupled to the hillslopes above, the impulse now spreads out over area, in other words it diffuses over space as a wave of aggression. Finally, right at the top of the slope one soil particle gradually moves away from the divide as a response to a 20 m sea level change! Space itself thus becomes a barrier to change. There is another consequence, as Schumm (1977) discovered. Area means a lot of sediment, which comes into the gorge. The stream cannot remove it and therefore valley fill takes place. The complex response scenario begins. From the viewpoint of this paper, it is clear that spatial scale is a vital element of landscape sensitivity because, with space, change becomes increasingly diversified and complex and prediction becomes less and less certain. This also means there are two types of sensitive landscape. Those that are the least complex in response choices can respond simply and directly if the impulse is of the right kind. Those that can move in many directions can recognise more impulse types and either bifurcate in different directions or develop a complex adjustment mechanism.

System-state Resistance

All systems have a history. This means that they have been subjected to impulses of change and formative events in a spatial pattern and temporal sequence that is unique for that system. Although there may be general patterns applied to neighbouring systems, it is not possible for two systems to receive exactly the same number, frequency, duration or magnitude of events. For example, the partial area rainfall-runoff concept means that one part of a system changes, the other does not. This is not a denial of the general principles of science because the response mechanisms still follow natural laws. All it means is that they are not called into play all over and all the time. In terms of sensitivity, some system components are further along the road toward an ability to respond to a given impulse. The sequence of events applied to them in the past have favourably predisposed them to change.

An important point to emphasise is that the historical events may also have done all the work required, A stable state may have been achieved and the system components may have relaxed to a characteristic form (Brunsden & Thornes 1979). Additionally, the historical event sequencing may mean that the previous process environment may have been very efficient. In the succeeding domain a harsh regime will be required for further change. Perhaps only the very high magnitude events can be successful. Such landscapes are overrelaxed and insensitive, and this is the basis of landscape persistence.

REFERENCES

Baker, V. R. and Twidale, C. R. (1991). The reenchantment of geomorphology. *Geomorphology*, **4**, 73−100.

Brunsden, D. (1987). Principles of hazard assessment in neotectonic terrains. *Memoirs of the Geological Society of China*, **9**, 305−334.

Brunsden, D. (1990). Tablets of stone: towards the Ten Commandments of Geomorphology. *Zeitschrift für Geomorphologie, Suppl.*, **79**, 1−37.

Brunsden, D. and Jones, D. K. C. (1980). Relative time scales and formative events in coastal landslide systems. *Zeitschrift für Geomorphologie*, **84**, 1−19.

Brunsden, D. and Thornes, J. B. (1979). Landscape sensitivity and change. *Transactions of the Institute of British Geographers*, **4**, 463–484.

de Boer, D. H. (1992). Hierarchies and spatial scale in process geomorphology: a review. *Geomorphology*, **4**, 303–318.

Iida, T. and Okunishi, K. (1983). The development of hillslopes due to landslides. *Zeitschrift für Geomorphologie*, **46**, 66–77.

Isaac, K. P. (1983). Tertiary laterite weathering in Devon, England and the Palaeogene continental environment of south west England. *Proceedings of the Geological Association*, **94**, 105–115.

King, L. (1953). Canons of landscape evolution, *Bulletin of the Geological Society of America*, **64**, 721–762.

Sanders, B. (1970). *An Introduction to the Study of the Fabric of Geological Bodies*. Permagon Press, Oxford.

Scheidegger, A. E. (1987). The fundamental principles of landscape evolution. *Catena*, **10**, 199–210.

Schumm, S. A. (1973). Geomorphic thresholds and complex responses of drainage systems. In Morisawa, M. (ed.), *Fluvial Geomorphology*, Publications in Geomorphology, Binghampton, pp. 299–310.

Schumm, S. A. (1977). *The Fluvial System*, Wiley, New York.

Schumm, S. A. (1979). Geomorphic thresholds: the concept and its applications. *Transactions of the Institute of British Geographers*, **4**, 485–575.

Schumm, S. A. (1985). Explanation and extrapolation in geomorphology: seven reasons for geologic uncertainty. *Transactions of the Japanese Geomorphology Union*, **6**(1), 1–18.

Trudgill, S. T. (1976). Rock weathering and climate: quantitative and experimental aspects. In Derbyshire, E. (ed.), *Geomorphology and Climate*, Wiley, London, pp. 59–99.

Wolcott, S. (1990). Flume studies of grave bed surface responses to flowing water. Unpublished Ph.D thesis, University of British Columbia.

Part 1

GEOMORPHIC PROCESSES, LANDSCAPE SENSITIVITY AND CLIMATIC CHANGE

3 The Sensitivity of River Channels in the Landscape System

PETER W. DOWNS and KEN J. GREGORY
Department of Geography, University of Southampton, UK

The river channel acts as the conduit for fluid flow in the landscape and should be seen as an integral part of the landscape system for three reasons. First, because present river networks provide a conveyor belt for sediment movement within the landscape system, and therefore the interaction between river channel forms and processes must be a central ingredient of the landscape system and merits research investigation. Secondly, because previous relationships between river channels and landscape systems may have been different from present and this is one *raison d'être* for palaeohydrological research. Thirdly, it is necessary to consider how the interrelationships between river channels and fluid flow may be different in future environments and ultimately to use this knowledge to complement and improve current approaches to integrated river basin management (Downs, Gregory & Brookes, 1991).

Research focused on each of these three aspects naturally involves the extent to which river systems are susceptible to change. The primary way in which a river system can change is by alterations to the quality and quantity of its water. Quality aspects embrace mechanical components as suspended sediment and bedload, chemical components involving solutes and pollutants, water temperature, and the biological content, particularly in the form of organic load. The river channel itself may also undergo change through a number of modes of self-adjustment (Hey, 1978) or degrees of freedom which represent the fundamental parameters of the channel morphology and processes that are able to adjust. Such adjustments are influenced by changes in the quantity or quality of fluid flow, by direct modifications of the channel or by alterations of the area adjacent to the channel. These adjustments may be autogenically or allogenically induced (Lewin, 1977). The precise number of degrees of freedom specified by Hey (1978, 1982, 1986, 1988), by Gregory (1980, 1987a) and by Knighton (1984) are all slightly different. Degrees of freedom have often been distinguished for both river processes and river channel morphology together rather than separately. Considering only the degrees of freedom of the river channel allows changes to be categorised (Gregory, 1987b) according to scale to include *in-channel* changes of sediment or of the pool-riffle sequence, changes of the *channel cross-section* involving width, depth and capacity, changes that apply to the *reach* involving channel planform and channel slope, and at a larger scale changes of the *channel network* or of the *basin* as a whole.

However, even when autogenic or allogenic changes occur in the quantity and quality of river flow, or as a result of modification of the channel perimeter, subsequent adjustments of river channel morphology will not necessarily occur. Whether a change occurs or not is a reflection of the sensitivity of the river channel to change. This chapter therefore outlines

Landscape Sensitivity. Edited by D. S. G. Thomas and R. J. Allison
© 1993 John Wiley and Sons Ltd

relevant approaches in geomorphology and approaches developed in other disciplines as a basis for establishing what is required in fluvial geomorphology and how sensitivity may be defined in relation to river channel adjustments. These discussions are a prelude to suggesting general and specific examples of approaches that can be used for the examination of river channel sensitivity and are capable of further application in relation to river channel management.

RIVER CHANNEL SENSITIVITY AND FLUVIAL GEOMORPHOLOGY

There have been comparatively few explicit reviews of sensitivity in relation to fluvial geomorphology but a number of developments are implicitly relevant. Variations of the equilibrium condition were acknowledged in Langbein & Leopold's (1964) review of quasi-equilibrium states applied to the fluvial system. Subsequently in a series of seminal papers Schumm explored the potential of river metamorphosis (1969) and considered thresholds (e.g. 1973) beyond which adjustments of the fluvial system could occur. Schumm (1985) explicitly identified sensitivity as one of the seven reasons for geological uncertainty and suggested that when combined with singularity there are instances where a minor input may have a major effect, but elsewhere in the landscape the same input may have no consequences. The implications for Quaternary investigations have been explored by Begin & Schumm (1984). Schumm (1985, p. 13) concluded that within a landscape composed of singular landforms there will be sensitive and insensitive landforms. Schumm (1991) also defined sensitivity as the propensity of a system to respond to minor external change, so that if a system is sensitive and close to a threshold it will respond to an external influence. Subsequently, Schumm (1988) considered recognition of sensitivity in relation to the identification and prediction of geomorphic hazards. Brunsden & Thornes's (1979) consideration of landscape sensitivity and change developed ideas that had been used in relation to hill slopes whereby a factor of safety approach devolved upon the relationship between the magnitude of the disturbing and resisting forces. They concluded (Brunsden & Thornes, 1979, p. 476) that

> The sensitivity of a landscape to change is expressed as the likelihood that a given change
> in the controls of a system will produce a sensible, recognizable and persistent response.
> The issue involves two aspects: the propensity for change and the capacity of the system
> to absorb the change.

Brunsden & Thornes (1979) were concerned with landscape sensitivity, and therefore they did not offer an explicit definition that could be applied to river channel sensitivity; four interpretations are thus given below (pp. 20). Other general approaches have also envisaged definitions of geomorphological sensitivity based on balance of forces and their periodicity. A comprehensive description is that of Chorley, Schumm & Sugden (1984, p. 7) who proposed that

> changes of inputs (i.e. process magnitudes) into systems produce changes in (sediment)
> output and changes in the forms or structures of the internal components of the system
> (i.e. *subsystems*). The speed with which changes in input are reflected in these resulting
> form changes is expressed by the so-called *relaxation time* of the system (Melton, 1958)
> . . . Thus landform response can be measured in terms of *sensitivity* (i.e. recurrence
> interval/relaxation time) . . . where *recurrence interval* is the average length of time
> separating events of a relevant magnitude.

In addition to these general approaches there have also been specific attempts to differentiate areas according to their susceptibility to change. Thus along river channels in the south-west of the United States, Graf (1979, 1982, 1988) demonstrated how analysis of shifts of river channel pattern allow river channels to be categorised according to degrees of instability. Such studies illustrate that some reaches of river channel are more sensitive than others, so that change is not equally distributed along the fluvial system. In a more general context Buckley (1991) has advocated environmental sensitivity maps to show the probable environmental impacts of different types of land use or development. He suggested that environmental sensitivity refers to the relationship between applied stresses and environmental responses. There are, however, many instances, even in the field of geomorphological contributions to environmental management, where the sensitivity of landscape has not been explicitly developed as a central theme (e.g. Toy & Hadley, 1987; Cooke & Doornkamp, 1990). Prior to suggesting how sensitivity may be related explicitly to river channel behaviour, it is useful to ascertain how sensitivity has been used in other disciplines.

CONNOTATIONS OF SENSITIVITY IN VARIOUS DISCIPLINES

'Sensitivity' is defined in the *Oxford English Dictionary* (1989) as: 'The degree to which a device, test or procedure responds to small amounts of, or slight changes in, that to which it is designed to respond; the ratio of the response of a device to the stimulus causing it' (Vol. 14, p. 986). Furthermore it is suggested by Smith (1944, in the *Oxford English Dictionary*, 1989) that the term 'sensitivity' exists at a variety of scales that are nested, i.e. the sensitivity of the system output is relative to the sensitivities of its component parts, which are themselves sensitive to their components. Therefore sensitivity can be mathematically described as the ratio of two differentials that express the response or induced output change consequent upon a stimulus or applied input change. Techniques of sensitivity analysis have therefore been developed in a number of disciplines in order to examine the extent to which a data series, instrument or model may be sensitive to specified factors. Table 3.1 illustrates the recent (1988−91) uses of sensitivity analysis in a wide range of disciplines, based on a search of 67 computerised literature databases. It demonstrates a bias towards the physical sciences and technology in particular, thus corresponding to the interpretation by Smith in the *Oxford English Dictionary*. Sensitivity analysis is also encountered in mathematics, in the study of water resources and in business applications where the technique may be used as a test of statistically or mathematically based financial estimates (Chaterjee & Hadi, 1988). A number of applications are suggested from databases, including geomorphology, but investigations suggest that many of these have been within ecological studies. Subsequent searches of the complete content of suitable databases centred on sensitivity analysis related to river channels or the drainage basin yielded few references directly applicable to geomorphology.

However, one discipline that may afford an important analogy for geomorphology in its use of sensitivity analysis is hydrology, particularly the field of physically based, distributed hydrological modelling of drainage basin discharge response. The use of sensitivity analysis is regarded by McCuen (1973) as 'a modeling tool that, if properly used, can provide a model designer with a better understanding of the correspondence between the model and the physical processes being modeled' (p. 38). Examples of sensitivity analyses of separate models are included in Cunningham & Sinclair (1979), Rodgers *et al.* (1985) and Bathurst (1986). Other

Table 3.1   Individual databases with more than 10 examples of sensitivity analysis, 1988–1991

Frequency	Individual databases; description of disciplines included
1463	all science and technology disciplines
1255	all engineering and technology disciplines
1231	physics, electronics, computers and control, information technology
548	all science and technology (French based)
516	United States Government sponsored research
377	dissertations awarded
260	mathematics
258	water resources science, planning and economics
219	business journals of trade, industry and commerce
209	business/industry management and administration
154	geography, geology, ecology and related disciplines
147	business news, developments and markets
97	chemistry
92	business journals of trade, industry and commerce
89	geosciences based around geology and geophysics
80	mechanical/civil engineering, production engineering and engineering management
79	transportation research (US basis)
47	science and management of marine and freshwater environments
45	economics
41	fluids engineering
39	all social sciences
31	computers, telecommunications and electronics
25	scholarly/general interest publications and the most popular academic journals
22	science and practice of metallurgy
17	marine related disciplines
13	material science and engineering (primarily non-metallic)
12	psychology and behavioural sciences
11	educational materials from Education Resources Information Center, US
11	geosciences based around geology and geophysics

uses of sensitivity within hydrology include, for example, its application to analyses of evapotranspiration (Coleman & De Coursey, 1976; Beven, 1979), groundwater recharge and flow (Chiew & McMahon, 1990; Prince, Reilly & Franke, 1989), runoff (Karl & Riebsame, 1989) and water quality (e.g. Brandt, Bergstrom & Sanders, 1987; Melack, Stoddard & Ochs, 1985). The parameters or functions of optimised models are subjected to sensitivity analysis in order to indicate which are the most important for model performance and output, for example, in Beven (1979) and Bathurst (1986). In these examples, and that of Cunningham & Sinclair (1979), the results are utilised to indicate which parameters require the most accurate estimation. Chiew & McMahon (1990) use sensitivity analysis to gauge whether streamflow records calibrated by a rainfall runoff model as an indicator of groundwater recharge are in fact sensitive to the same parameters as those that influence the recharge.

The function 'sensitivity' has a mathematical basis derived from a Taylor series expansion (McCuen, 1973; Beven, 1979), which is normally simplified to produce a general definition of sensitivity, S

$$S(F_i) = \frac{dF_o}{dF_i}$$

where F are factors being compared, o the output and i the input (McCuen, 1973). It is this general definition of sensitivity that forms the basis from which parametric and component sensitivities are derived (McCuen, 1973). Furthermore, in order to facilitate comparison of sensitivities, the measure of relative sensitivity, R_s, compensates for the input and output factor magnitudes as

$$R_s = \frac{dF_o/dF_o}{dF_i/F_i} = \frac{dF_o}{dF_i} \cdot \frac{F_i}{F_o}$$

Coleman & De Coursey (1976) note that, as partial derivatives are difficult to define, in practice sensitivity is more often based on a method of finite differences. A number of similar measures for absolute and relative parameter sensitivity have been proposed for the method of finite differences (e.g. McCuen, 1974; Coleman & De Coursey, 1976; Cunningham & Sinclair, 1979; Bathurst, 1986; Chiew & McMahon, 1990). Therefore, in the hydrological literature, sensitivity has been used as a term to describe a precise function. This function follows from the optimisation of a deterministic mathematical model and produces an index (normally relative) that describes how the output of the model, for instance the discharge, varies in relation to the variation in the factors contained within the model.

CATEGORIES OF RIVER CHANNEL SENSITIVITY

In geomorphology the first use of sensitivity has been as the ratio of the disturbing to the resisting forces, which indicates the propensity for change and may have a bearing upon the capacity of the system to absorb change. This indicates two distinct ways in which sensitivity has been employed, because whereas the ratio of the disturbing to the resisting forces is dimensionless and simply indicates a ratio value for a particular system, the notion of propensity for change endeavours to indicate for a particular system its proximity to particular thresholds and so introduces an element of probability in relation to the likelihood of change occurring. It is this proximity to thresholds that was seen as an important feature of sensitivity by Schumm (1991) who also argued that sensitivity could be considered as an aspect of singularity but in view of its significance justifies separate treatment. A further and third geomorphological interpretation is obtained by relaxation time divided into recurrence interval; this is really a measure of the landform response or recovery time and so represents an indication of the time required prior to change being accommodated.

Approaches to river channel sensitivity could mirror those used in physically based distributed hydrological modelling. This would involve consideration of river channel sensitivity as the extent to which morphological changes to the channel are dependent upon the change in any of the factors known to influence the morphology. However, response indeterminacy complicates the consideration of river channel change. Response indeterminacy primarily involves two aspects: first, the existence of thresholds in the landscape system which determine whether an event is effective (Wolman & Gerson, 1978; Newson, 1980); and secondly, the possibility of differences in the rate and duration of changes in time (Brunsden & Thornes, 1979; Chorley, Schumm & Sugden, 1984; Schumm, 1985) which contribute to the complex response identified by Schumm (1985). These problems of response indeterminacy may partially account for the imbalance between conceptual understanding and modelling ability in geomorphology identified by Anderson & Sambles (1988). In addition, the lack of practical modelling in geomorphology is seen by Anderson & Sambles (1988) to reflect uncertainties concerning model assumptions, the influence of large-magnitude, low-frequency events, the difficult in validating models, which relates to the difficulty of documenting the output of the model, and the fact that geomorphological models are likely to be unique to the time base for which the model is constructed. Faced with these difficulties in producing a 'standard' measure of sensitivity, one remaining possibility is to model the system using a coarse scale of physically based components which override a number of smaller-scale thresholds. This would allow a deterministic approach, based on physical concepts and experiment (Thomas & Huggett, 1980) to produce an empirical result that goes some way to overcoming the probabilistic nature of response.

Consideration of these geomorphological aspects in relation to sensitivity suggests several possible ways by which sensitivity can be applied to river channels. First, is the application of the simple ratio of disturbing to resisting forces. However, this simply indicates the possible influence of such disturbing forces so that a second interpretation associated with the propensity for change would additionally involve the proximity to threshold conditions. A third possible interpretation is the ability of a system to recover from change because river channel systems are such that the disturbing and the resisting forces are not always coincident in time. The full development of this third route involves the definition of a ratio of recurrence interval to relaxation time (Chorley, Schumm & Sugden, 1984). A fourth possible interpretation that integrates the previous three to some extent arises because it is possible to model adjustments and changes of river channels and to establish the sensitivity of model outputs to changes of model inputs. These four possible interpretations can be illustrated in terms of their application to river channels.

EXAMPLES OF RIVER CHANNEL SENSITIVITY

An immediate problem confronting the identification of river channel sensitivity centres upon definition of the river channel. There is no single way in which the river channel can be defined comprehensively and it has been suggested (Gregory, 1987a) that there are a series of scales at which the river channel can be viewed by hydrogeomorphologists. These range from the scale of in-channel characteristics, through the channel cross-section and channel geometry parameters, to the channel reach scale, and eventually to the entire drainage network of channels contained within a specific drainage basin. Sensitivity can be applied to any one of these scales and the interpretation may be slightly different according to the scale that is used.

A further distinction can be made between contemporary river channel systems and river channel change or adjustments. Although it is often expedient to distinguish between these two broad categories, inevitably they are not clearly differentiated and the relationship between the two varies, for example in the ways in which recovery can occur from large channel-modifying or from channel-forming events (Wolman & Gerson, 1978). The interpretation of contemporary river channel systems involves notions of equilibrium that have been applied to present systems and these have been clarified particularly by Graf (1988) when he proposed that equilibrium is a useful explanatory concept in humid regions but that in the case of dryland rivers more emphasis needs to be given to disequilibrium behaviour.

A way of expressing the sensitivity of river channels as a whole is offered in Table 3.2 by identifying the four major categories of sensitivity interpretation outlined above together with examples of applications to contemporary river channel systems and to changes of river channels. A number of examples can be quoted from the literature of the way in which these interpretations of sensitivity apply to contemporary channel systems. The ratio of disturbing

Table 3.2 Examples of river channel systems and river channel changes related to four interpretations of sensitivity

Interpretation of sensitivity	River channel systems: equilibrium conditions	River channel changes consequent upon			
		Dams/reservoirs	Land-use changes	Urbanisation	Channelisation
Ratio of disturbing to resisting forces.	Force resistance, e.g. Graf (1979) Impact of flood events, Baker, Kochel & Patton (1988)	Clear water erosion Sediment load reduced, e.g. Petts (1984)	Change in frequency of flows, e.g. Gregory & Madew (1982)	Changes in frequency of flows, and in sediment transport, e.g. Gupta (1984)	Impact of large events and increase in flow velocity, e.g. Brookes (1988)
Propensity for change shown as proximity to threshold conditions.	Change of channel pattern, e.g. proximity to single-thread/multi-thread threshold, e.g. Schumm (1979)	Possibility of multi-thread replaced by single thread channel, e.g. Gregory (1987b)	Single thread to multi-thread channels, e.g. Schumm (1979)	Possibility of incised channels, e.g. Nanson & Young (1981), or channel erosion, e.g. Whitlow & Gregory (1989)	Erosion producing larger channel according to high or low power sites, e.g. Brookes (1988)
Ability to recover from change	Recovery of channel systems from large flood events, e.g. Gupta & Fox (1974)	Channel capacities changed, e.g. Williams (1978); new distribution of rapids, e.g. Graf (1980)	Channel morphology changed, e.g. Starkel (1987)	Development of new channel morphology, e.g. Gregory (1977)	New channel morphology, e.g. Brookes (1990)
Modelling of change and its sensitivity	Use of form-process equations, e.g. Wharton et al. (1989) for sensitivity analysis	Use of model describing channel change for sensitivity analysis, e.g. using equation given by Williams & Wolman (1984) for changes downstream of dams			

to resisting forces is exemplified by the study of Graf (1979) of Central City in Colorado and also by the impact of large flood events on river channel systems (Baker, Kochel & Patton, 1988). The propensity for change shown through proximity to threshold conditions is exemplified, for example, in New Zealand as indicated by Schumm (1979) and for gully development as indicated in the United States by Graf (1988). The ability to recover from change is classically indicated by studies of the recovery of river channels from large events such as those instigated after hurricane Agnes on the east coast of the United States: the significance for channels is summarised by Gupta & Fox (1974). Against this background Table 3.2 incorporates specific examples of river channel adjustments according to changes associated with dams and reservoirs, land-use changes, urbanisation effects and channelisation. The ratio of disturbing to resisting forces in each case is reflected in the changing frequency of flows and in sediment transport, whereas the propensity for change can be reflected in various morphological adjustments such as the change from single thread to multithread channels, the possibility of channel enlargement or changes in channel character, for example according to high or low power sites as identified in the case of channelisation by Brookes (1988). In each case the ability to recover from change relates to the time in which channel adjustments can be completed or new features can be introduced and this can be exemplified by the rate law as developed by Graf (1977). For contemporary channels, as for changes over time involving morphological adjustments, it is possible to use sensitivity analysis to examine how changes in output may be affected by changes in input parameters.

SURROGATES FOR SENSITIVITY

The several interpretations of sensitivity as applied to river channels (Table 3.2) indicate a need for approaches to identification of the potential sensitivity of channel reaches, and this can be attempted in three major ways. First, and most effectively, by endeavouring to establish process response equations that describe the physical system and which could be employed to identify channels that are most sensitive to adjustment. This approach has been attempted by Hey (1986) and by other researchers but has usually had to focus upon a particular aspect of the channel system such as the planform or the channel geometry and is as yet very difficult to achieve. A second approach employs surrogates for this physically based method and a very good example is the way in which Patton & Schumm (1975) in the Piceance basin of Colorado used the relationship between drainage area and slope to differentiate gullied from ungullied basins. A third approach is to look at the spatial variation of channel adjustment as an indicator of variations in sensitivity and then to extend this approach by establishing why some areas are more sensitive than others. In this approach the amount or type of change summarises the balance of forces, resistance and proximity to thresholds. Although comparisons of maps of different dates (e.g. Graf, 1979; Burkham, 1972; Hooke & Kain, 1982) afford a means of comparing different reaches according to amount of change, this does not easily distinguish between autogenic and allogenic change. An alternative method is to use survey techniques to deduce spatial variations in adjustment and this is illustrated for two examples (Figure 3.1A).

The Monks Brook is a tributary of the River Itchen north of Southampton, draining an area of 41 km^2, and has been influenced by four major phases of urbanisation in the twentieth century (Figure 3.1B). The channel shows considerable signs of both direct modification and morphological adjustment and was mapped by the authors as two independent surveyors using

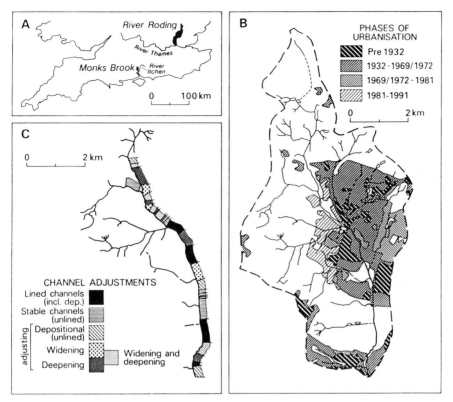

Figure 3.1 Spatial pattern of channel change along the Monks Brook, Hampshire. Location of the Monks Brook and the Roding (Figure 3.2) is shown in (A), the date of urbanisation is shown in (B), and the spatial pattern of channel change related to sensitivity is shown in (C)

a checklist scheme of vegetation, morphological and structural indicators that could locate areas of river channel adjustment. Previous attempts to utilise vegetation indicators (Gregory, 1992) and morphological and structural indicators of change (e.g. Lewin, Macklin & Newson, 1988) have not been extended for the systematic identification of the distribution of channel change. Therefore a detailed study (Gregory, Davis & Downs, 1992) was undertaken on the Monks Brook to see if the indicators could be employed to establish exactly where change was occurring. The results of field surveying of the Monks Brook channel (Figure 3.1C) showed that the channel could be categorised into lined sections that were not able to adjust, stable sections that did not show signs of change, and sections that were adjusting either as a result of deposition or erosion. In the case of the sections adjusting by erosion, it appeared that some were adjusting by widening and in other cases deepening was dominant, whereas there were some in which evidence of both types of change appeared. In the small basin of the Monks Brook the results obtained by two independent surveyors showed a high degree of agreement and of delimited homogeneous reaches in 95% of the channel length surveyed (Gregory, Davis & Downs, 1992). It was concluded that the mapping technique was sufficient to establish the spatial variation of channel adjustment, which could be taken to be a reflection of sensitivity. In the case of adjusting river channels it is now known (Gregory, 1987a,b) that extensive adjustments have occurred along river channels of the world but the degree

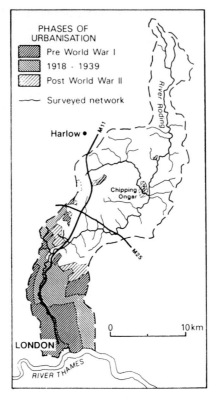

Figure 3.2 Land-use influences on the river channel of the River Roding. The characteristic channel adjustments are shown in Table 3.3 and the location of the basin is shown in Figure 3.1(A)

and exact location of adjustment can vary along a changed channel as illustrated along the Monks Brook, and is of significance to river channel management.

A second, and contrasting, example is provided from a survey of river channel adjustments for channels in the 366 km^2 Roding basin (Figure 3.2), which rises in central Essex and confluences with the estuarine River Thames to the east of London. The basin has been subject to a number of allogenic influences and subsequent channel changes may have been facilitated by the presence of easily eroded drift deposits in most of the basin. The lower part of the basin is urban and successive expansions of London have proceeded up valley from the mouth of the River Roding until the 1960s, when the new town status of Harlow deflected urban growth into the neighbouring Stort basin. With this expansion have been a number of associated channelisation measures including a number of modifications by large cut-offs which continue due to major road developments. In contrast the middle and upper basin have been influenced by direct and indirect modifications as a consequence of small town expansion and the construction of the M25 motorway. These channels may also have been influenced by hydrological and sedimentological changes brought about by intensive field drainage schemes, which often have included channel straightening, for the purposes of arable agriculture.

The channels were surveyed by one operator in late summer 1990 and a majority of reaches resurveyed by the same operator early in 1991 to take account of any seasonal differences

Table 3.3 Field observed indicators of channel adjustment in river channels: developed for lowland UK channels

Form of indication	Type[a]	Indicators
Morphological	1	bank erosion by undercutting/slumping/desiccation
	1	exposure of fresh materials in lower bank
	1	evidence of cutoffs/chutes and associated features
	2a	erosion of central channel to form compound cross-section
	2a	tributary bed level higher than main channel
	2a	accretion of continuous low bench
	2b	'fresh' gravels — concentrated at riffle locations or generally over whole bed
	2b	formation of step/pool sequence
	2b	accretion of point bar
	2b	silt deposition across whole bed
Vegetational	1	exposed tree roots
	1	undercut tree roots/turf root mat
	1	deformed tree
	1	protruding tree (erosion behind)
	1	trees in channel — bed or bench level
	2b	encroachment of (non-seasonal) vegetation onto channel bed or bench
	2b	complete cover of bed by seasonal vegetation — exceeding riffle length
Structural	1	erosion behind/under bank protection (leading to damage/collapse)
	1	structure protruding into channel
	1	structure set back from current channel bank
	1	erosion immediately downstream of structure
	2a	structure base not at original height relative to channel bed level

The particular combination of indicators observed determines which of the 10 channel conditions in Figure 3.3 is used to describe the reach.

[a] Types of indicator: (1) Bankside indicators — if indicators occur on opposite banks, then allogenic change is assumed; an occurrence on one bank may indicate autogenic change. (2) Bed indicators — (a) likely to represent allogenic changes; (b) may represent either allogenic or autogenic changes.

in sedimentological and vegetational evidence. The survey technique was similar in design to the one used for the Monks Brook and was implemented using the morphological, vegetational and structural indicators indicated in Table 3.3. The result of the field survey was to delimit 126 reaches of different length, each of which was homogeneous in character according to river channel adjustment. It was then necessary to classify the 126 reaches. To provide results suitable as input to river channel management, the adjustments observed were classified into a range of conditions designed to represent all possible adjustments of the channel, acknowledging (Mosley, 1987) that all classification schemes must reflect an intended purpose. The result, given in Figure 3.3, is a 10-fold classification based upon the notion that natural channel change involves channel cross-sections that are either reducing, enlarging, shifting (migration) or stable (no changes). The approach also allows, because of the influence of allogenic factors on channel adjustment, for any two of these conditions to exist simultaneously as river channels 'recover' to a capacity reflecting a possibly modified

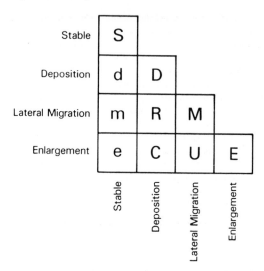

Figure 3.3   Basis for categories of channel adjustment. These are used for the Roding basin shown in Figure 3.2 and contained in Table 3.3 and explained in the text p. 25

hydrological regime. The result categories and the extent of surveyed channel of each type are given in Table 3.4. This shows that in the Roding basin, a broad spectrum of channel adjustments exist that may present a variety of problems for channel management, not least in their upstream and downstream interactions. The adjustments and their distribution should reflect the culmination of the varied influences on the river basin and the adjustments of the channels in the Roding basin, and in three other basins within the Thames basin, are being analysed in relation to various factors of rock type, channel gradient, land use and channel management to isolate the factors to which the river channels appear to be the most sensitive.

CONCLUSION

The first three categories of sensitivity identified in Table 3.2 in relation to river channel changes need to be identified in situations where river channels are adjusting. It is necessary to be able to detect adjusting river channels and to differentiate areas that are more or less sensitive to adjustment. This is particularly important as it is now known that river channel adjustments can be very extensive and Brookes & Gregory (1988) included reference to 137 studies in different parts of the world that demonstrate the magnitude and significance of river channel adjustments. Methods are therefore required to identify channels that are sensitive and this has been achieved for planform changes by using historical techniques which involve comparison of surveys of different dates. However, the examples of the Monks Brook and the Roding show how morphological, ecological and structural indicators can be employed to identify changes of channel cross-section, and by analysing such changes along reaches it is possible to demonstrate spatial patterns of river channel adjustment. Ideally the two approaches, namely planform change at the medium or long term and cross-section change at the short term, need to be combined.

River channel sensitivity is relevant to river channel management because it is necessary

Table 3.4   Characteristic channel adjustments in the Roding basin

Channel condition	Description	Number of reaches	Percentage length
S	no morphological activity	22	16.1
D	deposition on both banks and bed	9	5.4
M	migration of most bends	16	14.0
E	enlargement of both banks and bed	10	3.5
d	less intense deposition	18	19.1
m	less intense migration	20	20.0
e	less intense enlargement	7	7.8
R	migration and deposition	6	5.1
U	enlargement and migration	10	6.9
C	enlargement and deposition	3	1.3
culverts	channel not visible	5	0.9
Total	(length surveyed 111.45 km)	126	100.1

to know which channels are adjusting and which may be susceptible to adjustment in the future. For this purpose it is important to recognise the distinction between the different types of sensitivity summarised in Table 3.2. As river management increasingly works with the river and not against it (Winkely, 1972) and there is an increasing trend (e.g. Brookes, 1990) to design channels that imitate natural conditions, it is important to acknowledge that adjustment is now an integral part of the character of many river channels. It is therefore necessary to have an awareness of what may happen in particular sections of river channels as well as what is happening at the present time, and it could prove to be most effective to be aware of potential channel changes when river management schemes are planned and designed. The methods for mapping channel adjustment applied in this paper to the Monks Brook and to the river Roding afford techniques that can be employed to differentiate reaches of river channel according to their sensitivity to adjustment and, as such, may assist towards the greater input of hydrogeomorphological understanding in holistic river management.

ACKNOWLEDGEMENTS

One of the authors (P. W. Downs) is grateful to the NERC for the provision of a studentship CASE with the National Rivers Authority (Thames) when some of the research reported on the Roding was undertaken, and the assistance of Mr R. J. Davis with fieldwork for the Monks Brook is gratefully acknowledged.

REFERENCES

Anderson, M. G. and Sambles, K. M. (1988). A review of the bases of geomorphological modelling. In Anderson, M. G. (ed.), *Modelling Geomorphological Systems*, Wiley, Chichester, pp. 1–32.
Baker, V. R., Kochel, R. C. and Patton, P. C. (eds) (1988). *Flood Geomorphology*, Wiley/Interscience, New York.
Bathurst, J. C. (1986). Sensitivity analysis of the Système Hydrologique Européen for an upland catchment. *Journal of Hydrology*, **87**, 103–123.
Begin, Z. B. and Schumm, S. A. (1984) Gradational thresholds and landform singularity: significance for Quaternary studies. *Quaternary Research*, **31**, 267–274.

Beven, K. J. (1979). A sensitivity analysis of the Penman–Monteith actual evapotranspiration estimates. *Journal of Hydrology*, **44**, 169–190.

Brandt, M., Bergstrom, S. and Sanden, P. (1987). Environmental impacts of an old mine tailings deposit: modelling of water balance, alkalinity and pH. *Nordic Hydrology*, **18**, 291–300.

Brookes, A. (1987). River channel adjustments downstream from channelization works in England and Wales. *Earth Surface Processes and Landforms*, **12**, 337–351.

Brookes, A. (1988). *Channelized Rivers: Perspectives for Environmental Management*, Wiley, Chichester.

Brookes, A. (1990). Restoration and enhancement of engineered river channels: some European experiences. *Regulated Rivers: Research and Management*, **5**, 45–56.

Brookes, A. and Gregory, K. J. (1988). Channelization, river engineering and geomorphology. In Hooke, J. M. (ed.), *Geomorphology in Environmental Planning*, Wiley, Chichester, pp. 145–167.

Brunsden, D. and Thornes, J. B. (1979). Landscape sensitivity and change. *Transactions of the Institute of British Geographers*, **NS4**, 463–484.

Buckley, R. (1991). *Perspectives in Environmental Management*, Springer-Verlag, Berlin.

Burkham, D. E. (1972). Channels changes of the Gila river in Safford valley, Arizona, 1846–1970. *United States Geological Survey, Professional Paper*, 655G.

Chaterjee, S. and Hadi, A. S. (1988). *Sensitivity Analysis in Linear Regression*, Wiley, New York.

Chiew, F. H. S. and McMahon, T. A. (1990). Estimating groundwater recharge using a surface watershed model: sensitivity analysis, *Journal of Hydrology*, **114**, 305–325.

Chorley, R. J., Schumm, S. A. and Sugden, D. E. (1984). *Geomorphology*, Methuen, London.

Coleman, G. and De Coursey, D. G. (1976). Sensitivity and model variance analysis applied to some evaporation and evapotranspiration models. *Water Resources Research*, **12**, 873–879.

Cooke, R. U. and Doornkamp, J. C. (1990). *Geomorphology in Environmental Management: A New Introduction*, Oxford University Press, Oxford.

Cunningham, A. B. and Sinclair, P. J. (1979). Application and analysis of a coupled surface and groundwater model. *Journal of Hydrology*, **43**, 129–148.

Downs, P. W., Gregory, K. J. and Brookes, A. (1991). How integrated is river basin management? *Environmental Management*, **15**, 299–309.

Graf, W. L. (1977). The rate law in fluvial geomorphology. *American Journal of Science*, **277**, 178–191.

Graf, W. L. (1979). Mining and channel response. *Annals of the Association of American Geographers*, **69**, 262–275.

Graf, W. L. (1980). The effect of dam closure on downstream rapids. *Water Resources Research*, **16**, 129–136.

Graf, W. L. (1982). Spatial variation of fluvial processes in semi-arid lands. In Thorn, C. E. (ed.), *Space and Time in Geomorphology*, George Allen & Unwin, London, pp. 193–217.

Graf, W. L. (1988). *Fluvial Processes in Dryland Rivers*, Springer-Verlag, Berlin.

Gregory, K. J. (1977). Channel and network metamorphosis in Northern New South Wales. In Gregory, K. J. (ed.), *River Channel Changes*, Wiley, Chichester, pp. 389–410.

Gregory, K. J. (1980). River channels. In Gregory, K. J. and Walling, D. E. (eds), *Man and Environmental Processes*, Dawson, Folkestone, pp. 123–143.

Gregory, K. J. (1987a). River channels. In Gregory, K. J. and Walling, D. E. (eds), *Human Activity and Environmental Processes*, Wiley, Chichester, pp. 207–235.

Gregory, K. J. (1987b). Environmental effects of river channel changes. *Regulated Rivers: Research and Management*. **1**, 358–363.

Gregory, K. J. (1987c). Hydrogeomorphology of Alpine proglacial areas. In Gurnell, A. M. and Clark, M. J. (eds), *Glacio-Fluvial Sediment Transfer*, Wiley, Chichester, pp. 87–107.

Gregory, K. J. (1992). Vegetation and river channel process interactions. In Boon, P. J. Calow, P. and Petts, G. E. (eds), *River Conservation and Management*, Wiley, Chichester, pp. 255–269.

Gregory, K. J. and Madew, J. R. (1982). Land use change, flood frequency and channel adjustments. In Hey, R. D., Bathurst, J. C. and Thorne, C. R. (eds), *Gravel Bed Rivers: Fluvial Processes, Engineering and Management*, Wiley, Chichester, pp. 757–782.

Gregory, K. J., Davis, R. J. and Downs, P. W. (1992). Identification of river channel change due to urbanisation. *Applied Geography*, **12**, 299–318.

Gupta, A. (1984). Urban hydrology and sedimentation in the human tropics. In Costa, J. .E. and Fleischer, P. J. (eds), *Developments and Applications of Geomorphology*, Springer-Verlag, Berlin, pp. 240–267.

Gupta, A. and Fox, H. (1974). Effects of high-magnitude floods on channel form: a case study in Maryland Piedmont. *Water Resources Research,* **10**, 499−509.

Hey, R. D. (1978). Determinate hydraulic geometry of river channels. *Journal of Hydraulics Division, Proceedings of the American Society of Civil Engineers,* **104**, 869−885.

Hey, R. D. (1982). Gravel bed rivers: form and processes. In Hey, R. D., Bathurst, J. C. and Thorne, C. R. (eds), *Gravel Bed Rivers: Fluvial Processes, Engineering and Management,* Wiley, Chichester, pp. 5−13.

Hey, R. D. (1986). River mechanics. *Journal of the Institution of Water Engineers and Scientists,* **40**, 139−159.

Hey, R. D. (1988). Bar form resistance in gravel bed rivers. *Journal of Hydraulic Engineering, Proceedings of the American Society of Civil Engineers,* **114**, 1498−1508.

Hooke, J. M. and Kain, R. J. P. (1982). *Historical Change in the Physical Environment: A Guide to Sources and Techniques,* Butterworths, London.

Karl, T. R. and Riebsame, W. E. (1989). Impact of decadal fluctuations in mean precipitation and temperature on runoff: a sensitivity study over the United States, *Climatic Change,* **15**, 433−447.

Knighton, A. D. (1984). *Fluvial Forms and Processes,* Edward Arnold, London.

Langbein, W. B. and Leopold, L. B. (1964). Quasi-equilibrium states in channel morphology. *American Journal of Science,* **262**, 782−794.

Lewin, J. (1977). Channel pattern changes. In Gregory, K. J. (ed.), *River Channel Changes,* Wiley, Chichester, pp. 167−184.

Lewin, J., Macklin, M. G. and Newson, M. D. (1988). Regime theory and environmental change — irreconcilable concepts? In White, W. R. (ed.), *International Conference on River Regime,* Hydraulics Research Ltd, Wallingford.

McCuen, R. H. (1973). The role of sensitivity analysis in hydrologic modelling. *Journal of Hydrology,* **18**, 37−53.

McCuen, R. H. (1974). A sensitivity and error analysis of procedures used for estimating evaporation. *Water Resources Bulletin,* **10**, 486−498.

Melack, J. M., Stoddard, J. L. and Ochs, C. A. (1985). Major ion chemistry and sensitivity to acid precipitation of Sierra Nevada lakes. *Water Resources Research,* **21**, 27−32.

Melton, M. A. (1958). Correlation structure of morphometric properties of drainage systems and their controlling agents. *Journal of Geology,* **66**, 442−460.

Mosley, M. P. (1987). The classification and characterisation of rivers. In Richards, K. S. (ed.), *River Channels: Environment and Process,* Blackwell, Oxford, pp. 295−320.

Nanson, G. C. and Young, R. W. (1981). Downstream reduction of channel size with contrasting urban effects in small coastal streams of south eastern Australia. *Journal of Hydrology,* **55**, 239−255.

Newson, M. D. (1980). The geomorphological effectiveness of floods — a contribution stimulated by two recent events in mid-Wales. *Earth Surface Processes,* **5**, 1−16.

Oxford English Dictionary, 2nd edn (1989) Clarendon Press, Oxford.

Patton, P. C. and Schumm, S. A. (1975). Gully erosion, northern Colorado: a threshold phenomenon. *Geology,* **3**, 88−90.

Petts, G. E. (1984) *Impounded Rivers: Perspectives for Ecological Management,* Wiley, Chichester.

Prince, K. R., Reilly, T. W. and Franke, O. L. (1989). Analysis of the shallow groundwater flow system near Connetquot Brook, Long Island, New York. *Journal of Hydrology,* **107**, 223−250.

Rodgers, C. C. M., Beven, K. J., Morris, E. M. and Anderson, M. G. (1985). Sensitivity analysis, calibration and predictive uncertainty of the Institute of Hydrology Distributed Model. *Journal of Hydrology,* **81**, 179−191.

Schumm, S. A. (1969). River metamorphosis, *Journal of the Hydraulics Division, Proceedings of the American Society of Civil Engineers,* **95**, 255−273.

Schumm, S. A. (1973). Geomorphic thresholds and complex response of drainage systems. In Morisawa, M. (ed.), *Fluvial Geomorphology,* Binghampton Publications in Geomorphology, SUNY, pp. 299−309.

Schumm, S. A. (1979). Geomorphic thresholds: the concept and its applications. *Transactions of the Institute of British Geographers,* **NS4**, 485−515.

Schumm, S. A. (1985). Explanation and extrapolation in geomorphology, seven reasons for geological uncertainty. *Transactions of the Japanese Geomorphological Union,* **6-1**, 1−18.

Schumm, S. A. (1988). Geomorphic hazards — problems of prediction. *Zeitschrift für Geomorphologie,*

67, 17−14.

Schumm, S. A. (1991). *To Interpret the Earth: Ten Ways to be Wrong*, Cambridge University Press, Cambridge.

Starkel, L. (ed.) (1987). *Evolution of the Vistula Valley during the last 15000 years, Vol. II: Geographical Studies*, Polish Academy of Sciences, Institute of Geography and Spatial Organisation, Warsaw, Special Issue no. 4.

Thomas, R. W. and Huggett, R. J. (1980). *Modelling in Geography: A Mathematical Approach*, Harper & Row, London.

Toy, T. J. and Hadley, R. F. (1987). *Geomorphology and the Reclamation of Disturbed Lands*, Academic Press, London.

Wharton, G., Arnell, N. W., Gregory, K. J. and Gurnell, A. M. (1989). River discharge estimated from channel dimensions. *Journal of Hydrology,* **106**, 365−376.

Whitlow, J. R. and Gregory, K. J. (1989). Change in urban stream channels in Zimbabwe. *Regulated Rivers: Research and Management,* **4**, 27−42.

Williams, G. P. (1978). The case of the shrinking channels — the North Platte and Platte Rivers in Nebraska, *United States Geological Survey, Circular* 781.

Williams, G. P. and Wolman, M. G. (1984). Downstream effects of dams on alluvial rivers. *United States Geological Survey, Professional Paper* 1286.

Winkley, B. R. (1972). River regulation with the aid of nature. *International Commission on Irrigation and Drainage*, Eighth Congress Verna Transactions, Vol. V, 29.1, pp. 433−437.

Wolman, M. G. and Gerson, R. A. (1978). Relative scales of time and effectiveness of climate in watershed geomorphology. *Earth Surface Processes,* **3**, 189−208.

4 Use of Caesium-137 Measurements to Investigate Relationships between Erosion Rates and Topography

TOM A. QUINE and DES E. WALLING
Department of Geography, University of Exeter, UK

ABSTRACT

The costs associated with both the on-site and off-site impacts of soil loss from agricultural land highlight the importance of the identification of landscape sensitivity to erosion. Sensitivity of the landscape to erosion at the micro-scale has been addressed in the present study by assessment of topographic controls upon the processes of erosion and deposition. The caesium-137 (^{137}Cs) technique is particularly suited to this task because of the point-specific nature of the long-term erosion rate estimates obtained and the high degree of spatial resolution that is achievable. In this study the pattern of ^{137}Cs-derived soil redistribution rates has been compared with the site topography for five arable fields on a range of soil types from southern UK. Topographic attributes (elevation, slope angle, profile curvature, planform curvature, upslope distance and upslope area) for individual sampling points were extracted from a digital terrain model of each site. In order to identify the topographic controls upon the processes of erosion and deposition, data sets of point erosion rates and point deposition rates were separately regressed against the selected topographic attributes for the individual sampling points. The feasibility of producing multivariate erosion prediction equations for each site, based upon a selection of topographic variables, was examined by regressing point soil redistribution rates (including both erosion and deposition) against the topographic attributes. Finally the results of the regression analysis were used to identify topographically-defined sensitive areas on the soil types studied.

THE CONTEXT

Growing awareness of both the on-site and off-site problems associated with soil loss from agricultural land (e.g. Crosson & Stout, 1983; Clark, Havercamp & Chapman, 1985) has been accompanied by a recognition of the need to predict those areas of the landscape that are most sensitive or susceptible to erosion. The identification of such sensitive areas is an essential prerequisite for the development of effective land management and soil conservation strategies for maintaining soil productivity and reducing downstream sediment yields. Identification of sensitive areas may be approached at several spatial scales. At the macro-scale, attention may focus on those areas within a particular region that are most susceptible to erosion under specified land use practices, as a result of regional variations in soil, topographic, hydrological and climatic conditions (cf. Boardman and Burt, Heathwaite & Trudgill this volume). At the meso-scale, concern may be with identifying the relative erosion

Landscape Sensitivity. Edited by D. S. G. Thomas and R. J. Allison
© 1993 John Wiley and Sons Ltd

risk associated with individual fields or management units in response to local topography and soil types. At the micro-scale, it may be necessary to identify those areas *within* a specific field or small area that are particularly susceptible to erosion. At this scale, soil conditions will exhibit a greater degree of spatial uniformity and topographic position could be expected to exert a primary control. Soil loss prediction procedures such as the USLE (Wischmeier & Smith, 1978) provide a means of identifying erosion-susceptible areas at the macro- and meso-scale. However, these are less effective at the micro-scale, since the topographic factors employed are rather crude, providing only a measure of average slope length and steepness. At this scale it is necessary also to consider variables such as the profile and planform curvature of the slope and the upslope area, in order to locate the zones most likely to erode. Other approaches are required.

The two approaches that have been most frequently employed to date in order to investigate micro-scale sensitivity to erosion are field observation and model-based prediction. Field monitoring of carefully selected control plots can provide a basis for developing relationships between topographic parameters and erosion incidence, but extrapolation from these data and relationships to the complex topography of the agricultural landscape presents many problems (cf. Roels, 1985). More direct observations or surveys of the spatial distribution of erosion on cultivated fields, such as those undertaken by Evans (1988) and Boardman (1990), provide data that are more readily applicable to the assessment of within-field controls on soil erosion. However, such surveys rely on visual identification of eroded areas and visual assessment of spatial variations in severity of erosion. They are most useful for delimiting areas subject to erosion rather than defining spatial variation in erosion rates, and the low resolution of the resultant data limits their use in detailed analysis of spatial patterns of erosion rates and their topographic control. A further problem associated with both field measurement approaches is the need for repeat or long-term measurements, in order to distinguish essentially random variations in the spatial incidence of erosion from those reflecting consistent topographic control. Such long-term measurements may be expensive and difficult to maintain and it may be necessary to undertake these measurements in a variety of local areas in order to provide sufficient information for generalisation.

The problems of spatial resolution and expense associated with the field measurement approaches outlined above may to some extent be overcome by the use of model-based prediction. The application of theoretical relationships between erosion rates and topographic parameters allows the identification of sensitive areas to be based on surrogate variables that are relatively easy to measure or derive. Moore, O'Loughlin & Burch (1988) have, for example, developed a predictive procedure based on a topographic model coupled with an assessment of the transport capacity of surface runoff which demonstrates the potential of this approach to generate high-resolution spatial patterns of erosion susceptibility from easily measured topographic data. However, the patterns predicted by these models are only as reliable as the functions and relationships used in their derivation and the need for field calibration and verification remains an important constraint in their effective use. The problems associated with field measurements noted above provide a situation where it is difficult, if not impossible, to obtain field data of sufficient spatial resolution for model verification.

Recent advances in the use of ^{137}Cs measurements for documenting rates and patterns of soil erosion on agricultural land (cf. Loughran, Campbell & Walling, 1987; Walling & Quine, 1990) may provide a means of overcoming many of the limitations of existing field measurement techniques for identifying erosion-sensitive areas at the micro-scale and in turn for providing data suitable for more effective verification and calibration of theoretical models.

For example, Pennock & De Jong (1987) have demonstrated the potential for using ^{137}Cs-derived erosion rates in the examination of the influence of slope curvature on soil erosion and deposition. This paper explores this approach further by presenting the results of a preliminary assessment of the potential for using ^{137}Cs measurements to identify topographic controls on within-field variations in erosion rates in several areas of arable cultivation in lowland Britain.

THE CAESIUM-137 TECHNIQUE

Background and Potential

Caesium-137 was produced in significant quantities as a by-product of the atmospheric testing of nuclear weapons during the period from the mid-1950s until the mid-1970s. As a result of such weapons tests, ^{137}Cs was distributed globally in the stratosphere and deposited as fallout. The value of the radionuclide as an environmental tracer of sediment movement lies in the fact that in most environments fallout arriving at the land surface is rapidly and strongly adsorbed by soil particles, and particularly the clay fraction (Bachhuber *et al.*, 1982; Livens & Baxter, 1988; Squire & Middleton, 1966). Experimental studies, for example, have shown that clay minerals extract caesium very efficiently from solutions with 0.001 M concentration (Livens & Loveland, 1988; Klobe & Gast, 1970; Sawnhey, 1964, 1966). Because the concentrations of ^{137}Cs in rainfall during the main period of atmospheric deposition of weapons-test fallout were much lower than those employed in the experimental studies, very efficient fixing of atmospherically deposited ^{137}Cs can be assumed. After its initial deposition, subsequent redistribution of radiocaesium is therefore associated with the erosion and transport of soil particles. Direct evidence for the particle-associated movement of ^{137}Cs is provided by profile distributions of the radionuclide collected from cultivated sites. Where elevated ^{137}Cs inventories exist, the ^{137}Cs profile is invariably found to extend below the current plough depth at comparable concentrations to those found within the plough layer. In contrast, at sites where the ^{137}Cs inventories are indicative of erosion or stability, the concentration of ^{137}Cs in the soil declines rapidly to zero below the plough depth (Walling & Quine, 1990). This situation is consistent with the proposition that elevated inventories of ^{137}Cs reflect the deposition of ^{137}Cs-bearing sediment, rather than lateral movement of ^{137}Cs prior to particle association.

The potential for the use of ^{137}Cs in the study of sediment movement was identified as early as 1965 by Rogowski & Tamura (1965) and the technique has subsequently become established as a powerful tool in erosion research (Loughran, 1989). Recent discussions of the application of the ^{137}Cs technique to the study of erosion on arable fields in the UK (Walling & Quine, 1990, 1991) have highlighted some of the advantages of the approach. In the present context the most important aspects are as follows:

1. The erosion rate estimates provided are point-specific.
2. The erosion rate estimates represent long-term averages for the period extending from 1954 to the time of sampling.
3. Erosion rates can be estimated retrospectively on the basis of a single site visit.
4. The degree of spatial resolution is constrained only by the sampling density, not by the technique.

With its potential for estimating long-term (~ 35 years) average erosion rates with a high degree of spatial resolution on the basis of a single site visit, the ^{137}Cs technique must be seen as possessing considerable potential as a tool in the study of within-field variation in susceptibility to erosion and in the identification of major controlling variables (Pennock & De Jong, 1987). This potential may be demonstrated by analysis of data collected in a recent survey of soil erosion on arable land in the UK.

Erosion Rate Data Derived from Caesium-137 Measurements

During the course of a reconnaissance survey of soil erosion on arable land in the UK, intensive ^{137}Cs sampling was undertaken on 13 fields. The approach employed has been described by Walling & Quine (1990, 1991) and the erosion rate estimates from all the sites have been summarised elsewhere (Quine & Walling, 1991). For each site investigated, two data sets are available. First, approximately 100 point estimates of soil redistribution rate were obtained by applying calibration relationships which relate the degree of departure of the point ^{137}Cs inventory from the local fallout reference inventory to the rate of erosion or aggradation. The calibration relationships employed in the present study take into account spatial variation in the nature of erosion, the frequency of erosion events and the potential for particle-size differentiation and variation in the ^{137}Cs content of eroded and deposited sediment. The resultant point estimates of soil redistribution rate may be used to establish the pattern of erosion and aggradation within the study fields. Secondly, at the time that the soil cores needed for the ^{137}Cs measurements were collected, spot height data were obtained for each sampling point. These data can be used to produce a digital terrain model (DTM) for each study area. Combination of these two data sets provides a basis for examining topographic controls upon within-field variation in soil erosion and aggradation rates, and thereby micro-scale sensitivity to erosion. In order to investigate these controls, the data from a sample of five of the 13 sites have been subject to further analysis, which is described in the following discussion. The sites were selected to represent a range of soil textures, extending from sands to clays (Table 4.1).

Table 4.1   Study sites — soil types and redistribution ranges

Site and county	Soil association and type	^{137}Cs-derived soil redistribution rates (tonnes ha^{-1} year^{-1}	
		Range	Inter-quartile range
Rufford Nottinghamshire	551b Cuckney 1 Typical Brown Sand	$-62-+58$	$-18--3$
Dalicott Shropshire	551a Bridgnorth Typical Brown Sand	$-72-+57$	$-14--2$
Lewes Sussex	343h Andover 1 Brown Rendzina	$-36-+34$	$-6-+2$
Fishpool Gwent	571b Bromyard Typical Argillic Brown Earth	$-37-+25$	$-7-+4$
Keysoe Bedfordshire	411d Hanslope Typical Calcareous Pelosol	$-16-+12$	$-3-+2$

   The analysis undertaken may be conveniently subdivided into three stages: (i) topographic characterisation; (ii) examination of relationships between soil redistribution and topography; and (iii) development of multivariate predictive equations. All statistical analyses were performed using SPSS/PC+.

TOPOGRAPHIC CHARACTERISATION

Topographic characterisation involved both selection and derivation of topographic attributes that may exercise control over rates of soil redistribution. If soil erosion by water is considered to be controlled by the processes of detachment and transport (cf. Meyer & Wischmeier, 1969), then the role of topography will primarily reflect its influence on the detachment and transport capacity of surface runoff and therefore on runoff amount and energy. This is reflected in the incorporation of slope length and steepness in meso- and macro-scale predictive procedures such as the USLE (Wischmeier and Smith, 1978). The attributes slope angle and upslope distance were therefore also selected for this study. However, at the micro-scale, runoff amount may be better reflected by the area upslope of the point under study, so this variable was also included. Visual examination of the pattern of measured erosion rates from study fields identified slope crests as areas susceptible to erosion, a finding consistent with field observations by Boardman & Robinson (1985) on the South Downs. In order to highlight areas such as these, which are characterised by a high rate of change of slope angle, both profile and planform curvature were added to the list of variables. Profile curvature is the rate of change of slope along the line of greatest slope angle, and planform curvature is the rate of change of slope along a line perpendicular to the greatest slope. Downslope and cross-slope convexities are characterised by high positive profile curvature and high negative planform curvature, respectively. The final list of topographic attributes selected was as follows:

- upslope distance;
- upslope area;
- slope angle;
- profile curvature;
- planform curvature.

Within this list, the first two parameters may be broadly identified with runoff amount and the remaining three with the runoff energy, though this is clearly a somewhat artificial distinction. It was possible to derive all of the topographic parameters directly from the DTM for each site using software developed by Zevenbergen & Thorne (1987).

   It would have been desirable to examine the influence of grid size on the spatial distribution of topographic variables within the study fields, in order to define the optimum grid size for the analysis. However, the grid spacing was predefined by the field sampling programme carried out previously. The spacing had been selected as providing adequate definition of the local topography and was therefore considered to be appropriate in the current analysis. The grid spacing was 20 m for all sites except Lewes, where a 25 m spacing was employed. The relationships between the topographic variables are presented in Table 4.2, which lists the range of correlation coefficients between the pairs of variables found at the five study sites. The clearest feature is the high degree of positive correlation between the upslope area and upslope distance. As might be expected, these variables also show a fairly consistent

Table 4.2 Correlation matrix for topographic attributes; ranges of correlation coefficients for five sites

Attribute	Elevation (m)	Slope angle (degrees)	Profile curvature	Planform curvature (degrees)/100 m	Upslope area (m²)	Upslope distance (m)
Elevation		−0.005 to −0.74	0.43 to 0.81	−0.20 to −0.48	−0.43 to −0.51	−0.45 to −0.78
Slope angle			0.02 to −0.004 to −0.14	0.21 to 0.05 to −0.18	0.30 to 0.01 to −0.29	0.27 to 0.06 to −0.11
Profile curvature				−0.24 to −0.74	−0.12 to −0.37	−0.15 to −0.52
Planform curvature					0.16 to 0.53	0.09 to 0.58
Upslope area						0.66 to 0.90
Upslope distance						

negative correlation with elevation. The range of the correlation coefficients shown by the other pairs of parameters evidences substantial differences between sites in the degree of variable interdependence. This must be recognised when examining the extent of topographic explanation of soil redistribution provided by individual parameters.

TOPOGRAPHIC CONTROLS

Correlation with Redistribution Rate

In this analysis the topographic attributes of the sampling points were correlated with the soil redistribution rates for the same points. Negative redistribution rates are indicative of erosion and positive rates denote aggradation. Those relationships where a significance level of 90% was surpassed for a two-tailed test are listed in Table 4.3. The pattern of correlations reveals a number of important features. The first and clearest feature is a contrast between sites. The site at Keysoe on the clay soil differs from the other sites in the absence of correlation with either profile or planform curvature, and also in the sign of the correlations with upslope area and upslope distance. The second feature is the occurrence of a statistically significant correlation between soil redistribution rate and planform and profile curvature and upslope area at four of the five sites. Finally, and in contrast, slope angle and upslope distance, which have often been used as variables in erosion assessment studies, show statistically significant correlations at only two and three sites, respectively.

The implications of the correlation analysis, and particularly the features noted above, must be considered. First, if the intensity of erosion was proportional to the runoff amount, then soil redistribution, as defined above, should be negatively correlated with the upslope parameters. This is only the case at Keysoe on the clay soil. In contrast, a positive correlation is found between redistribution rate and upslope area at Fishpool, Rufford and Lewes, and with upslope distance at both of the latter sites. This suggests that at these sites points with maximum upslope area and distance are typified by aggradation rather than erosion, and that erosion takes place at the top of the slopes, implying that runoff amount is not a limiting factor.

Table 4.3 Statistically significant correlations with soil redistribution rate at the 95% level; figures in parentheses are significant at the 90% level

Soil and site	Elevation (m)	Slope angle (degrees)	Profile curvature	Planform curvature	Upslope area (m^2)	Upslope distance (m)	No.
			(degrees/100 m)				
Sandy							
Rufford	−0.47	−0.32	−0.52	0.34	0.48	0.51	117
Dalicott	−0.34		−0.62	0.45			83
Chalky							
Lewes	−0.35	−0.21	−0.44	0.27	0.27	0.33	99
Silty							
Fishpool			(−0.19)	0.32	(0.19)		99
Clay							
Keysoe					−0.30	(−0.18)	96

Turning attention to those variables broadly associated with runoff energy, the most striking feature is the occurrence of statistically significant correlations between profile and planform curvature and soil redistribution rate at all four sites on the sandy, chalky and silty soils. The signs of the correlation coefficients for both parameters reflect a pattern of erosion from convexities and aggradation in concavities, both down and across the slope. Under conditions of overland flow, profile convexities should be associated with increasing rates of flow downslope. Planform convexities are thought to be associated with flow dispersion, and therefore potential reduction in downslope flow, but increase in rate of lateral flow. In contrast the planform concavities are associated with flow convergence and therefore increasing runoff amounts. Erosion from both profile and planform convexities suggests that increase in rate and energy of runoff, whether downslope or lateral, is more important than runoff amount. The high correlation of erosion with convexities, which are situated toward the top of the slope, may partially explain the lower correlation with slope angle. High rates of erosion at the top of the slope may result in the transport capacity of the runoff being exceeded above the areas of highest slope angle, with consequent lower rates of erosion from such zones than might be expected on the basis of predicted runoff energy.

The relationships identified between the topographic variables and soil redistribution rates have provided some insight into the processes operating at the individual study sites and have demonstrated that different controls operate on soils of different textures. This is seen most clearly in the contrast between the site at Keysoe on the clay soil where erosion is preferentially located in depressions, characterised by high upslope area and distance, and the remaining sites on the sandy, silty and chalky soils where depressions are associated with aggradation and maximum erosion occurs on slope convexities, characterised by a high degree of slope curvature. Only at Keysoe do rates of erosion appear to be controlled by runoff amount. At all the other sites runoff energy appears to be more important. However, the conclusions must be tentative because of the small number of fields examined. Nevertheless it may be possible to obtain further evidence of the influence of topography by separating eroding and aggrading points. This will be considered in the following section.

Erosion and Aggradation — Controlling Variables

If different sets of processes promote erosion and aggradation, then it might be expected that they will be reflected in control by different parameters. Each set of eroding and aggrading points was therefore separately correlated with the topographic attributes and the results are summarised in Table 4.4. An additional indication of the dominant controlling topographic variables and the degree of explanation provided by the topographic parameters was obtained by subjecting the separate data sets to stepwise multiple regression, with erosion rate and aggradation rate as the dependent variable. At Lewes no independent variables were included using the stepwise approach with the aggradation data set, and at Fishpool no independent variables were included with either the erosion or the aggradation data sets. In these cases the equations were produced using the backward method, which imposes less stringent demands upon variable inclusion. The beta coefficients and adjusted R^2 values for the equations are listed in Table 4.5. The adjusted R^2 statistic is used as this attempts to correct the overestimate of goodness of fit that may be produced by use of multiple independent variables. The results listed in Tables 4.4 and 4.5 tend to confirm the notion of contrasting controlling variables for erosion and aggradation, as no variable is found in both regression equations at one site. However, a number of aspects merit specific comment.

Table 4.4 Statistically significant correlations with soil erosion and aggradation rates at the 95% level; figures in parentheses are significant at the 90% level

Soil and site	Elevation (m)	Slope angle (degrees)	Profile curvature	Planform curvature (degrees/100 m)	Upslope area (m^2)	Upslope distance (m)	No.
Sandy							
Rufford							
Erosion	−0.34	−0.26	−0.43	0.36	(0.19)	0.27	103
Aggradation					0.77	0.76	14
Dalicott							
Erosion			−0.49	0.35			67
Aggradation	−0.53						16
Chalky							
Lewes							
Erosion	−0.27		−0.47			(0.21)	68
Aggradation							31
Silty							
Fishpool							
Erosion				(0.21)			63
Aggradation					(0.30)		36
Clay							
Keysoe							
Erosion					−0.29	(−0.23)	52
Aggradation		(0.28)	0.35				44

First, examination of the variables controlling erosion on the sandy, chalky and silty sites reaffirms the importance of slope curvature. A statistically significant corelation between erosion rates and profile curvature is found at both sandy sites and the chalky site, and between erosion rates and planform curvature at both sandy sites and the silty site (Table 4.4). No other parameters show statistically significant correlations with erosion rates at more than two of these sites, and slope angle is correlated with erosion rates at Rufford alone. Inclusion of variables in the multiple regression equations and their beta coefficients (Table 4.5) confirm the dominant role of profile curvature in controlling erosion rates at the sandy (Rufford and Dalicott) and chalky (Lewes) sites. The adjusted R^2 values indicate that variation in profile curvature at these sites explains 17−23% of the variation in erosion rates, and only at Rufford is a significant improvement in explanation, from 17 to 27%, achieved by inclusion of slope angle and planform curvature. This strengthens the previous suggestion that on sandy and chalky soils increase in runoff energy in areas of high profile curvature is more important than runoff amount.

The dominance of planform curvature at Fishpool reflects the nature of the local topography, which is characterised by downslope spurs separating within-field depressions. Such areas display a high degree of variation in planform curvature, with relatively low levels of profile curvature. However, the recognition of planform convexities as the areas most susceptible to erosion reaffirms the suggestion that on this soil texture also, runoff energy is more important than runoff amount. Despite this, the low adjusted R^2 values for both erosion and aggradation at Fishpool demand consideration. One possible explanation is the presence of a number of

Table 4.5 Beta coefficients for variables included in stepwise multiple regression equations; figures in parentheses refer to the backward multiple regression procedure. The dependent variables are erosion and aggradation rates

Soil and site	Elevation (m)	Slope angle (degrees)	Profile curvature	Planform curvature	Upslope area (m²)	Upslope distance (m)	Adjusted R²
			(degrees/100 m)				
Sandy							
Rufford							
Erosion		−0.27	−0.34	0.20			0.27
Aggradation					0.77		0.55
Dalicott							
Erosion			−0.49				0.23
Aggradation	−0.53						0.22
Chalky							
Lewes							
Erosion			−0.47				0.21
Aggradation		(−0.21)					(0.01)
Silty							
Fishpool							
Erosion				(0.21)			(0.03)
Aggradation					(0.30)		(0.07)
Clay							
Keysoe							
Erosion					−0.29		0.07
Aggradation			0.35				0.10

high erosion and aggradation rates at Fishpool which introduce a degree of non-linearity into the relationship.

Secondly, the data presented in Tables 4.4 and 4.5 reaffirm the distinction in controlling variables, and by implication processes, between the Keysoe site and the others. At Keysoe, the only statistically significant relationships between erosion rates and the topographic variables are negative correlations with upslope area and upslope distance, and the former is the only variable selected in the regression equation. Again this confirms the pattern noted in the previous section and suggests that at Keysoe the processes of erosion may be topographically controlled and located in zones of high upslope area and distance, where runoff amounts may be expected to be maximised. This could, however, also reflect runoff generation from saturated areas in the lowest lying ground.

Finally, it is necessary to consider the factors influencing aggradation. There is a wide range in the adjusted R^2 values for the multiple regression equations, from 0.01 to 0.55, but no clear textural pattern is apparent either in the magnitude of these values or the inclusion of variables in the regression equations. However, examination of Table 4.4 reveals a positive correlation between aggradation rate and upslope area at Fishpool, and with both upslope parameters at Rufford. At Dalicott only elevation shows a statistically significant correlation (negative) with aggradation rate. No statistically significant correlation between aggradation rate and the topographic parameters is found at Lewes, but Table 4.3 shows a positive correlation between soil redistribution rate and both upslope area and upslope distance. All these relationships point toward a concentration of aggradation at the base of slopes, in valley-bottoms or within-slope depressions, at these sites. The range of adjusted R^2 values in part reflects the incidence of such locations within the study sites. For example, the site at Rufford enclosed a well-defined valley bottom, which doubtless contributed to the high degree of explanation. In addition, on the sandy soils at Rufford and Dalicott the transported sediment will be dominated by coarse material, which will be readily deposited in response to a reduction in flow transport capacity in valley bottom or slope base locations. This may explain the higher degree of explanation of aggradation for these sites, than for those on other textures.

Although a tentative association between aggradation and valley bottoms may be proposed at Lewes, the most striking feature about this site is the very low degree of explanation afforded by the topographic variables. This suggests that at this site, on chalky soil, aggradation is controlled by factors other than topography. A further complicating factor may be the occurrence of episodic erosion of the valley floor in extreme events, observed at chalky sites such as this. These episodes may lead to dispersal of the deposited sediment and so mask the topographic controls upon earlier aggradation.

As in the case of controls on erosion, the Keysoe site shows a marked contrast with the other four sites in its pattern of aggradation. Statistically significant correlations are found between aggradation rates and both slope angle and profile curvature. In both cases the correlations are positive, indicating a tendency for aggrading areas to be situated on convexities on relatively steep slopes. Explanation of this pattern is problematic. However, the large fields now found in this area are the result of the amalgamation of smaller units. Sediment deposition may have taken place at the downslope edge of the previous smaller units, when separated by hedges. The line of such former hedges would therefore be associated with both long-term aggradation and slight slope discontinuities, characterised by relatively high profile curvature. However, this explanation is only tentative and requires further investigation.

Redistribution Patterns and the Evidence of Residuals

The regression equations developed for erosion and aggradation may be used to calculate the variation in the rates of soil redistribution that may be attributed to these topographic controls. If the ^{137}Cs-derived distribution of eroding and aggrading sites is employed it is possible to select the appropriate equation to be used at each sample point, and thereby produce a plot of predicted soil erosion and aggradation rates based upon the topographic attributes. This may be compared with the ^{137}Cs-derived rates to provide a visual impression of the goodness of fit to complement the adjusted R^2 values discussed above. Figures 4.1(a) and (b) illustrates the ^{137}Cs-derived and predicted distributions for the site at Rufford.

The difference between the two distributions may be quantified by the calculation of residuals. Figure 4.1(c) illustrates the spatial distribution of these residuals for the Rufford site and two features are immediately apparent. First, most of the high negative residuals (underestimation of erosion or overestimation of aggradation) coincide with areas of peak erosion on the plot of ^{137}Cs-derived rates (Figure 4.1(a)) but also lie in areas of high erosion on the plot of predicted rates (Figure 4.1(b)). This indicates that though the regression equation correctly identifies the ranking of erosion, there is a consistent underestimation of peak rates. Secondly, many of the high positive residuals (overestimation of erosion or underestimation of aggradation) are located along the eastern flank of the field. This may be indicative of cross-slope wind erosion and aggradation along this flank, which would not be readily correlated with the topographic attributes subject to analysis. Examination of residuals may therefore identify the impact of additional processes, which if quantifiable may improve the degree of explanation achievable.

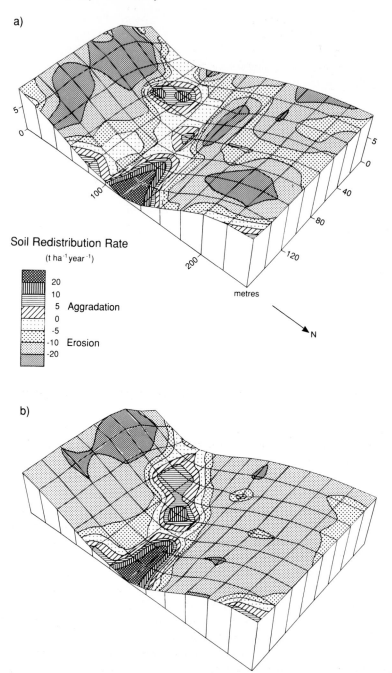

Soil Redistribution Rate

(t ha⁻¹ year⁻¹)

20
10
5   Aggradation
0
-5
-10 Erosion
-20

metres

N

Figure 4.1 Patterns of soil erosion and aggradation and residuals superimposed upon isometric projections of the field topography, for the Rufford site, Nottinghamshire. (a) ^{137}Cs-derived rates of soil erosion and aggradation. (b) Predicted rates of soil erosion and aggradation. (c) Residual rates (^{137}Cs-derived − predicted). Positives residuals indicate overestimates of erosion or underestimates of aggradation; negative residuals indicate underestimates of erosion or overestimates of aggradation

c)

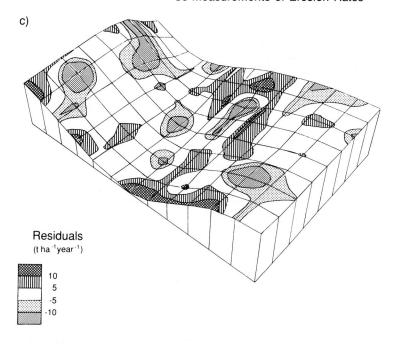

Residuals
(t ha^{-1}year^{-1})

10
5
-5
-10

DEVELOPMENT OF MULTIVARIATE PREDICTION EQUATIONS

The maps of predicted rates of soil redistribution, produced using the multiple regression equations for erosion and aggradation, provide a visual impression of the degree of explanation that may be attributed to the topographic parameters employed. However, these equations cannot be used to predict redistribution rates since their use requires prior knowledge of the distribution of eroding and aggrading points. If prediction of redistribution rates is required, it is necessary to produce a single equation that may be applied to both erosion and aggradation. In order to examine the feasibility of developing such site-based prediction relationships, stepwise multiple regression analyses were performed for each site with soil redistribution rate as the dependent variable.

Table 4.6 lists the variables included in the resulting equations, their respective beta coefficients and the adjusted R^2 values. In assessing the adjusted R^2 values it is important to recognise that the regression equations are not only predicting the point rates but also defining the spatial distribution of erosion and aggradation. Consequently, although the regression equation for Rufford explains less than half of the variation in soil redistribution rate, there is a good visual agreement between the ^{137}Cs-derived and the predicted pattern of erosion and aggradation (Figure 4.2, cf. Figure 4.1(a)). A similarly good visual fit is obtained for the field at Dalicott Farm, also on sandy soil. However, Table 4.6 shows a clear decline in adjusted R^2 values from the sandy, through silty, to clay soils in the study fields. This decline in adjusted R^2 mirrors a reduction in the range of ^{137}Cs-derived soil redistribution rates. Local variation in measured rates unrelated to topography could be expected to have a larger impact upon the goodness of fit where the range of rates is lower.

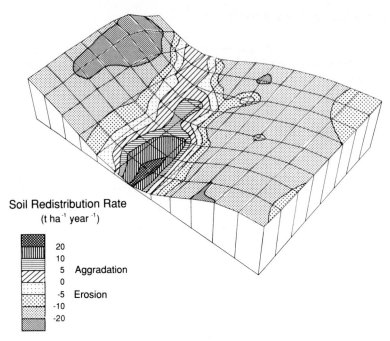

Soil Redistribution Rate
(t ha⁻¹ year⁻¹)

20
10
5   Aggradation
0
-5   Erosion
-10
-20

Figure 4.2   The pattern of predicted soil redistribution rates superimposed upon an isometric projection of the field topography for the Rufford site, Nottinghamshire

In terms of predicting the areal distribution of areas susceptible to erosion and aggradation, the regression equations for Rufford and Dalicott provide useful tools despite explaining only half of the variation in soil redistribution rates. At the other sites, the regression equations provide a useful first estimate of the distribution of sensitive areas, but the lower adjusted R^2 values are indicative of the high degree of generalisation and the role of factors other than site topography in defining the pattern of soil redistribution.

Table 4.6   Beta coefficients for variables included in stepwise multiple regression equations. Dependent variable is the soil redistribution rate

Soil and site	Elevation (m)	Slope angle (degrees)	Profile curvature	Planform curvature	Upslope area (m²)	Upslope distance (m)	Adjusted R^2
			(degrees/100 m)				
Sandy							
Rufford		−0.31	−0.36			0.31	0.43
Dalicott			−0.62				0.38
Chalky							
Lewes		−0.28	−0.48				0.25
Silty							
Fishpool				0.32			0.09
Clay							
Keysoe					−0.30		0.08

IDENTIFYING SENSITIVE AREAS

Using only topographic indices, it has been possible to explain between 8 and 43% of the variation in the rates of soil redistribution within five study fields on a range of soil types. It has been suggested that these figures mask the capacity of the predictive equations to identify the sensitive areas as the goodness of fit refers to the estimation of specific point values of erosion and aggradation rate rather than the more general delineation of the main features of the two distributions. For example, in the case of the Rufford site the predictive equation correctly allocated 85% of points to eroding and aggrading categories, and 44% of points had predicted point rates within 5 tonnes ha^{-1} year^{-1} of the ^{137}Cs-derived rates. None the less, a greater degree of explanation would be desirable. When eroding sites alone were examined, at three of the five sites between 21 and 27% of the variation in erosion rates could be explained by reference to the chosen topographic attributes. These figures highlight the need to consider factors other than topography alone in the estimation of erosion rate, even when considering long-term trends. Although the topography may determine the runoff energy, it is important to consider the potential impact of variation in soil properties upon runoff generation and sediment entrainment. For example, spatial variation in soil texture and structure may lead to variation in minimum detachment energy and therefore in sediment mobilisation. These properties may also be significant in determining variation in infiltration rate, with the consequent influence on runoff generation. The controlling effect of soil properties on erosion rates at the macro- and meso-scale has been recognised in the past, in the inclusion of soil parameters in prediction equations. However, the results of the analysis presented here suggest that micro-scale variation in soil properties may exert more control over within-field patterns of soil redistribution than has been recognised previously.

Table 4.7(a) Erosional significance of correlation between soil redistribution rate and topographic attributes

Positive correlation with topographic attributes:	Slope angle	Profile curvature	Planform curvature	Upslope area and distance
Location of eroding zones:	low angle	concavity	convexity	upper slope
Negative correlation with topographic attributes:	Slope angle	Profile curvature	Planform curvature	Upslope area and distance
Location of eroding zones:	high angle	convexity	concavity	lower slope

Table 4.7(b) Classification of sensitive areas

Soils	Zones	Processes
Sandy Chalky Silty	crests and spurs, towards top of slope, often high angle	sheet and rill/inter-rill
Clay	valley bottom	concentrated valley flow

a)

b)

c)

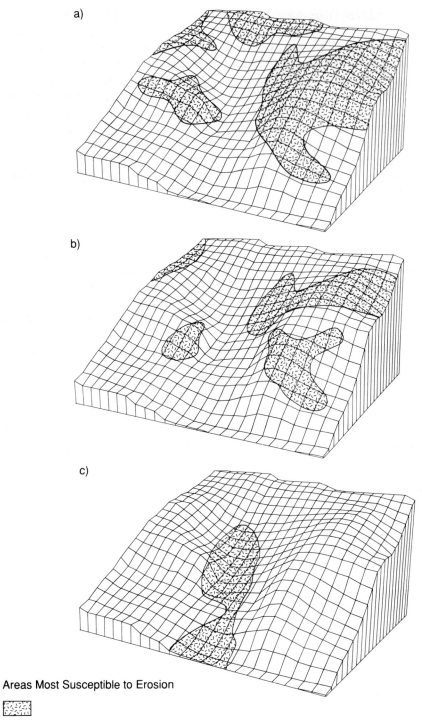

Areas Most Susceptible to Erosion

Figure 4.3 Tentative prediction of the spatial distribution of areas which are susceptible to erosion, based upon topographic attributes of a hypothetical landscape. (a) Sandy soils. (b) Silty soils. (c) Clay soils

Despite these constraints upon topographic delineation of zones sensitive to erosion, it is possible to make a tentative classification of susceptible landscape features on the basis of the analysis described above. Table 4.7(a) summarises the erosional significance of positive and negative correlations with the topographic attributes examined, and in Table 4.7(b) these are used to produce a tentative classification. These topographically defined, sensitive areas are illustrated in diagrams of hypothetical landscapes in Figure 4.3. The identification of slope crests and spurs as the most sensitive areas on sandy, chalky and silty soils suggests that the significance of slope length may have been overstated in previous attempts to explain propensity to erosion. The pre-eminence of these sites also suggests that sheet and rill/inter-rill erosion on slopes may be of greater significance than is often recognised. This classification suggests that it is only on the clay site in this sample that erosion is dominated by concentrated flow in the valley floor.

CONCLUSION

The point-specific measurement of soil erosion at high spatial resolution provided by the ^{137}Cs technique has permitted the examination of topographic controls upon landscape sensitivity to erosion at the within-field or micro-scale. The data have highlighted textural distinctions in the degree and nature of topographic control upon erosion, which may be related to process. The analysis has also revealed the importance of considering factors other than topography in the prediction of spatial patterns and rates of erosion. This is of particular significance to topographic models of erosion that assume constant soil properties, such as the contour-based model proposed by Moore, O'Loughlin & Burch (1988). The importance of infiltration has been recognised as of fundamental importance in hydrologic modelling (Smith & Herbert, 1979; DeCoursey, 1988) and in a recent analysis of the 'ANSWERS' model De Roo, Hazelhoff & Burrough (1989) identified infiltration as one of the most sensitive parameters in the model. The recognition of factors other than topography that may exercise control over the spatial patterns of soil erosion and aggradation reinforces the need to carry out field verification of predictive models (Pennock & De Jong, 1987). Use of ^{137}Cs-based erosion rate measurements may provide a practical means of achieving such verification.

ACKNOWLEDGEMENTS

The erosion rate measurements used in this paper were undertaken as part of the NERC/NCC 'Agriculture and the Environment' initiative, and the analysis reported was undertaken while one of the authors (T. A. Quine) was in receipt of a Lloyd's of London Tercentenary Foundation Fellowship. This financial support is gratefully acknowledged. Particular thanks are due to Professor C. R. Thorne for supplying the landscape analysis software, and to Dr R. Evans and Dr J. Boardman for comments regarding the ^{137}Cs-derived erosion rates. We also thank Mr J. Grapes for assistance with the gamma spectrometry analysis, and the various landowners who granted access to their properties and permitted soil sampling.

REFERENCES

Bachhuber, H., Bunzl, K., Schimmack, W. and Gans, I. (1982). The migration of ^{137}Cs and ^{90}Sr in multilayered soils: results from batch, column, and fallout investigations. *Nuclear Technology*, **59**, 291−301.

Boardman, J. (1990). Soil erosion on the South Downs: a review. In Boardman, J., Foster, I. D. L. and Dearing, J. A. (eds), *Soil Erosion on Agricultural Land*, Wiley, Chichester, pp. 87−105.

Boardman, J. and Robinson, D. A. (1985). Soil erosion, climatic vagary and agricultural change on the Downs around Lewes and Brighton, Autumn 1982. *Applied Geography*, **5**, 243−258.

Clark, E. H., Havercamp, J. A. and Chapman, W. (1985). *Eroding Soils: The Off-farm Impacts*, The Conservation Foundation, Washington, DC.

Crosson, P. R. and Stout, A. T. (1983). *Productivity Effects of Cropland Erosion in the United States*, Resources for the Future, Washington, DC.

DeCoursey, D. G. (1988). A critical assessment of hydrologic modelling. In *Modelling Agricultural, Forest and Rangeland Hydrology: Proceedings of the 1988 International Symposium*, ASAE Publication 07−88, pp. 478−493. St Joseph, Michigan.

De Roo, A. P. J., Hazelhoff, L. and Burrough, P. A. (1989). Soil erosion modelling using 'ANSWERS' and Geographical Information Systems. *Earth Surface Processes and Landforms*, **14**, 517−533.

Evans, R. (1988). *Water Erosion in England and Wales 1982−1984*, report for Soil Survey and Land Research Centre, Silsoe.

Klobe, W. D. and Gast, R. G. (1970). Conditions affecting cesium fixation and sodium entrapment in hydrobiotite and vermiculite. *Soil Science Society of America Proceedings*, **34**, 746−750.

Livens, F. R. and Baxter, M. S. (1988). Chemical associations of artificial radionuclides in Cumbrian soils. *Environmental Radioactivity*, **7**, 75−86.

Livens, F. R. and Loveland, P. J. (1988). The influence of soil properties on the environmental mobility of caesium in Cumbria. *Soil Use and Management*, **4**, 69−75.

Loughran, R. J. (1989). The measurement of soil erosion. *Progress in Physical Geography*, **13**, 216−233.

Loughran, R. J., Campbell, B. L. and Walling, D. E. (1987). Soil erosion and sedimentation indicated by caesium-137: Jackmoor Brook catchment, Devon, England. *Catena*, **14**, 201−212.

Meyer, L. D. and Wischmeier, W. H. (1969). Mathematical simulation of the process of soil erosion by water. *Transactions of the American Society of Agricultural Engineers*, **12**, 754−759.

Moore, I. D., O'Loughlin, E. M. and Burch, G. J. (1988). A contour-based topographic model for hydrological and ecological applications. *Earth Surface Processes and Landforms*, **13**, 305−320

Pennock, D. J. and De Jong, E. (1987). The influence of slope curvature on soil erosion and deposition in hummock terrain. *Soil Science*, **144**, 209−217.

Quine, T. A. and Walling, D. E. (1991). Rates of soil erosion on arable fields in Britain: quantitative data from caesium-137 measurements. *Soil Use and Management*, **7**, 169−176.

Roels, J. M. (1985). Estimation of soil loss at a regional scale based on plot measurements — some critical considerations. *Earth Surface Processes and Landforms*, **10**, 587−595.

Rogowski, A. S. and Tamura, T. (1965) Movement of cesium-137 by run-off, erosion and infiltration on the alluvial Captina silt loam. *Health Physics*, **11**, 1333−1340.

Sawhney, B. L. (1964). Sorption and fixation of microquantities of cesium by clay minerals: effect of saturating cations. *Soil Science Society of America Proceedings*, **28**, 183−186.

Sawhney, B. L. (1966). Kinetics of cesium sorption by clay minerals. *Soil Science Society of America Proceedings*, **30**, 565−569.

Smith, R. E. and Herbert, R. H. N. (1979). A Monte Carlo analysis of the hydrologic effects of spatial variability of infiltration. *Water Resources Research*, **15**, 419−429.

Squire, H. M. and Middleton, L. J. (1966). Behaviour of ^{137}Cs in soils and pastures — a long term experiment. *Radiation Botany*, **6**, 413−423.

Walling, D. E. and Quine, T. A. (1990). Use of caesium-137 to investigate patterns and rates of soil erosion on arable fields. In Boardman, J., Foster, I. D. L. and Dearing, J. A. (eds), *Soil Erosion on Aricultural Land*, Wiley, Chichester, pp. 33−53.

Walling, D. E. and Quine, T. A. (1991). The use of caesium-137 measurements to investigate soil erosion on arable fields in the UK: potential applications and limitations. *Journal of Soil Science*, **42**, 147−165.

Wischmeier, W. H. and Smith, D. D. (1978). *Predicting Erosion Losses — A Guide to Conservation Planning*, Agricultural Handbook 537, Science and Education Administration, USDA, Washington, DC.

Zevenbergen, L. W. and Thorne, C. R. (1987). Quantitative analysis of land surface topography. *Earth Surface Processes and Landforms*, **12**, 47−56.

5 Landscape Sensitivity and Change on Dartmoor

A. JOHN W. GERRARD
School of Geography, University of Birmingham, UK

ABSTRACT

Landform change is brought about either by alterations in the controlling processes or by internal changes within the landform systems. The speed with which change occurs is a function of landscape sensitivity. This chapter examines a number of ways of assessing landscape sensitivity on the granite upland of Dartmoor, south-west England. Sensitivity is assessed over the whole of Dartmoor by means of an examination, within kilometre squares, of values of specific relief, maximum slope angle, stream density and frequency and stream order. This allows sensitivity gradients to be established. Sensitivity is also examined at the basin and valley-side slope scale. The synthesis allows a number of statements to be made concerning the way in which the Dartmoor landscape has changed since the Tertiary Period as a result of changes in the operation of geomorphological processes.

INTRODUCTION

Much attention has been directed, over the last 20–30 years, to identifying the agents of change in the physical landscape. Change is brought about by those formative events (Wolman & Gerson, 1978) that are effective in shaping the landscape. The effectiveness of such events depends upon the force exerted, the return period of the events, the sensitivity of the landscape and the magnitude of the constructive or restorative processes that occur during the intervening intervals. It has been suggested that, during stable periods, with relatively constant processes, characteristic forms will be produced. Thus, for any given set of environmental conditions, it is anticipated that there will be a tendency, over time, to produce a set of characteristic landforms (Brunsden & Thornes, 1979). The logical extension of these ideas is that landscape evolution is episodic, with long periods of stability being separated by shorter periods of instability. Landscape change will mostly take place when the efficiency of formative processes is increased or when the resistance of the landscape is decreased or both. An increase in the efficiency of formative processes can be brought about by increasing the rate of operation of such processes or by replacing one set of processes by another, more effective group.

Environmental conditions have varied enormously over the approximately 1.5 million years of the Quaternary, even more so if the 65 million years of the Tertiary Period are included. The sequence of larger-scale changes in the British Isles has been examined by Brown (1979) and a number of suggestions have been made concerning geomorphological processes and

Landscape Sensitivity. Edited by D. S. G. Thomas and R. J. Allison

landform development. Such syntheses raise important questions, such as how individual landscapes respond to these changes, the speed of response and the persistence of any changes. Parsons (1988), in reviewing information concerning hillslope process—form relationships, came to three conclusions. First, relict hillslope forms will be common where substantial changes in hillslope processes have occurred during the last 100 000 years; secondly, the morphological effects of intermittent, though currently acting processes, will persist for substantial periods of time; and thirdly, processes that operated in the past may act to constrain or modify those presently active. This implies that any landscape is a mosaic of relic and currently forming landforms. Thus, it is important to be able to establish the environmental conditions that have occurred in the past and to attempt to predict the dominant processes involved and the response in terms of landform development.

THE DARTMOOR LANDSCAPE

Dartmoor is the largest and most prominent of the granitic uplands of south-west England. It is basically an upland plateau, varying in height from 400 to 600 m and deeply dissected by generally southward flowing streams with relative reliefs of up to 200 m (Figure 5.1).

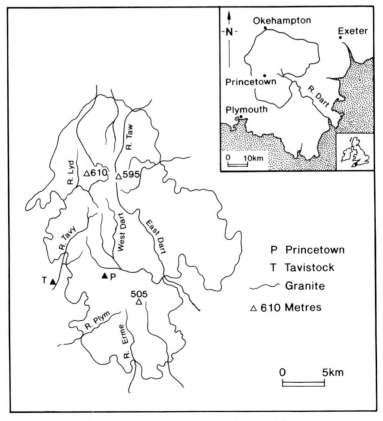

Figure 5.1   General nature and location of Dartmoor

The ridge tops are often flat or gently sloping, occasionally crowned by tors and fringed by rock buttresses. Various aspects of the geomorphology of Dartmoor have been examined. These have included the construction of denudation chronologies (e.g. Waters, 1960; Brunsden, 1963; Orme, 1964), examination of the weathering of granite (Waters, 1957; Brunsden, 1964; Fookes, Dearman & Franklin, 1971; Dearman & Baynes, 1978), the evolution of tors (Linton, 1955; Palmer & Neilson, 1962; Waters, 1964; Eden & Green, 1971; Gerrard, 1974, 1978) and assessment of the impacts of former periglacial processes (Waters, 1964; Gerrard, 1988a, 1989a). The comparatively level upper surfaces and accordant summit heights are thought to represent relict erosion surfaces. Most workers have regarded these surfaces as sub-aerial Tertiary peneplains but modified pediplanation may also have been involved. Dartmoor has never been glaciated but has been affected considerably by periglacial processes during the cold phases of the Pleistocene period. The landscape at present appears to be remarkably stable with little indication of soil erosion or mass movement. Water flow on slopes is dominated by throughflow (Williams, Ternan & Kent, 1984), with overland flow occurring during periods of soil saturation. The rivers flow in generally steep-sided valleys and contain much coarse material which is only moved in the most extreme floods.

LANDSCAPE SENSITIVITY

Geomorphological systems are continually subject to disturbances that arise because of changes in the external conditions of the system or from internal structural instabilities. External shocks to landscape systems have been classified as either ramped or pulsed inputs. Ramped inputs include major changes of climate, vegetation, land use and base level, whereas pulsed inputs are the low frequency—high magnitude events such as extreme, but short-lived climatic events. The response in the landscape to such events can be complex and different parts will respond differently. Brunsden & Thornes (1979) have outlined three basic patterns of spatial response. First, a change may be ubiquitous because of instantaneous changes in a widely acting process. Weathering is one such process and on Dartmoor is represented by the change to dominant physical weathering during the cold phases of the Quaternary Period. This change would have occurred over the entire landscape, though its effectiveness would have been constrained by the pre-existing topography and materials. Secondly, change may be propagated linearly along sensitive erosional axes, such as river channels. Changes of base level will be felt along river channels, and physical weathering may have been directed along major joint systems in the granite. Thirdly, many models of landscape change argue for the propagation of change as diffuse waves of aggression away from river channels or linear axes. A change in river activity from aggradation to incision may lead to change in adjoining valley-side slope processes. A major factor in the Dartmoor landscape is the alluvial infill that exists in many of the major valleys, which means that currently rivers rarely impinge on the base of slopes. It can also be argued that some tors are the manifestations of waves of aggression of physical weathering. Granite outcrops can be attacked by frost action and gradually retreat across the landscape, leaving a pile of boulders at the base or perhaps an altiplanation terrace. Such processes will possess greatest potential in those locations where jointing is most favourable and where material can be removed from the base of tors, e.g. steep slopes, at break of slope or on summits.

This introduces the idea of sensitivity to change. The sensitivity of a landscape to change can be expressed as the likelihood that a given change in the controlling factors of the system will produce a recognisable and persistent change of landscape form. Some parts of the landscape will be more sensitive to change than others. Exposures of well-jointed granite on steep slopes will be very sensitive to change, whereas exposures of poorly jointed granite on gentle slopes will be less sensitive. Areas near river channels will be more sensitive to change than the interfluves, which lie far from boundary changes. Interfluves and plateau areas will be affected by ubiquitous changes, such as the influence of climate on weathering, but they will generally be far removed from the transmission of basal changes. Landscape sensitivity is scale-dependent. The sensitivity of an entire landscape will be different from the sensitivity of a portion of an individual slope. This is because different processes operate over different spatial scales and because the organised adjustment of a whole landscape will take longer than for individual components. The challenge for the geomorphologist is to devise ways of assessing such sensitivity. This is now explored by examining the sensitivity of the Dartmoor landscape at three spatial scales: the entire landscape, the drainage basin and individual slopes.

SENSITIVITY AT THE LANDSCAPE SCALE

In analysing large-scale landscape sensitivity, several approaches are possible. Field mapping of landforms and materials is one such approach — with care taken to ensure that the basic categories mapped can be related to the concepts involved in sensitivity. The drawback with such an approach is the time and effort involved. The approach adopted here is to use basic geomorphological concepts to derive indices of sensitivity that can be obtained relatively easily from topographic maps. As outlined earlier, landscape sensitivity can be related to a number of fundamental aspects of the landscape. Material will be removed from steep slopes quicker than from gentle slopes. Thus, measures of slope form and steepness are required. Rivers act both as erosional and transporting agents. Landscapes with a large number of drainage channels are likely to be more sensitive to change than areas with fewer drainage channels. This means that landscape sensitivity will need to take account of the nature and number of drainage channels. These measures would seem to concentrate on axial transmission of change. However, for ubiquitous change to be effective material has to be transported. To this end three slope variables and three drainage variables were chosen to represent landscape sensitivity on Dartmoor. The variables have been obtained for each of the 541 km squares that cover the upland granite area of Dartmoor at a scale of 1:25 000. It is recognised that the results will be scale-dependent. However, this is not meant to be a definitive study but an exploration of possibilities. Slope variables measured were specific relief (difference between maximum and minimum elevation in each square), maximum slope (as assessed by the distance between the closest spaced contours with a contour interval of 10 m) and number of contours crossing the edges of each square. Drainage variables were drainage density, stream frequency and maximum Strahler stream order, all measured from the blue line network. Drainage density provides an inverse indication of the length of slope profiles affected by stream action, stream frequency is a spatial index of debris evacuation and stream order provides some indication of stream power.

The spatial distribution of specific relief highlights the contrast between interior plateau areas and deeply dissected northern, eastern and southern fringes (Figure 5.2). The spatial distribution of number of contour crossings is very similar to that for specific relief (Figure 5.2). The contrast between the plateau areas and dissected fringes is also evident on the map

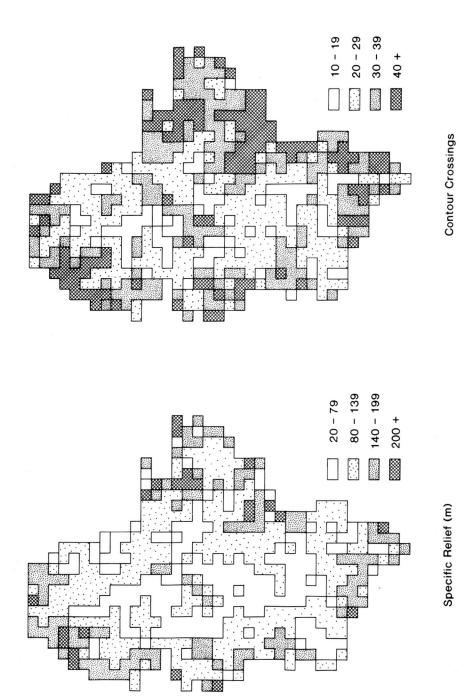

Specific Relief (m)

Contour Crossings

Figure 5.2   Spatial distributions of specific relief and number of contour crossings

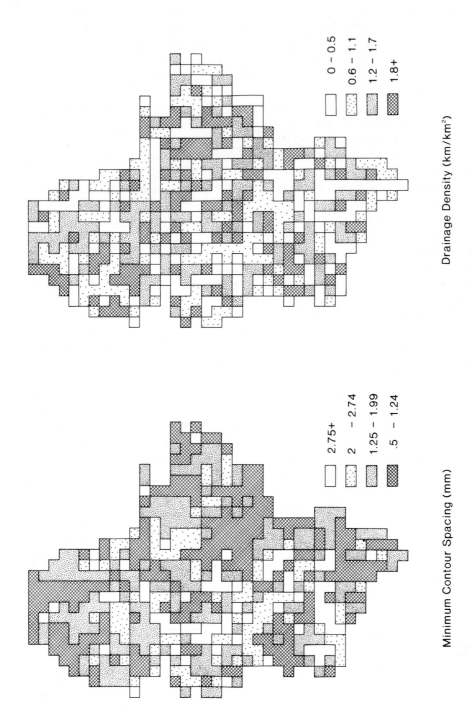

Minimum Contour Spacing (mm)

	2.75+
	2 − 2.74
	1.25 − 1.99
	.5 − 1.24

Drainage Density (km/km^2)

	0 − 0.5
	0.6 − 1.1
	1.2 − 1.7
	1.8+

Figure 5.3 Spatial distributions of minimum contour spacing and drainage density

of minimum contour spacing (Figure 5.3). This analysis emphasises that the distributions of slope characteristics, while similar in some respects, also possess marked differences. Patterns of drainage density (Figure 5.3) and stream frequency (Figure 5.4) appear more random but the map of stream order demonstrates the way in which many of the rivers combine to form the River Dart system, in the central area (Figure 5.4). The map of drainage density is similar to that of Gardiner (1971).

A cursory investigation of the spatial distributions suggests that there are relationships between some of the variables. This is not surprising as it has already been suggested that river action and slope form may be related. Also specific relief and number of contour crossings are essentially indicators of mean slope angle. Such relationships become clear in the correlation matrix (Table 5.1). The very high coefficient between specific relief and number of contour crossings demonstrates that essentially the same factor is being measured. Minimum contour spacing is negatively related to specific relief, number of contour crossings and drainage density. Drainage density seems unrelated to specific relief and number of contour crossings.

Maps of single variables provide an interesting insight into the possible spatial distribution of landscape sensitivity if it is accepted that they can be used as surrogates for the operation of general landscape processes. The broad detail substantiates the general statements made earlier with the central upland areas of north and south Dartmoor possessing low landscape sensitivity and the periphery and areas adjacent to the major rivers possessing high sensitivity. But a number of more subtle patterns emerge. There appear to be isolated areas of low sensitivity surrounded by highly sensitive areas. This might suggest that, were conditions to change, such insensitive areas would be affected by changes propagated through the more sensitive regions. It is much rarer for high sensitivity areas to be surrounded, and therefore protected, by areas of low sensitivity. Thus, the gradients portrayed in the maps are useful indications of the way in which change may be propagated through the landscape. The distribution also implies that such change is not always propagated along stream channels. There may also be anisotropy in the gradients possibly reflecting a granitic structural control. This analysis, though requiring refinement, has provided an insight into landscape-scale sensitivity. It has also suggested a number of lines for future research.

SENSITIVITY AT THE BASIN SCALE

It is at the basin scale that detailed relationships between stream activity and slope development can be investigated. A number of studies have shown that soil, slopes and associated processes vary systematically within drainage basins. Such relationships have been examined in the basin of the River Cowsic, central Dartmoor. The River Cowsic, a fourth-order stream on

Table 5.1 Degree of correlation between landscape parameters

	Specific relief	Minimum contour spacing	Contour crossings	Drainage density
Specific relief	1.0	−0.52	0.96	−0.27
Minimum contour spacing		1.00	−0.58	+0.43
Contour crossings			1.00	−0.12

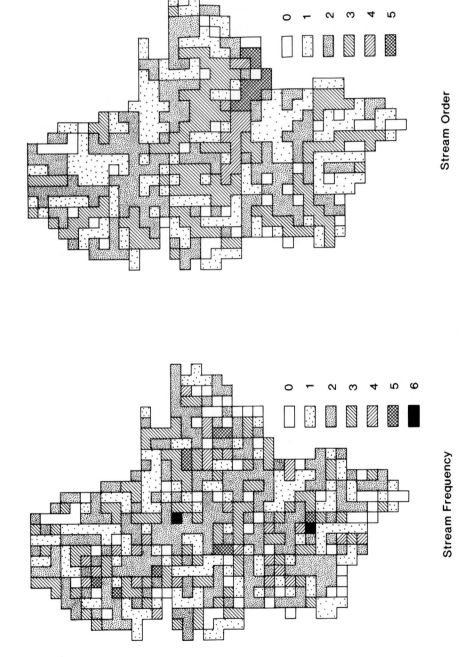

Figure 5.4 Spatial distributions of stream frequency and stream order

Table 5.2 Descriptive parameters for slope profiles in the basin of the River Cowsic grouped according to stream order (after Gerrard, 1988b)

Slope order	Mean mean angle	Mean maximum angle	Mean length (m)	Number of slopes
1	10.8	13.9	13.8	17
2	8.7	17.1	126.3	11
3	10.9	26.9	352.2	13
4	7.0	17.4	333.0	14

the Strahler (1952) system of stream ordering using the blue line network, rises in the high central mass of northern Dartmoor and flows south to meet the River West Dart at Two Bridges. Fifty-five slope profiles, located using a stratified random sample, were measured with a 3 m recording interval. The means of particular slope parameters for individual slope profiles are shown in Table 5.2. Fourth-order slopes possess the lowest mean values and the difference between third- and fourth-order slopes is most marked, as is the difference between second- and third-order slopes. Systematic trends occur in that mean and maximum angles and length of slopes increase up to third order and then decrease. Differences in overall slope form also exist in that fourth-order slopes possess a greater proportion of rectilinear segments (Gerrard, 1982).

These results reinforce the link between slope and stream activity alluded to earlier. If one accepts the ergodic principle, it enables possible changes to slope form to be predicted if there is a change in stream activity such as rejuvenation or climate and land use changes that might alter the drainage density and value of stream order.

There are also differences in soils and materials on slopes leading off streams of different orders (Gerrard, 1990). Fourth-order slopes are dominated by iron pan podzols and generally have quite thick humic horizons (average 31 cm). Soil on third-order slopes exhibits greater variability, is stonier and humic horizons are thinner (14 cm on average). Steeper third-order slopes are generally occupied by brown podzolic soils, or a transition between iron pan podzols and brown earths. Soil on second-order slopes is shallower and stonier. This has repercussions for slope sensitivity in that any change in the balance between throughflow or overland flow or between mass movement processes and water action will be affected by the nature of the soil profile.

As a link between the basin and entire landscape scales, the spatial grouping of drainage basins has been examined. The blue line network on 1:25 000 scale OS maps produces 27 third-order basins, wholly or partly draining the granitic area of Dartmoor. A hierarchical cluster analysis using a variety of morphometric values allows groupings of basins to be identified (Figure 5.5). The patterns of basins and the way the pattern simplifies at higher similarity levels bears some resemblance to the pattern of the morphologic indicators. The groupings at a similarity coefficient of 1.1 have been listed in Table 5.3 with some summary statistics shown in Table 5.4. Basins in group 1 are characterised by small size, high relief ratio and are sandwiched between larger basins at the edge of the upland. The second group are characterised by the intermediate size of the basins and are mostly peripheral to the main upland mass and clustered in the northeastern part of Dartmoor. Group 3 is the largest cluster and includes many of the basins that drain the high northern and southern areas. Group 4 contains the five largest basins and those with the greatest average relative relief.

The way in which the basins group together at different levels helps explain how the drainage

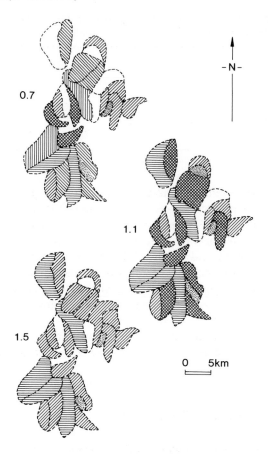

Figure 5.5   Grouping sequence of drainage basins (reproduced by permission from
Gerrard, 1989b)

Table 5.3   The major groups of drainage basins at a similarity coefficient of 1.1 (after
Gerrard, 1989b)

Group 1	Group 2	Group 3	Group 4	Ungrouped basins
Broadaford	Forder	Swincombe	Erme	Wallabrook
Ballabrook	Plall	Avon	Plym	Bovey
Blackaton	Becka	Yealm	Meavy	West Dart
	Cowsic	S. Teign	East Dart	
	Burramoor	N. Teign	W. Okement	
	W. Webburn	E. Okement		
		Blackabrook		
		Cherrybrook		
		E. Webburn		
		Glazebrook		

Table 5.4 Mean values of morphometric properties of the major groups of Dartmoor drainage basins (after Gerrard, 1989b)

Morphometric property	Group 1	Group 2	Group 3	Group 4
Total number of streams	8.0	15.2	13.6	31.0
Area of third-order basins (sq km)	5.7	11.9	16.8	32.6
Drainage density of third-order basins (km/sq km)	1.2	1.2	1.1	1.1
Total basin relief (m)	284.4	304.4	333.6	419.9
Basin relief ratio	411.5	294.4	236.8	195.2

pattern of Dartmoor has evolved. The majority of basins that join at an early stage in the clustering process have streams whose headwaters start on the upper plateau surfaces. Long profiles of streams and ridge profiles suggest that Dartmoor has been uplifted several times, forcing the rivers to cut new valleys inside their old ones. It appears that the earliest rivers flowed east and removed the cover rocks of the Dartmoor granite (Green, 1985). During the mid-Tertiary earth movements, Dartmoor was tilted to the south and the south-flowing Plym, Avon, Yealm and Erme were initiated as well as the long north bank tributaries of the Dart. The drainage was rejuvenated intermittently throughout the Tertiary and early Pleistocene Periods. The changing conditions of the Quaternary will have affected characteristics such as drainage density, but the overall pattern was probably little affected. This suggests that the large-scale sensitivity of the landscape was established by the early phases of the Quaternary Period.

SENSITIVITY AT THE SLOPE SCALE

The nature of the slope deposits, clitter, benched hillslopes and tors indicate that the Dartmoor slopes have been affected to a considerable extent by former periglacial processes (Gerrard, 1988a). The effect of the periglacial phases seems to have been to increase the efficiency of certain processes and speed up the development of landforms rather than to create distinct slope forms. However, characteristic slope angles in the range 8–12° (Gerrard, 1987) are mantled with solifluction material, and high pore water pressures associated with such periglacial slope processes may well have produced slopes that have been little affected by processes acting over the majority of the Holocene Period. Fossil slope forms are likely to be created when environments change from a high energy to a lower energy state (Gerrard, 1991). This would imply that periglacial processes have desensitised the Dartmoor slope systems. However, this does not mean that changes have not occurred on the slopes since the past periglacial phase, as a close examination reveals a series of infilled gullies on many slopes (Gerrard, 1988c). One of the distinctive features of the gullies is the presence of layered infill, consisting of alternating layers of coarse sand and fine sand or silt. The gullies cut through the solifluction deposits and represent a later phase of landscape instability. The infill might represent a phase of instability at the start of the Holocene or soil erosion initated by vegetation changes during the Bronze and Iron Ages. Soils on these and neighbouring slopes appear comparatively youthful and are out of equilibrium with slope form (Gerrard, 1990), also suggesting that changes have occurred. The general conclusion is that the periglacial effect has not been obliterated but has been altered subtly.

SYNTHESIS

These deliberations imply a number of consequences for the Dartmoor landscape. In the insensitive areas morphoclimatic characteristic forms will not be produced unless the environmental conditions remain constant for long periods. These areas will tend to be dominated by structurally controlled elements. This implies that the inner plateau areas have changed relatively little in form, certainly throughout the Quaternary Period, and owe their extent and relative position to geological factors. A corollary of this is that the mobile elements, such as river channels, are relatively constant in location as they cannot overcome the intrinsic barriers. Slopes linking river channels with interfluves are intermediate in terms of sensitivity to change. Some change is propagated, but not as much as lower down the slopes and in river channels. Some of the major changes take place in the transition stage between one climatic type and another or one process domain and another.

These ideas can be incorporated into a suggested scheme for the changes that have taken place on Dartmoor (Table 5.5). It is very difficult to suggest what the landscape was like at the end of the Tertiary Period, but the major upper plateau areas were probably very similar to their present form and the major lineaments of the rivers were established. A variable, but probably quite shallow, thickness of weathered material was probably created. Towards the end of the Tertiary Period the climate became colder and chemical weathering would have become less important. Slope processes and river action would be dominant.

The difficulty with assessing periglacial changes is that there is only evidence of what was presumably the last phase. The thickness of soliflucted material is not excessive but stream channels would have received more material, lower slopes become accumulation zones and other areas stripped to reveal bare rock. Some valley-side tors were created and others destroyed. Modification of slope form undoubtedly occurred with stepped hillslopes and some gullying. Changes during the Holocene have been more subtle and difficult to establish. However, pollen analysis and archaeological remains provide some information (Simmons,

Table 5.5 Phases in the evolution of the Dartmoor landscape (items in bold indicate the dominant processes) (after Gerrard, 1982)

Period	Process domains	Landform effects
Early/mid-Tertiary	**chemical weathering** slope processes, river action	**erosion surfaces** **possible etchplains** **major lineaments established**
Late Tertiary	**chemical weathering** **slope processes** **river action**	**slope modification** **rock exposed in high energy situations**
Pre-Glacial Pleistocene	**river action** **slope processes**	as above
Quaternary-Glacial	**physical weathering** **solifluction** **wind erosion** rill action	**benched hillslopes** **tor formation** **gullying, lower slope accumulation**
Quaternary-Interglacial	chemical weathering **soil creep** **throughflow**	**infilled gullies** **soil development**

Table 5.6 Suggested chronology of landscape change on Dartmoor during the Holocene Period (from Gerrard, 1988c)

Period (BP)	Geomorphological activity
10250–9450	some solifluction, gullying, soil erosion
9450–8450	landscape stabilisation, soil and vegetation development
8450–7450	extensive soil development (brown earths?)
7450–4450	soil and vegetation changes, start of peat development
4450–3500	peat accumulation on high ground, forest clearance, some erosion
3500–2450	quite extensive erosion, soil modification
2450–1000	gradual landscape stabilisation, podzol development
1000–present	localised effects due to tinning, enclosure and improvement, generally a stable phase

1964a, 1964b, 1969; Staines, 1979; Maguire, Ralph & Fleming, 1983) on which to erect a chronology (Table 5.6).

There is little information on which to base the significance of extreme, but short-lived, pulsed inputs on Dartmoor. A number of floods on Dartmoor streams have been recorded but there is little evidence of the effects of such events. Some of these early floods have been summarised by Worth (1967). During one storm in the West Okement on 17 August 1917, a shallow channel, approximately 1.5 m wide and 1 m deep, draining an old road to the Forest Mine, was excavated to a depth of 5 m. The Redaven Brook carried boulders and stones, blocking its former channel and creating a new channel. All such accounts relate to stream activity and there is no indication that any slope failures occurred.

In conclusion, the Dartmoor landscape can be subdivided into three components on the basis of rapidity of landscape change and the residue from earlier phases still in existence (Figure 5.6). The scale of most major Dartmoor valleys matches the 1 km scale adopted earlier. The upper plateau areas (zone A) have suffered some surface lowering, solifluction and tor formation. The major slopes (zone B) have suffered retreat and lowering in their upper portions and accumulation towards their bases. Tors have been created on the steeper sections and towards the slope crests and a stepped slope form has evolved in some localities. This zone essentially reflects the former influence of periglacial processes and minor modification in

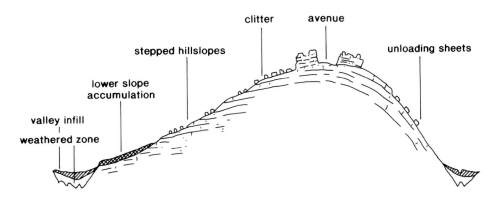

Figure 5.6 Main elements in the Dartmoor landscape (after Gerrard, 1988a)

historic times, and is intermediate in sensitivity between the upper plateau areas and the more rapidly changing zone of the stream channels, with alternating cut and fill. However, little is known of the detailed changes in stream regimes during the Holocene Period. The nature of the alluvial infill indicates that changes have occurred but the detailed sequence has still to be elucidated. Many of these ideas must remain tentative but hopefully they will serve to stimulate future work to help resolve some of the uncertainties.

REFERENCES

Brown, E. H. (1979). The shape of Britain. *Transactions of the Institute of British Geographers,* **NS4**, 449–462.

Brunsden, D. (1963). The denudation chronology of the River Dart. *Transactions of the Institute of British Geographers,* **32**, 49–63.

Brunsden, D. (1964). The origin of decomposed granite. In Simmons, I. G. (ed.), *Dartmoor Essays,* Devonshire Association for the Advancement of Science, Exeter, pp. 97–116.

Brunsden, D. and Thornes, J. B. (1979). Landscape sensitivity and change. *Transactions of the Institute of British Geographers,* **NS4**(4), 463–484.

Dearman, W. R. and Baynes, F. J. (1978). A field study of the basic controls of weathering patterns in the Dartmoor granite. *Proceedings of the Ussher Society,* **4**, 192–203.

Eden, M. J. and Green, C. P. (1971). Some aspects of granite weathering and tor formation on Dartmoor. *Geografiska Annaler,* **53A**, 92–99.

Fookes, P. G., Dearman, W. R. and Franklin, J. A. (1971). Some engineering aspects of rock weathering with field examples from Dartmoor and elsewhere. *Quarterly Journal of Geology,* **4**, 139–185.

Gardiner, V. (1971). A drainage density map of Dartmoor. *Transactions of the Devonshire Association,* **103**, 167–180.

Gerrard, A. J. (1974). The geomorphological importance of jointing in the Dartmoor granite. *Institute of British Geographers Special Publication,* no. 7, pp. 39–51.

Gerrard, A. J. (1978). Tors and granite landforms of Dartmoor and eastern Bodmin Moor. *Proceedings of the Ussher Society,* **4**, 204–210.

Gerrard, A. J. (1982). *Slope form, soil and regolith characteristics in the basin of the River Cowsic, Central Dartmoor, Devon.* Unpublished Ph.D. thesis, University of London.

Gerrard, A. J. (1987). Slope form in the basin of the River Cowsic, Dartmoor, England, Part I: characteristic and threshold slope angles. *National Geographer,* **22**, 95–115.

Gerrard, A. J. (1988a). Periglacial modification of the Cox Tor–Staple Tors area of western Dartmoor, England. *Physical Geography,* **9**, 280–300.

Gerrard, A. J. (1988b). Slope form in the basin of the River Cowsic, Dartmoor, England, Part II: spatial variation in slope angles. *National Geographer,* **23**, 1–14.

Gerrard, A. J. (1988c). Partially infilled gully systems on Dartmoor. *Proceedings of the Ussher Society,* **7**, 86–89.

Gerrard, A. J. (1989a). The nature of slope materials on the Dartmoor granite. *Zeitschrift für Geomorphologie,* **33**, 179–188.

Gerrard, A. J. (1989b). Drainage basin analysis of the granitic upland of Dartmoor, England. *Geographical Review of India,* **51**, 1–17.

Gerrard, A. J. (1990). Soil variations on hillslopes in temperate environments. *Geomorphology,* **3**, 225–244.

Gerrard, A. J. (1991). The status of temperate hillslopes in the Holocene. *The Holocene,* **1**, 86–90.

Green, C. P. (1985). Pre-Quaternary weathering residues, sediments and landform development: examples from southern Britain. In Richards, K. S., Arnett, R. R. and Ellis, S. (eds), *Geomorphology and Soils,* George Allen & Unwin, London, pp. 58–77.

Linton, D. L. (1955). The problem of tors. *Geographical Journal,* **121**, 470–487.

Maguire, D., Ralph, N. and Fleming, A. (1983). Early land use on Dartmoor — palaeobotanical and pedological investigations on Holne Moor. In Jones, M. (ed.), *Integrating the Subsistence Economy,* BAR International Series no. 181, pp. 57–105.

Orme, A. R. (1964). The geomorphology of southern Dartmoor. In Simmons, I. G. (ed.), *Dartmoor Essays*, Devonshire Association for the Advancement of Science, Exeter, pp. 31–72.

Palmer, J. and Neilson, R. A. (1962) The origin of granite tors on Dartmoor, Devonshire. *Proceedings of the Yorkshire Geological Society*, **33**, 315–340.

Parsons, A. J. (1988). *Hillslope Form*, Routledge, London.

Simmons, I. G. (1964a). An ecological history of Dartmoor. In Simmons, I. G. (ed.), *Dartmoor Essays*, Devonshire Association for the Advancement of Science, Exeter, pp. 193–216.

Simmons, I. G. (1964b). Pollen diagrams from Dartmoor. *New Phytologist*, **63**, 165–180.

Simmons, I. G. (1969). Environment and early man on Dartmoor, Devon, England. *Proceedings of the Prehistoric Society*, **35**, 203–219.

Staines, S. J. (1979). Environmental change on Dartmoor. In Maxfield, V. (ed.), *Prehistoric Dartmoor in its context*, Devon Archaeological Society, Jubilee Conference Proceedings, pp. 21–47.

Shrahler, A. N. (1952). Dynamic basis of geomorphology. *Bulletin of the Geological Society of America*, **63**, 923–938.

Waters, R. S. (1957). Differential weathering and erosion on oldlands. *Geographical Journal*, **123**, 503–509.

Waters, R. S. (1960). The bearing of superficial deposits on the age and origin of the upland plain of East Devon, West Somerset and South Somerset. *Transaction of the Institute of British Geographers*, **28**, 89–97.

Waters, R. S. (1964). The Pleistocene legacy to the geomorphology of Dartmoor. In Simmons, I. G. (ed.), *Dartmoor Essays*, Devonshire Association for the Advancement of Science, Exeter, pp. 73–96.

Williams, A. G., Ternan, J. L. and Kent, M. (1984). Hydrochemical characteristics of a Dartmoor hillslope. In Burt, T. P. and Walling, D. E. (eds), *Catchment Experiments in Fluvial Geomorphology*, Geobooks, Norwich, pp. 379–398.

Wolman, M. G. and Gerson, R. (1978). Relative scales of time and effectiveness of climate in watershed geomorphology. *Earth Surface Processes*, **3**, 189–208.

Worth, R. H. (1967). *Dartmoor*, David & Charles, Newton Abbot.

6 Landscape Stability and Biogeomorphic Response to Past and Future Climatic Shifts in Intertropical Africa

NEIL ROBERTS and PHILIP BARKER
Department of Geography, Loughborough University of Technology, UK

ABSTRACT

In low-latitude regions such as intertropical Africa, arid and humid phases have alternated over time as the monsoonal circulation system has been weakened and strengthened. Empirical palaeoclimatic data further indicate that past dry−wet shifts have frequently been abrupt and step-like in character. As a result of climatic change, the semi-arid zone of maximum bioclimatic erosion potential has expanded and contracted across much of intertropical Africa. In this chapter we investigate the impact of previous dry-to-wet shifts in climate on earth surface biotic and physical systems, and compare them with that predicted to occur under the scenario of a warmer, wetter climate. The last important dry-to-wet climatic transition in intertropical Africa occurred ~ 12 500 BP at the end of the late Pleistocene arid phase, and according to Knox's (1972) biogeomorphic response model, catchment erosion and sediment flux would have peaked immediately after this. We present data on lake sediment accumulation rates from cores spanning this dry-to-wet transition with a view to testing the Knox model empirically. While a number of problems hinder the translation of core accumulation data into quantitative values of clastic sediment flux, palaeoenvironmental records clearly highlight the role that vegetation can play in mediating the erosional response to climate. It is concluded that the terminal Pleistocene climatic transition may provide a valuable means of calibrating the future response of tropical landscapes to the climate projected to occur on an Earth warmed by greenhouse gases.

INTRODUCTION

One of the most important aspects of any future greenhouse gas induced atmospheric warming will be its effect upon the landscape. Landscape stability or instability is reflected in many and different ways, but one of the clearest, simplest indications is the intensity of erosion and sediment flux. This is important, not only because understanding how earth surface systems function and respond to forcing is a basic aim of geomorphology *per se*, but also because land degradation through accelerated soil loss is a pressing environmental problem from a human perspective. To understand better the sensitivity of landscapes to such external forcing as rapid climate change, it is important to know how far present-day rates of erosion are typical of the long-term geological norm, to what degree they are climatically controlled,

Landscape Sensitivity. Edited by D. S. G. Thomas and R. J. Allison
© 1993 John Wiley and Sons Ltd

how far they have been anthropogenically accelerated, and if so over what time period they have been changed (Douglas, 1967, 1976). In this endeavour a Holocene, or even longer, timescale of analysis is essential (Knox, 1984; Eybergen & Imeson, 1989). In this paper we review evidence of landscape response to past climatic changes in intertropical Africa, with many of the issues raised also having relevance in other low- and mid-latitude regions of the world.

MODELLING THE EROSIONAL RESPONSE TO CLIMATIC FORCING

One of the most effective means of understanding landscape sensitivity is through a coupled empirical/model-based approach in which models, either conceptual or numerical, are tested and calibrated against real-world data sets. In the case of soil erosion on slopes, predictive models such as the Universal Soil Loss Equation (USLE), or regional variants of the formula, such as the Soil Loss Estimation for Southern Africa (SLEMSA), have been developed on the basis of the major controlling parameters (Wischmeier & Smith, 1978; Elwell, 1984). These may be tested against observed data from experimental erosion plots on individual slopes or from small catchment monitoring studies (e.g. Roberts & Lambert, 1990).

At a broader level, Langbein & Schumm (1958) attempted to predict the relationship between catchment sediment yield and climate. Following the not unreasonable assumption that soil loss is increased when protective vegetation cover is depleted but rainfall erosivities remain high, they argued that catchment sediment yield should peak where effective precipitation is approximately 300 mm year^{-1}, i.e. under a semi-arid regime. Subsequent authors (e.g. Walling & Webb, 1983) have been unable to verify the predicted semi-arid erosion peak and have questioned the validity of its existence. Certainly there are other strongly seasonal climates, such as the monsoonal regimes of southern Asia, where climatically controlled erosion rates may also be high. However, analysis of global data sets for catchment sediment yield cannot easily disaggregate the effects of climate from other parameters such as tectonics and human impact. It is evident that the 'natural' bioclimatic erosion zones of Langbein & Schumm have been modified by human action in many parts of the world. Fuelwood collection and grazing have reduced vegetation cover in much of Africa's sub-humid woodland savanna zone, while conversion to agricultural land has often led to complete vegetation clearance. The consequence of these has been to push Langbein & Schumm's erosion peak from the semi-arid zone into areas of wetter climate, and hence even higher erosion potential. At appropriate spatial scales and in the absence of human disturbance, the link between climate, vegetation and erosion therefore appears to remain valid.

Palaeoclimatic research has demonstrated that low-latitude climatic belts have been far from constant during the late Quaternary (Street-Perrott, Roberts & Metcalfe, 1985). In regions such as intertropical Africa, the dominant climatic signal over the last glacial−interglacial cycle has been effective precipitation — precisely the prime control over erosion identified by Langbein & Schumm. Arid and humid phases have alternated as the monsoonal circulation system has been weakened and strengthened (Kutzbach, 1983). As a result, the postulated semi-arid zone of maximum bioclimatic erosion potential would have expanded and contracted across much of intertropical Aftica. Thus at times when the climate was wetter than today's in northern intertropical Africa (e.g. early Holocene) the zone of greatest erosion potential would have moved into presently arid areas such as the Sahara. By contrast, at times of drier climate (e.g. late Pleistocene) this zone would have lain in areas now covered by tropical moist woodland (Hamilton, 1982). It is evident that climate change had a differential effect on erosion across Africa, with the same climatic change — e.g. to wetter conditions — increasing erosion rates in one area but reducing them in another.

The Langbein—Schumm model is essentially a static one which assumes that equilibrium conditions exist (or existed) between climate, vegetation and erosion. Clearly this is not always the case, and it is unlikely to be true for the erosional response to environmental forcing over, say, the next century. For periods of climatic transition, dynamic models will be of greater relevance, and in this study the biogeomorphic response model proposed by Knox (1972) will be employed. According to Knox's model, shifts in effective precipitation can

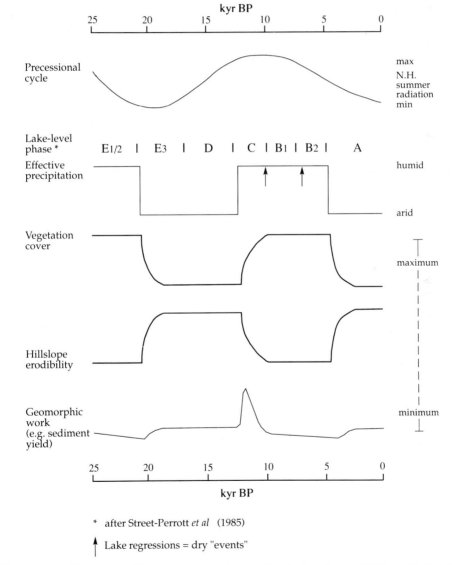

Figure 6.1 Biogeomorphic response model (modified after Knox, 1972) applied to the semi-arid zone of intertropical Africa

lead to periods of temporary disequlibrium between climate, vegetation and hillslope erodibility. In particular, an abrupt shift from a dry to a wet climate would generate high rainfall erosivities while hillslope erodibilities were also still high, before a protective vegetation cover had been established over the landscape. Under these conditions of landscape instability, intense geomorphological work in the form of erosion and sediment flux can take place (Figure 6.1).

The response of the precipitation−evaporation ($P-E$) ratio in low-latitude regions to future global warming is uncertain (Rind, 1990), but we can examine General Circulation Models (GCMs) of atmospheric circulation to see the likely direction of change in the parameters which control erosion. While there is considerable regional variation between individual GCMs, there is an overall consensus that tropical precipitation will increase under a 2 × CO_2 atmosphere (Henderson-Sellers, 1992). All other things being equal this will cause greater rainsplash impact and soil particle detachment. Despite this absolute increase in P, a decrease in $P-E$ and hence in soil moisture status is projected for most areas because of higher evapotranspiration. The future consequences for vegetation cover are less clear. On the one hand, the higher CO_2 levels will tend to stimulate plant growth, but on the other, lower soil moisture may put greater stress on vegetation, especially in the dry season, reducing winter ground cover in savanna-type ecosystems. Soil erodibilities may therefore be enhanced at the onset of the monsoonal rains. In many areas, of course, vegetation cover has already been depleted by grazing pressure and deforestation. Changes in these parameters for northern intertropical land areas according to the UKMO GCM experiment are shown in Figure 6.2(b). According to this model, precipitation will be higher on average by ∼15%, but up to 25% (200 mm year^{-1}) in regions such as East Africa, while runoff will increase during the wet summer season.

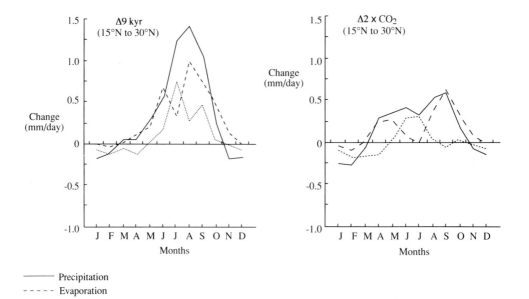

Figure 6.2 UKMO GCM derived estimates for precipitation, evaporation and runoff at 9 kyr and 2 × CO_2 in the northern hemisphere tropics (after Mitchell, 1990)

These changes broadly correspond to those of the dry-to-wet shift in Knox's model, i.e. greater rainfall erosivities without adequate compensatory adjustment in protective vegetation cover. The timescale over which greenhouse gas-induced climate change may take place is also subject to considerable uncertainty. Projections give envelope dates between 2015 and 2050 for a doubling of CO_2 equivalent from pre-industrial levels (Warrick & Farmer, 1990). Either of these time periods is sufficiently rapid for there to be a period of disequilibrium between parameters such as soil moisture and vegetation adjustment, and certainly when viewed on a Holocene timescale future changes are likely to approximate to the stepwise climatic shifts of the biogeomorphic response model. This does not mean that the increase in geomorphic work predicted by Knox's model will actually take place under a warmer Earth scenario. Like Langbein & Schumm's (1958) linkage of climate and sediment yield, the biogeomorphic response model remains conceptually attractive but empirically unproven.

PALAEOENVIRONMENTAL DATA SOURCES

One way of improving our confidence in models is to test them against field data, which in the case of the Knox model means time-series, and hence delving into environmental history. Over a decadal timescale it is possible to draw on documented records of climate and hydrogeomorphological response. In East Africa a significant and abrupt meterorological shift took place during the early 1960s (Lamb, 1966). Following a decade of relatively low rainfall, there was a sharp rise in precipitation between 1959 and 1961 (Figure 6.3(a)). This evoked a clear hydrological response, with the water levels of major lakes rising dramatically (Figure 6.3(b)) and major rivers such as the White Nile experiencing large increases in discharge. The difficulty of modelling this climate−hydrology coupling numerically for Lake Victoria through conventional water balance methods was interpreted by Kite (1981) as the result of underestimation of rainfall over the lake surface. Alternately, the hydrological response may have been amplified by much greater than predicted surface runoff. Certainly the geomorphological response in major gauged catchments like the Tana River was markedly non-linear, where sediment yield peaked abruptly during the dry-to-wet transition (Figure 6.3(c)). These data are far from ideal for model testing, coming as they do from different catchments, but they do serve to emphasise how dramatic the regional landscape response to climate shifts can be even within recent times.

For a more major and permanent period of transient adjustment between climate and earth surface systems it is necessary to examine longer-term sequences which occurred over late Quaternary time. In northern intertropical Africa, the last major shift from an arid to a humid climate took place between ~ 13 000 and ~ 9000 years ago, in other words during the period of the last glacial−interglacial transition (Roberts, 1990). The humid ('pluvial') climate of the early Holocene has been suggested by some authors (e.g. Butzer, 1980a) as analogous to that which may occur in the near future in low latitudes. Mitchell (1990) has recently convincingly argued that from a climatic point of view, the early Holocene does not make a good analogue for a greenhouse gas-warmed Earth. Comparing the UKMO GCMs for 9000 years ago (9 kyr) and $2 \times CO_2$ reveals the underlying mechanisms of change to be quite different, namely, greater seasonality at 9 kyr associated with increased receipt of solar radiation during the northern summer compared with increased mean temperatures but no change in seasonality for $2 \times CO_2$. Although their meteorological causes are not the same, the consequences of these two climatic scenarios for rainfall and hydrology in the northern

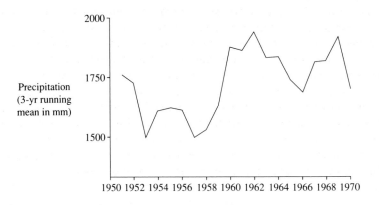

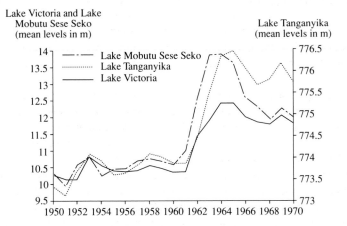

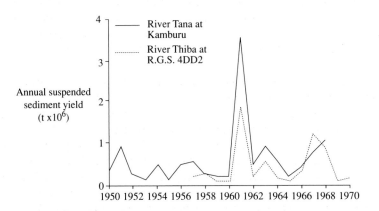

Figure 6.3 Precipitation, lake levels and sediment yield from East Africa, 1950−70. (a) Precipitation over Lake Victoria (after Kite, 1981); (b) water level change in selected East African lakes (after Kite, 1981); (c) sediment transport by two Kenyan rivers (after Walling, 1984)

hemisphere tropics are not dissimilar (Figure 6.2). For example, both the 9 kyr and 2 ×
CO_2 simulations show a significant increase in summer precipitation over East Africa and
South Asia. Examining the environmental record of the terminal Pleistocene and early
Holocene may thus prove informative for future scenarios in terms of the response of earth
surface biotic and physical systems to the onset of a warmer, wetter climate.

For the late Pleistocene—early Holocene, proxy methods of environmental reconstruction
have to be employed. In the context of soil erosion histories, this involves calculating the
changing rate and character of sedimentation in dated colluvial, alluvial and lacustrine
sequences. These methods cannot achieve the precision of observed or gauged data but they
do encompass changes of an order of magnitude similar to those that may occur in a future
world warmed by anthropogenic increases in greenhouse gases. A sediment-based approach
to erosional history has been attempted explicitly in a number of small temperate zone lakes
where a meso-scale of analysis has avoided the worst of the problems of sediment storage
and of spatial representativeness. In sites such as Frains Lake and Braederroch Loch,
accumulation rates typically increase towards the present-day as a result of human-induced
vegetation clearance and soil erosion (Davis, 1976; Edwards & Rowntree, 1980). On the
other hand, it is easier to attempt quantitative reconstruction of lake catchment sediment flux
for recent centuries than it is for more distant time periods like the last glacial—interglacial
transition. A number of further difficulties present themselves if data are to be meaningful.

1. *Chronology*: ideally sequences should be dated by a continuous chronology through
   laminated sediments or magnetic declination/inclination. If not, close interval ^{14}C dating
   is essential, and even then the results may be problematic (e.g. Bogoria I core; Tiercelin
   & Vincens, 1987). Sequences may also include unconformities that, if not properly
   identified, will grossly distort accumulation rates.
2. *Core representativeness and changing depo-centres*: accumulation rates vary spatially
   across the bed of a lake, so that single cores/sections may not reflect mean sedimentation
   rates. Ideally deposits should be studied from multiple cores (Dearing, 1986), but until
   very recently multiple coring has been restricted to the larger African lakes (e.g. Lake
   Turkana; Barton & Torgensen, 1988). Seismic profiles have revealed that these lakes
   usually have complex patterns of sedimentation related to their tectonic origin (e.g. Scholz
   & Rosendahl, 1988). Large lakes and deltas, in particular, may be subject to major shifts
   in the focus of deposition (or depo-centre) as a result of seismic activity or water level
   fluctuations.
3. Not all lake sediments are of *allochthonous* origin in the form of clastics eroded from
   the catchment. Deposits of biochemical origin partly reflect the flux of solutes from the
   catchment, but also respond to purely authigenic processes determined by parameters
   such as water temperature. In principle, therefore, sediment accumulation data should
   be corrected for biogenic silica (primarily diatom frustules and sponge spicules), for
   organic matter of algal origin and for authigenic carbonate, if clastic influx is to be
   calculated. An important feature of many tropical lakes is that they are hydrologically
   closed, i.e. they lack a surface outlet. It is the constant adjustment of these lakes' surface
   area and water depth to changing water balance that makes them valuable indicators of
   the regional $P - E$ ratio. On the other hand, closed lakes become ionically concentrated,
   and as water salinity increases so sedimentation becomes dominated by a predictable
   series of authigenic carbonate minerals (Eugster & Kelts, 1983).

There are few existing data from intertropical Africa that fully overcome all of these methodological difficulties. But while none can be truly considered to provide rigorous, quantitative data on past erosion rates, a number of sequences do show clear indications of late Quaternary shifts in the rate and type of sediment accumulation at times of climatic change and vegetational readjustment.

LATE QUATERNARY RECORDS OF SEDIMENT FLUX

Importantly, empirical palaeoclimatic and palaeohydrologic data, notably from ^{14}C-dated changes in the water level and chemistry of non-outlet lakes, indicate the late Quaternary shifts in climate were abrupt in character, as Knox's model requires (Street-Perrott & Perrott, 1990). In some cases, such as the lakes of the central and northern Ethiopian rift, the most important climatic transition may have been at the onset of the Holocene, ~ 10 000 BP (Gasse & Street, 1978). In a long core from one of these lakes (Abiyata), the sediment accumulation rate was exceptionally high between 10 300 and 9800 BP before slowing down during the early Holocene high-stands of the lake ~ 8000 BP (Lezine & Bonnefille, 1982) (Figure 6.4).

On the other hand, in most areas of intertropical Africa the last major dry-to-wet shift occurred during the terminal Pleistocene, around 12 500 BP. Although the period 12 500−11 000 BP did not mark the lake-level maximum in most basins, it may have had greater erosional impact than the early Holocene humid phase because it was preceded by a longer phase of arid climate (Street-Perrott, Roberts & Metcalfe, 1985). The lakes of East Africa's western rift are fed by rivers whose catchments are today located in the humid tropical montane zone, but which were replaced by much more open vegetation formations at and after the last glacial maximum (Vincens, 1989; Bonnefille, Roeland & Gruiot, 1990). Cores from several of these lake basins, including Tanganyika, Kivu and Mobutu Sese Seko, show enhanced sediment accumulation between about 13 000 and 11 000 BP (Figure 6.4). Comparison of pollen and diatom records from Tanganyika core MPU 12 indicates a time-lag of several centuries in the vegetation response to the onset of a wetter climate as reflected in rising lake levels, around 12 500 BP (Gasse et al., 1989; Vincens, 1989). There is also evidence of disturbance in the headwater zone of these East African lake catchments during the terminal Pleistocene. For instance, at Muchoya Swamp in the Rukinga highlands of southwest Uganda (2260 m), a 4 m thick clay band of fluvial or mass movement origin was sandwiched between peat (Taylor, 1990). This clay was deposited rapidly ~ 11 000 BP at a time when pollen indicate a replacement of montane ericaceous vegetation by moist forest.

Not all East African lakes show significant changes in sediment accumulation during the late Quaternary. Where autochthonous sediments of biochemical origin (e.g. diatomite) have predominated, the sedimentation rate has been relatively insensitive to erosional changes which occurred within the lake catchment. Lake Victoria, for instance, possesses sediments that are rich in organic matter and biogenic silica, and shows no significant change in accumulation rate during the late Pleistocene and Holocene, reflecting the dominant influence of atmospheric inputs into this lake (Kendall, 1969; Stager, 1984).

The influence of catchment conditions upon lake sediment records is well illustrated by the basins of the southern Gregory Rift (Figure 6.5). Magadi, which lies at the lowest part of the rift, is today a hypersaline lake fed almost exclusively by groundwater. On the other hand, as the climate passed from dry to wet during the Pleistocene−Holocene transition, Magadi's modest surface catchment expanded massively, receiving not only runoff from areas

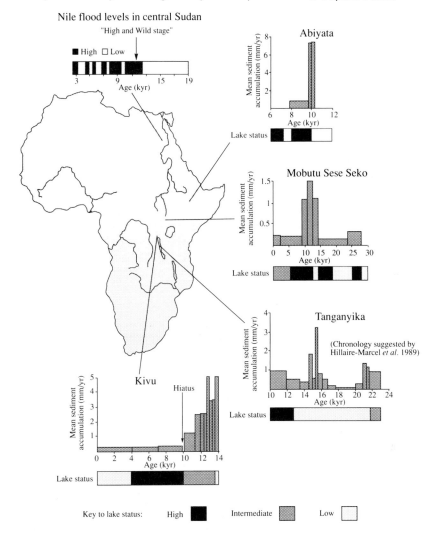

Figure 6.4 Palaeoenvironmental records of runoff and sedimentation at selected East African sites during the late Pleistocene and Holocene. Data sources — Abiyata: Lézine & Bonnefille (1982), Perrott (1979); Kivu: Degens & Hecky (1974); Mobutu Sese Seko: Harvey (1976); Nile: Adamson et al. (1980); Tanganyika: Gasse et al. (1989), Hillaire-Marcel et al. (1989)

presently characterised by ephemeral, spatially unconnected stream networks, but also overflow from adjacent catchments, which created a cascading hydrological system (Figure 6.6). Although the precise chronology of events has yet to be established, it is likely that overflows from Lake Naivasha passed through the Kedong valley, into the now-dry Koora basin and thence into Magadi (Washbourn-Kamau, 1975; Richardson & Dussinger, 1986; Baker, 1986). Certainly, evidence from a single ring of dated stromatolites shows that Magadi and Natron were jointed together as one deep lake between ~ 12 500 and ~ 9000 BP (Hillaire-Marcel & Casanova, 1987). It is even possible that Manyara overflowed northwards via Engaruka

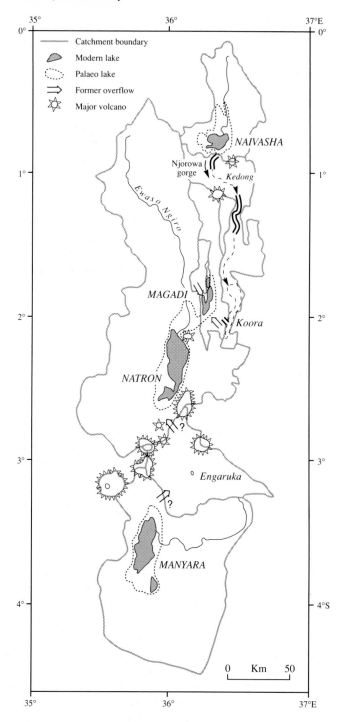

Figure 6.5   Past and present lake basins of the southern Gregory rift, East Africa

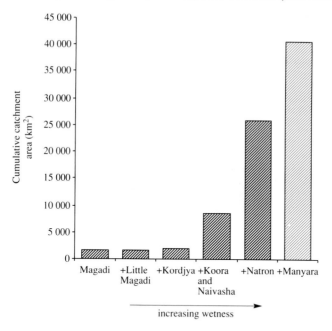

*N.B. An overflow from Manyara northward to lake Natron via the
Engaruka basin is yet to be confirmed.

Figure 6.6 Expansion in area of the Magadi catchment with increasing precipitation—
evaporation ratio

into Magadi—Natron. In sediment cores from Magadi, diatoms indicate an abrupt change
in lake water level and salinity at the beginning of the terminal Pleistocene wet phase ~
12 500 BP (diatom zone 1C/2 boundary: Barker *et al.*, 1990). By contrast, organic matter
above 2.78 m depth shows a lagged response, and in the intervening transient phase there
is a short but marked increase in magnetic susceptibility suggesting catchment erosion, before
it stabilises at lower levels (Taieb *et al.*, 1991; Williamson, 1991) (Figure 6.7).

Other African fluvial systems showed a similar expansion in drainage area during the arid-
to-humid transition at the end of the Pleistocene. During the arid phase around the time of
the last glacial maximum, the Nile's catchment shrank as headwater lakes like Victoria no
longer overflowed. But with the onset of a wetter climate its catchment area was enlarged
beyond its modern limits to include, for example, the Lake Turkana/Omo watershed (Adamson
et al., 1980). Nilotic sediments afford only limited evidence of sediment flux because most
of the Nile's sediment load has ended in the delta, but they do indicate flood levels and hence
catchment runoff. In this regard the highest flood levels recorded for the whole late Quaternary
are marked by the Sheikh Hassan silts (= Darau sub-stage), deposited above and beyond
the former floodplain. In upper Egypt these deposits lie up to 30 m above the floodplain and
have been described as belonging to a 'high and wild' stage of the River Nile (Butzer, 1980b).
Most importantly they date not to the period of highest lake levels in East Africa (9000—6000
BP) but to between 12 500 and 11 500 BP (Paulissen, 1986). Maximum runoff, it appears,
was generated not by maximum rainfall but by the onset of wet conditions before arboreal
vegetation had expanded, intercepting rainfall and releasing much of it back to the atmosphere

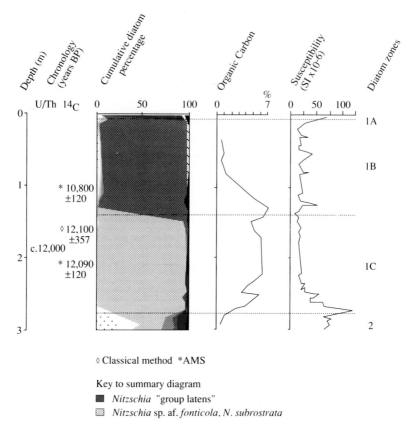

◊ Classical method   *AMS

Key to summary diagram
▦ *Nitzschia* "group latens"
▨ *Nitzschia* sp. af. *fonticola, N. subrostrata*

Figure 6.7 Diatom, organic carbon and magnetic susceptibility data for uppermost 3 m of Magadi core NF1 (data from Barker *et al.*, 1990; Taieb *et al.*, 1991; Williamson, 1991)

as evapotranspiration. Similar conclusions were obtained by Foucault & Stanley (1989) for sediment yield from the Ethiopian highland region drained by the Blue Nile and Atbara based on the Amphibole/Pyroxene ratios in Nile delta sediment cores.

In many respects, however, East Africa and the Nile does not represent an ideal setting for testing deterministic erosional models, because of the topographic complexity created by the Rift Valley system. A simpler climatic gradient is found in West Africa where bioclimatic zones are not interrupted by areas of high relief, but are aligned as broad latitudinal belts (Lezine, 1989). There are fewer lake sites available for study (or at least studied so far), making it difficult to use lake sedimentary records to assess changing erosional flux in different bioclimatic zones. However, the record from the small crater lake of Bosumtwi in Ghana is informative. This lake occupies a lowland location (+100 m above sea-level) and today lies in moist, semideciduous forest ~65 km south of the forest—savanna boundary. Core sediments covering the late Quaternary comprise fine clastics interlaminated with authigenic minerals, notably primary and diagenetic carbonates (Talbot *et al.*, 1984; Talbot and Kelts, 1986). The biogenic component is less important except for an algal sapropel mud deposited during the mid-Holocene. The onset of this phase of organic deposition at ~9000 BP was accompanied by a decrease in mean deepwater sedimentation from 0.66 to 0.39 m/1000

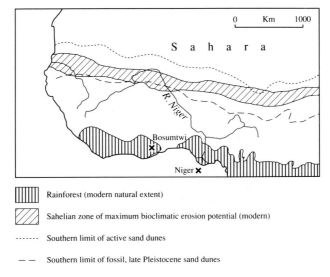

|||||| Rainforest (modern natural extent)

/////// Sahelian zone of maximum bioclimatic erosion potential (modern)

------- Southern limit of active sand dunes

— — Southern limit of fossil, late Pleistocene sand dunes

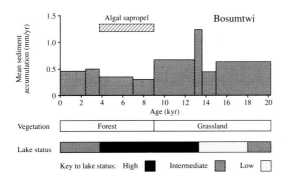

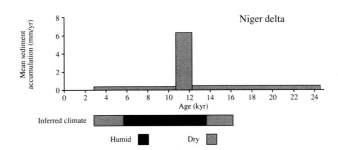

Figure 6.8 Late Quaternary changes in climate, vegetation and sediment flux in West Africa. Data sources — Lezine (1989); Bosumtwi: Talbot *et al*. (1984); Niger: Pastouret *et al*. (1978)

years (Figure 6.8). Pollen analyses indicate that these changes coincided with the replacement of wooded grassland by forest vegetation in the catchment, the latter significantly reducing the supply of clastic sediment to the lake. The highest late Quaternary accumulation rates in both central cores and around the basin margin occurred at and after a major rise in lake level ~ 12 500 BP (Talbot & Delibrias, 1980), corresponding to the period of 3000−4000 years between the first 'wetting up' and the readvance of forest. High rates of sediment flux are thus associated with a markedly seasonal climate (indicated by the deposition of clastic laminae) and a non-arboreal vegetation, with maximum values occurring during the dry-to-wet climatic transition at the end of the Pleistocene. The lowest flux rates took place during the phase between 9000 and 3750 BP when a woodland cover became established and when a low-seasonality climatic regime prevailed.

Evidence of terminal Pleistocene landscape instability in West Africa also comes from the sedimentary record of the River Niger. Much of the Niger catchment lies in the semi-arid Sahelian zone, and fossil dune fields testify to the fact that this zone shifted further south and, as with the Nile, reduced the river's catchment during the late Pleistocene. With the onset of wetter conditions ~ 12 500 BP the Niger discharge increased, eroding and transporting large volumes of sediment — such as dune sand — that were available (Talbot & Williams, 1979; Talbot, 1980). Extensive coarse-grained fluviatile sediments were laid down in the river floodplain or carried to the delta. A long well-dated core from the outer delta shows dramatic changes in sediment accumulation — being relatively low during both Pleistocene dry conditions and Holocene wet conditions, but very high during the transition between the two (Pastouret et al., 1978). Sedimentation rates increased more than 10 times between 11 500 and 10 900 BP, as the flux of sediment from the river basin reached a peak (Figure 6.8).

CONCLUSION

In most of intertropical Africa, the last major dry-to-wet climatic transition occurred 13 000−11 000 BP. Evidence of landscape instability at this time comes from the sedimentary records of African lake basins, and fluvial and deltaic sequences in the Nile and Niger catchments. The increase in sediment flux predicted by Knox's (1972) biogeomorphic response model is consequently given a measure of empirical support. However, not all of this increase was necessarily due to catchment soil erosion. Much of the material eroded was of aeolian origin, which itself had been deflated from dry lake basins during the preceding phase of arid climate. This recycling of the same sediment during alternating arid−humid cycles is not easily accommodated within the watershed catchment framework for measuring sediment flux. Sediment storage within catchments will also have led to a time lag between the period of hillslope erosion or atmospheric influx and the period when sediment was evacuated by the fluvial system, a lag which may have been substantial in larger drainage basins, such as the Nile and the Niger. In effect, the onset of a wetter climate ~ 12 500 BP saw a change from erosion limited by sediment transport (by water) to one limited by sediment supply, but in the interim high rates of sediment flux prevailed.

The Knox model may help to explain empirically derived data on late Quaternary sediment flux in intertropical Africa, but it does not seek to model the relationship between soil stability, rainfall and vegetation in any more than a schematic way. The vegetational response to increased precipitation, for example, is much more complex than this model allows. Biotic adjustment involves both a quick response by herbaceous ground cover, and slower compositional change. In moisture-limited habitats, the latter potentially includes the expansion of woodland at the expense of non-arboreal vegetation. The late Quaternary record from sites such as Bosumtwi confirms the historical importance of tree cover in reducing runoff and surface erosion. Not only are trees effective 'scavengers' in the competitive interaction for water between erosion and vegetation (Thornes, 1990), but their greater resilience to short-term drought helps maintain continuity of ground cover, whereas grassland offers very poor protection when degraded. Nor does the Knox model take into account all the relevant factors that have determined palaeohydrological conditions. A fully comprehensive model would also incorporate the role of hydrological pathways other than runoff (e.g. infiltration to groundwater) in which pedogenetic adjustment to climatic change plays a critical role (Fontes *et al.*, 1991).

While similarities do exist between the terminal Pleistocene—early Holocene transition from a drier, cooler to a warmer, wetter climate and that which is predicted to occur over the course of the next century as a result of increases in greenhouse gases, it would be wrong to draw any parallels too closely. Former periods of climatically induced landscape instability can never be true analogues for the environmental future. None the less, the rate of adjustment of biogeomorphic systems during the abrupt onset of a wetter climate ~ 12 500 BP gives us some measure of the timescale over which landscapes may be expected to readjust. Above all, the long-term record of erosion suggests that it will be during the transient phase of any future climatic change that landscape instability will be greatest, and that during this transition the response of earth surface systems to climatic forcing will be complex and non-linear.

ACKNOWLEDGEMENTS

The authors are pleased to thank Maurice Taieb, Michel Icole and David Williamson for permission to reproduce some of the data used in Figure 6.7, to referees for helpful comments, to Erica Millwain for cartographic assistance and to Gwynneth Barnwell for putting the bibliography in the correct format.

REFERENCES

Adamson, D. A., Gasse, F., Street, F. A. and Williams, M. A. J. (1980). Late Quaternary history of the Nile. *Nature*, **287**, 50—55.

Baker, B. H. (1986). Tectonics and volcanism of the southern Kenya Rift Valley and its influence on rift sedimentation. In Frostick, L. E. *et al.* (eds), *Sedimentation in the African Rifts*, Blackwell, Oxford, pp. 45—57.

Barker, P., Gasse, F., Roberts, N. and Taieb, M. (1990). Taphonomy and diagenesis in diatom assemblages: a Late Pleistocene palaeoecological study from Lake Magadi, Kenya. *Hydrobiologia*, **214**, 267—272.

Barton, C. E. and Torgersen, T. (1988). Palaeomagnetic and ^{210}Pb estimates of sedimentation in Lake Turkana, East Africa. *Palaeogeography, Palaeoclimatology, Palaeoecology*, **68**, 53—60.

Bonnefille, R., Roeland, J. C. and Gruiot, J. (1990). Temperature and rainfall estimates for the past 40 000 years in equatorial Africa. *Nature*, **346**, 347—349.

Butzer, K. W. (1980a). Adaptation to global environmental change. *Professional Geographer*, **32**, 269—278.

Butzer, K. W. (1980b). Pleistocene history of the Nile valley in Egypt and Lower Nubia. In Williams, M. A. J. and Faure, H. (eds), *The Sahara and the Nile*, Balkema, Rotterdam, pp. 253—280.

Davis, M. B. (1976). Erosion rates and land use history in southern Michigan. *Environmental Conservation*, **3**, 139—148.

Dearing, J. A. (1986). Core correlation and total sediment influx. In Berglund, B. E. (ed.), *Handbook of Holocene Palaeoecology and Palaeohydrology,* Wiley, Chichester, pp. 247—272.

Degens, E. T. and Hecky, R. E. (1974). Palaeoclimatic reconstruction of Late Pleistocene and Holocene based on biogenic sediments from the Black Sea and a Tropical African Lake. *Colloques Internationaux du CNRS,* **219**, 13—24.

Douglas, I. (1967). Man, vegetation and the sediment yield of rivers. *Nature,* **215**, 25—28.

Douglas, I. (1976). Erosion rates and climate: geomorphological implications. In Derbyshire, E. (ed.), *Geomorphology and Climate,* Wiley, Chichester, pp. 269—287.

Edwards, K. J. and Rowntree, K. M. (1980). Radiocarbon and palaeoenvironmental evidence for changing rates of erosion at a Flandrian stage site in Scotland. In Cullingford, R. A., Davidson, D. A. and Lewin, J. (eds), *Timescales in Geomorphology,* Wiley, Chichester, pp. 207—223.

Elwell, H. A. (1984). Soil loss estimation: a modelling technique. In Hadley, R. F. and Walling, D. E. (eds), *Erosion and Sediment Yield: Some Methods of Measurement and Modelling,* Geobooks, Norwich, pp. 15—36.

Eugster, H. P. and Kelts, K. (1983). Lacustrine chemical sediments. In Goudie, A. S. and Pye, K. (eds), *Chemical Sediments and Geomorphology,* Academic Press, London, pp. 321—368.

Eybergen, F. A. and Imeson, A. C. (1989). Geomorphological processes and climatic change. *Catena,* **16**, 307—319.

Fontes, J.-Ch., Andrews, J. N., Edmunds, W. M., Guerre, A. and Travi, Y. (1991). Palaeorecharge by the Niger River (Mali) deduced from groundwater geochemistry. *Water Resources Research,* **27**, 199—214.

Foucoult, A. and Stanley, D. J. (1989). Late Quaternary palaeoclimatic oscillations in East Africa recorded by heavy minerals in the Nile Valley, *Nature,* **339**, 44—46.

Gasse, F. and Street, F. A. (1978). Late Quaternary lake level fluctuations and environments of the Northern Rift Valley and Afar Region (Ethiopia and Djibouti). *Palaeogeography, Palaeoclimatology, Palaeoecology,* **24**, 279—325.

Gasse, F., Ledée, V., Massault, M., Fontes, J.-C. (1989). Water-level fluctuations of Lake Tanganyika in phase with oceanic changes during the last glaciation and deglaciation. *Nature,* **342**, 57—59.

Hamilton, A. C. (1982). *Environmental History of East Africa,* Academic Press, London.

Harvey, T. J. (1976). The palaeolimnology of Lake Mobutu Sese Seko, Uganda—Zaire: the last 28 000 years. Ph.D. Dissertation, Duke University.

Henderson-Sellers, A. (1993). Numerical modelling of global climates. In Roberts, N. (ed.), *The Changing Global Environment,* Blackwell, Oxford, in press.

Hillaire-Marcel, C. and Casanova, J. (1987). Isotopic hydrology and palaeohydrology of the Magadi (Kenya)—Natron (Tanzania) basin during the late Quaternary. *Palaeogeography, Palaeoclimatology, Palaeoecology,* **58**, 155—181.

Hillaire-Marcel, C., Aucour, A.-M., Bonnefille, R., Riollet, G., Vincens, A. and Williamson, D. (1989). ^{13}C/Palynological evidence of differential residence times of organic carbon prior to its sedimentation in East African rift lakes and peak bogs. *Quaternary Science Reviews,* **8**, 207—212.

Kendall, R. L. (1969). An ecological history of the Lake Victoria basin. *Ecological Monographs,* **39**, 121—176.

Kite, G. W. (1981). Recent changes in level of Lake Victoria. *Hydrological Sciences Bulletin,* **26**, 233—243.

Knox, J. C. (1972). Valley alluviation in southwestern Wisconsin. *Annals of the Association of American Geographers,* **62**, 401—410.

Knox, J. C. (1984). Responses of river systems to Holocene climates. In Wright, H. E. (ed.), *Late Quaternary Environments of the United States,* vol. 2: *The Holocene,* Longman, London, pp. 26—41.

Kutzbach, J. (1983). Monsoon rains of the Late Pleistocene and early Holocene: patterns, intensity and possible causes of changes. In Street-Perrott, A., Beran, M. and Ratcliffe, R. (eds), *Variations in the Global Water Budget,* Reidel, Dordrecht, pp. 371—389.

Lamb, H. H. (1966). Climate in the 1960s. *Geographical Journal,* **132**, 184—212.

Langbein, W. B. and Schumm, S. A. (1958). Yield of sediment in relation to mean annual precipitation. *Transactions of the American Geophysical Union,* **39**, 1079—1084.

Lezine, A.-M. (1989). Late Quaternary vegetation and climate of the Sahel. *Quaternary Research,* **32**, 317—334.

Lezine, A.-M. and Bonnefille, R. (1982). Diagramme Pollinique Holocéne d'un sondage du Lac Abiyata (Éthiopie, 7°42′ nord). *Pollen et Spores*, **24**, 463−480.

Mitchell, J, F. B. (1990). Greenhouse warming: is the Holocene a good analogue? *Journal of Climate*, **3**, 1177−1192.

Pastouret, L., Charnley, H., Delibrias, G., Duplessy, J.-C. and Thiede, J. (1978). Late Quaternary climatic changes in western tropical Africa deduced from deep-sea sedimentation off the Niger Delta. *Oceanologia Acta*, **1**, 217−232.

Paulissen, E. (1986). Characteristics of the 'Wild' Nile stage in Upper Egypt. *Changements globaux en Afrique*. Proceedings of the INQUA Dakar Symposium, pp. 367−369.

Perrott, F. A. (1979). Late Quaternary lakes in the Ziway−Shala basin, southern Ethiopia. Ph.D. thesis, University of Cambridge.

Richardson, J. L. and Dussinger, R. A. (1986). Paleolimnology of mid-elevation lakes in the Kenya Rift Valley. *Hydrobiologia*, **143**, 167−174.

Rind, D. (1990). Puzzles from the tropics. *Nature*, **346**, 317−318.

Roberts, N. (1990). Ups and downs of African lakes. *Nature*, **346**, 107.

Roberts, N. and Lambert, R. (1990). Degradation of dambo soils and peasant agriculture in Zimbabwe. In Boardman, J., Foster, I. D. L. and Dearing, J. A. (eds), *Soil Erosion from Agricultural Land*, Wiley, Chichester, pp. 537−558.

Scholz, C. A. and Rosendahl, B. R. (1988). Low stands of Lakes Malawi and Tanganyika, east Africa, delineated with multi-fold seismic data. *Science*, **240**, 1645−1648.

Stager, J. C. (1984). The diatom record of Lake Victoria: the last 17 000 years. In Mann, D. G. (ed.), *Proceedings of the 7th International Symposium on Living and Fossil Diatoms*, Philadelphia, pp. 475−476.

Street-Perrott, F. A. and Perrott, R. A. (1990). Abrupt climatic fluctuations in the tropics: the influence of Atlantic Ocean circulation. *Nature*, **343**, 607−612.

Street-Perrott, F. A., Roberts, N. and Metcalfe, S. (1985). Geomorphic implications of Late Quaternary hydrological and climatic changes in the Northern Hemisphere tropics. In Douglas, I. and Spencer, T. (eds), *Environmental Change and Tropical Geomorphology*. Allen & Unwin, London, pp. 165−183.

Taieb, M., Hillaire-Marcel, C., Barker, P., Bonnefille, R., Damnati, B., Gasse, F., Goetz, C., Icole, M., Lafont, R., Roberts, N., Vincens, A. and Williamson, D. (1991). Histoire paléohydrologique du lac Magadi (Kenya) au Pléistocène supérieure. *Comptes Rendues de l'Academie des Sciences, Paris*, Vol. 313. Série II, 339−346.

Talbot, M. (1980). Environmental responses to climatic change in the West African Sahel over the past 20 000 years. In Williams, M. A. J. and Faure, H. (eds), *The Sahara and the Nile*, Balkema, Rotterdam, pp. 37−62.

Talbot, M. and Delibrias, G. (1980). A new late Pleistocene−Holocene water-level curve for Lake Bosumtwi, Ghana. *Earth Planetary Science Letters*, **47**, 336−344.

Talbot, M. R. and Kelts, K. (1986). Primary and diagenetic carbonates in the anoxic sediments of Lake Bosumtwi, Ghana. *Geology*, **14**, 912−916.

Talbot, M. and Williams, M. A. J. (1979). Cyclic alluvial fan sedimentation on the flanks of fixed dunes, Janjari, Central Niger. *Catena*, **6**, 43−62.

Talbot, M., Livingstone, D. A., Palmer, P. G., Maley, J., Melack, J. M., Delibrias, G. and Gullicksen, S. (1984). Preliminary results from sediment cores from Lake Bosumtwi, Ghana. *Palaeoecology of Africa*, **16**, 173−192.

Taylor, D. M. (1990). Late Quaternary pollen records from two Ugandan mires: evidence for environmental change in the Rukinga highlands of southwest Uganda. *Palaeogeography, Palaeoclimatology, Palaeoecology*, **80**, 283−300.

Thornes, J. B. (1990). The interaction of erosional and vegetational dynamics in land degradation: spatial outcomes. In Thornes, J. B. (ed.), *Vegetation and Erosion*, Wiley, Chichester, pp: 41−53.

Tiercelin, J.-J. and Vincens, A. (1987). Le demi-graben de Baringo-Bogoria Rift Gregory, Kenya. *Bulletin du Centre de Recherche Exploration et Production Elf- Aquitaine*, **11**(2), 249−540.

Vincens, A. (1989). Les forêts claires zambéziennes du bassin Sud-Tanganyika. Évolution entre 25 000 et 6 000 ans BP. *Comptes Rendues de l'Academie des Sciences Paris*, **308**, série II, 809−814.

Walling, D. E. (1984). The sediment yields of African rivers. In Walling, D. E., Foster, S. S. D.

and Wurzel, P. (eds), *Challenges in African Hydrology and Water Resources*, Proceedings of the Harare Symposium, July 1984, IAHS Publ. 144, pp. 265–283.

Walling, D. E. and Webb, B. W. (1983). Patterns of sediment yield. In Gregory, K. J. (ed.), *Background to Palaeohydrology*, Wiley, Chichester, pp. 69–100.

Warrick, R. and Farmer, G. (1990). The greenhouse effect, climatic change and rising sea level: implications for development. *Transactions of the Institute of British Geographers, NS15*, 5–20.

Washbourn-Kamau, C. K. (1975). Late Quaternary shorelines of Lake Naivasha, Kenya. *Azania,* **10**, 77–92.

Williamson, D. (1991). Étude paléomagnetique de séquences sédimentaires de Méditerranée et d'Afrique intertropicale: paléoenvironnements, stratigraphie des dépôts et variation séculaire au Pléistocène supérieur. Ph.D. thesis, University Aix Marseille II.

Wischmeier, W. H. and Smith, D. D. (1978). Predicting rainfall erosion losses — a guide to conservation planning. *USDA ARS Conservation Research Report,* **35**, Washington, DC.

7 Geomorphological Processes, Environmental Change and Landscape Sensitivity in the Kalahari Region of Southern Africa

PAUL A. SHAW
School of Geology, Luton College of Higher Education, UK

and

DAVID S. G. THOMAS
Department of Geography, University of Sheffield, UK

ABSTRACT

Kalahari Group stratigraphy covers some 2.5 million km^2 of southern Africa. These sediments, of which the Kalahari Sand is the most ubiquitous, are of post-Cretaceous age, and have low fossil and organic contents — characteristics that have impeded geological interpretation in the past. In the last two decades geomorphological studies, focused on the semi-arid region (the 'Kalahari Desert') south of the Zambezi River, have provided contemporary and past geomorphological processes on the landscape. Research on caves, lakes, pans and fluvial links have indicated widespread humid episodes in the late Quaternary that are asynchronous with those in other sub-tropical arid regions, and which have been magnified, particularly in the Okavango Delta and adjacent parts of northern Botswana, by tectonic activity. Variations in groundwater recharge and chemistry have led to the formation of complex duricrust suites up to 100 m in thickness, many of which are associated with surface landforms. The most enigmatic landforms are vast 'fossil' linear dune fields, which cannot be related to present or past climatic parameters, and for which new explanations are being sought. Importantly, correct palaeoenvironmental interpretations require recognition and understanding of the sensitivities of individual landscape units to environmental changes. This in turn requires not just the application of modern analogues but recognition of regional hydrological complexities and responses to local, as well as more general, geomorphic controls.

INTRODUCTION

The postwar era has seen great advances in the understanding of tropical and subtropical landscapes at a variety of temporal and spatial scales. Significantly research has tended to concentrate on process studies, and on the reconstruction of past environments, mostly within the range of radiocarbon dating. In both directions there has been a tendency to emphasise erosional landscapes and well-defined climatic environments.

Landscape Sensitivity. Edited by D. S. G. Thomas and R. J. Allison
© 1993 John Wiley and Sons Ltd

Large areas of the low-latitude continents, however, experience semi-arid climates characterised by a range of temperature and precipitation conditions, including high inter-annual and intra-annual precipitation variability. Such regions are not only climatically unpredicatable, but also occupy positions between the tropical and temperate circulations that render them liable to dramatic change on the Quaternary timescale. Such environments require patient study; there is sufficient vegetation to mask surface landforms, while in process studies the wait for specific fluvial or aeolian events can be a frustrating experience.

The Kalahari is an extensive semi-arid depositional environment that has, over the past two decades, yielded much information on the rates and processes of landscape change. Some of these changes are specific to the Kalahari, while others have wider application. This paper summarises the results of this research.

THE KALAHARI

The initial impression of the Kalahari suggests that the landscape has recorded past episodes of both greater and lesser precipitation than the present mean, which varies from 250 mm year^{-1} in the south-west, to 700 mm year^{-1} on the Zambezi. Large dry valleys and extensive lake basins suggest the former, while massive linear dune fields covering most of the Kalahari Desert suggest greater aridity, based on the assumption that the dunes are now fixed by a vegetation cover. The problem then becomes one of identifying the extent and chronology of these climatic episodes, using a suite of sediments that are singularly unhelpful. Most of the Kalahari Group comprises sands, duricrusts, marls and conglomerates of Cretaceous to Recent age, which are non-fossiliferous and have, in the case of the ubiquitous sand, ambiguous sedimentary characteristics (Thomas, 1987). Attempts to define the stratigraphy of the Kalahari have been made since the beginning of the century (Passarge, 1904), but have been constantly thwarted by the inability to distinguish genuine strata from post-depositional alteration to duricrusts, which reach up to 100 m in thickness (Bruno, 1985). Although the interpretation problem has been recognised for some time (e.g. Rogers, 1936), there are still many geologists who persist in applying strict stratigraphic terminology to the Kalahari sequence, which, of course, varies spatially too (e.g. Wright, 1978; Malherbe, 1984).

The problem is compounded by the nature of the duricrust suite. Although research has covered many aspects of calcretes (Goudie, 1971) and silcretes (Smale, 1973; Summerfield, 1982, 1983), detailed site studies (e.g. Gwosdz and Modisi, 1983) show a bewildering range of intermediate duricrust forms over short distances, which could be termed sil-calcretes or cal-silcretes, depending on their major constituent. So great is this variety that Mazor et al. (1977) suggest the word 'crete' be used as a blanket term. Likewise, the thickess of the duricrusts has led to the assumption (e.g. Goudie, 1973, Summerfield, 1983) that they must be at least of Tertiary age. Such antiquity would render meaningless any climatic interpretation; Rust, Schmidt & Dietz (1984) suggest that climatic variability even within the last 20 000 years would render such an exercise pointless, though many attempts have been made (e.g. Heine, 1978, 1982).

From the tectonic viewpoint, the Kalahari Basin has been gradually evolving as a depositional setting since the break-up of Gondwanaland, with gradual epeirogenesis at the Kalahari rim. This stability has allowed Pickford (1990) to suggest, on the basis of the study of Pliocene cave floors, that the overall rate of rock erosion may be of the order of 0.1−0.2 mm year^{-1}. However, the extension of the East African Rift into the Middle Kalahari via the Gwembe

Trough and the Okavango Delta introduces an element of neo-tectonism that has had a profound effect on the geomorphology of the area (Cooke, 1980) and its sensitivity to climatic changes, and which persists in the year to year functioning of the Okavango Delta (Shaw, 1985).

'WET' LANDFORMS — CAVES, LAKES AND PERENNIAL RIVERS

The most persuasive evidence of greater precipitation comes from caves and lacustrine landforms in the Middle Kalahari (Figure 7.1). Drotsky's Cave in the Gcwihaba Hills of north-west Ngamiland has provided 26 ^{14}C dates from sinter deposits which represent periods in the late Quaternary during which local precipitation may have reached 300% of the present mean (Cooke and Verhagen, 1977; Cooke, 1984). Some of this carbon dating has been carried out in tandem with uranium/thorium dating on the same samples, while the latter method has recently been used to probe the environmental history back to 300 000 BP (Brook, Burney

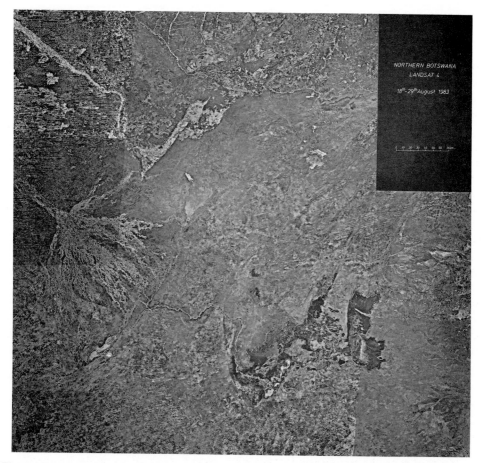

Figure 7.1 Landsat mosaic of the Middle Kalahari, northern Botswana. The Okavango Delta can be seen centre left linked, by the Boteti River, to the Makgadikgadi basin in the bottom right. Over much of the area, vegetated linear dunes appear as parallel striping, especially to the north and west of both the Delta and Makgadikgadi

86

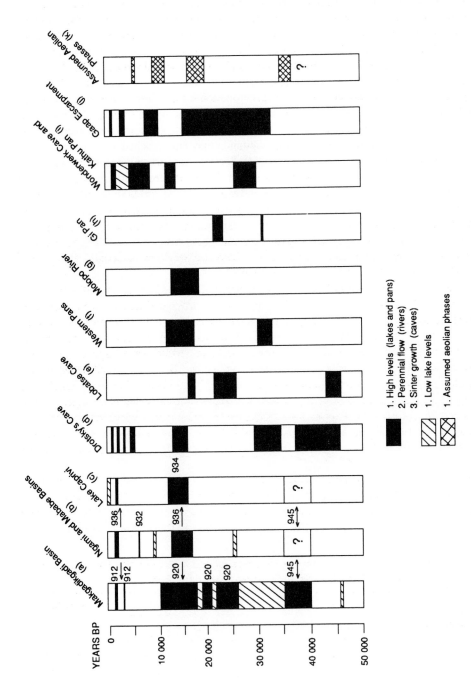

Figure 7.2 Summary of the palaeoclimatic data for the Kalahari. Sources — as per text

& Cowart, 1990). During the past 20 000 years wetter episodes are indicated at 16 000–13 000 BP, and on a number of occasions in the late Holocene (Figure 7.2).

The Okavango Delta is currently by far the most dynamic landform system in the Kalahari, with an annual flood cycle varying the inundated area from some 6000 to 13 000 km^2. Considerable variations have been noted in the distribution of floodwater throughout the Delta during the past 150 years (Wilson, 1973; Shaw, 1985), and on an inter-annual basis, depending on factors ranging from precipitation and tectonism through to fire and the activities of hippopotami! Approximately 400 000 m^3 year^{-1} of bedload sediment is deposited within the delta (UNDP/FAO, 1977), while probably three times that amount of solutes are deposited, the salt precipitation contributing greatly to the micromorphology of the islands and terraces in the upper delta (McCarthy & Metcalfe, 1991).

On a larger scale the Okavango (including 22 000 km^2 of contemporary and fossil sediments) is joined laterally to the Zambezi and Chobe Rivers, and downstream to the Makgadikgadi, Mababe and Ngami lake basins by a series of fault-controlled fluvial links. The configuration and evolution of the stages of the resulting massive palaeolake (Figure 7.3) have been described in a number of papers (see Shaw, 1988, for a recent summary), while calcretes and associated shell deposits on strandlines have provided 48 ^{14}C dates on which a chronology can be based (Cooke, 1984; Shaw, 1985; Shaw & Cooke, 1986; Shaw, Cooke & Thomas, 1988; Shaw & Thomas, 1988).

Two major stages have been identified. The higher, Lake Palaeo-Makgadikgadi stage (Grey & Cooke, 1977), at 845 m asl, lies at the limits of radiocarbon dating, and, with a lake area covering at least 60 000 km^2, has a hypothetical hydrological budget that can only be satisfied by the diversion of the Zambezi into the lake. Its interpretation is thus problematic, but cannot be viewed solely in climatic terms. The Lake Thamalakane stage at 936 m asl, however, would have been a shallow lake of some 7000 km^2 overflowing to the Makgadikgadi, with inputs from both the Angolan Highlands via the Okavango, Chobe and Zambezi Rivers and from local precipitation. Revised estimates of the hydrological budget suggest that only a 100% increase in mean annual precipitation would be sufficient to maintain this stage, as the flat terrain, finely adjusted fluvial links between lake basins, and the amplifier effect inherent in the hydrology of the Okavango Delta would lead to a high degree of sensitivity in the system.

The lake Thamalakane stage seems to have been coeval with the wet periods indicated in Drotsky's Cave. The subsequent desiccation of the lake led to widespread calcification of adjacent sediments, probably associated with the fall in the lake-associated water-table. This can be inferred from sets of paired dates associated with the 936 m level, in which aquatic

Table 7.1 Paired radiocarbon dates for sites associated with the Lake Thamalakane stage (variation expressed as a percentage of the 'real-time' age)

Site	Material	Date (BP)	Lab. no	Variation (%)
Ngwezumba River,	mollusca	15 570 ± 220	GrN 14788	
Mababe (936 m)	calcrete	13 070 ± 140	GrN 12623	16
Serondela Terrace,	mollusca	15 380 ± 140	GrN 13192	
Chobe River (934 m)	calcrete	11 550 ± 110	GrN 13191	25
Gidikwe Ridge,	mollusca	14 070 ± 150	GrN 14786	
Makgadikgadi (~920 m)	CO$_3$ reed casts	11 980 ± 130	GrN 15536	15

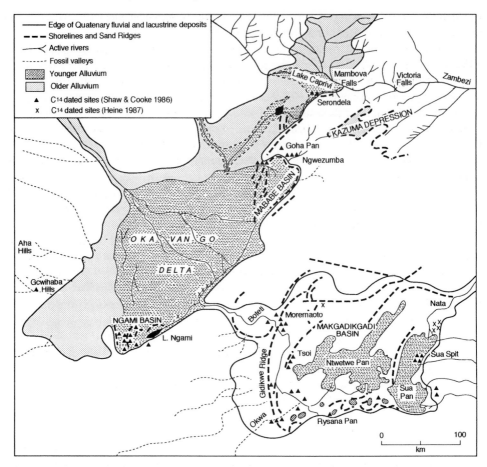

Figure 7.3   The drainage of the Middle Kalahari. After Shaw & Cooke (1986) and Heine (1987)

molluscs have been dated alongside their calcrete matrices, or associated calcareous forms (Table 7.1). Although the chances of cross-contamination between organic and inorganic carbonate are high, there appears to be a consistent variation between the 'real-time' indicator and the subsequent development of the matrix. This would suggest that much of the soft, alluvial calcrete associated with the palaeolakes is of recent origin, as are some of the duricrusts encountered on the lake beds, for example, the silcretes of Sua Pan (Shaw, Cooke & Perry, 1991).

GROUNDWATER LANDFORMS — PANS AND DRY VALLEYS

Pans and dry valleys (termed *mekgacha* in Botswana) are two major Kalahari landforms associated with extensive duricrust development. Until recently the evolution of the former has been associated with aeolian deflation of sediments, as suggested by the presence of fringing lunette dunes (e.g. Lancaster, 1978), while the latter have been attributed to fluvial action during periods of wetter climate. Detailed examinations of the morphology of both landforms, particularly in relation to underlying geological and hydrogeological conditions, suggests that both are a function of groundwater weathering and duricrust formation.

A detailed three-dimensional study of two pans in the south-east Kalahari (Butterworth, 1982; Farr *et al.*, 1982) indicates, in both cases, weathering and duricrust formation to the limits of drilling at 30 m, and a strong association with sub-Kalahari lineations. This link to zones of groundwater convergence has also been demonstrated in the southern Kalahari (Arad, 1984). Changes in the zonation of fresh and saline groundwaters in the pan, which would be significant in silcrete formation, have been explained by upward flushing of the less dense saline water during episodes of local recharge (Bruno, 1985).

Although lunette dunes are common on the leeward side of Kalahari pans, there are problems in linking them to pan evolution. Not least of these is the fact that the sediments comprising the lunettes are much coarser than those encountered in the pans themselves (Goudie & Thomas, 1986). Until this aspect has received further study it is essential to consider pans as polygenetic forms, with aeolian activity modifying pan morphology.

The dry valleys of the Kalahari can be divided into two categories: an endoreic system draining towards the Okavango Delta or the Makgadikgadi Basin, and a network of valleys associated with the Atlantic drainage of the Molopo River. The former, which includes the massive Okwa-Mmone network, with a potential catchment of 70 000 km^2, rise at the margins of the Kalahari Basin, where bedrock is either exposed, or lies close to the surface, and show a variety of distinctive valley forms, including an incised head-valley section associated with a wide range of duricrust types. Within the Kalahari itself the valley loses relief and adopts a dambo form of linked shallow depressions. The valleys lack active channels, and only a few records of sporadic flow exist, associated with high intensity rainfall events. Valleys of this type show deep weathering along pre-Kalahari (and in some cases, pre-Karoo) lineations, and Shaw & De Vries (1988) suggest that they are compatible with groundwater erosion along preferential flow paths, a hypothesis supported by the presence of near-surface groundwater levels until the early years of the 20th century.

The Molopo and its tributaries differ from these true *mekgacha* by displaying, in varying degrees, some of the characteristics of ephemeral hardveld rivers, including some active channel development. Floods have been recorded in all four major rivers (the Molopo, Nossop, Auob and Kuruman) at approximately 20 year intervals, while the Kuruman is perennial in its headwater section as a result of spring activity. These characteristics offer the opportunity to estimate the magnitude of past rainfall events by comparison of flood deposits, for the flood that extended the whole length of the Kuruman in February 1988 in response to a regional precipitation event of an estimated return time of 1 in 100 years is a useful measure of extreme conditions.

'DRY' LANDFORMS — THE KALAHARI ERG

The Kalahari erg is dominated by linear dunes, which, in the area south of the Zambezi, have been divided into three major dune fields on the basis of dune morphology and orientation. Linear dunes account for the vast majority of all dune forms in the Mega Kalahari and some 85% of dunes in the southern dunefield (e.g. Fryberger & Goudie, 1981), where annual preciptation is lowest. The pattern of the dunefields forms a 'wheelround' reflecting the anticyclonic circulation over the subcontinent (Figure 7.4). Less common dune forms include transverse dunes (barchans, barchanoid ridges) and parabolic dunes (lunettes, nested

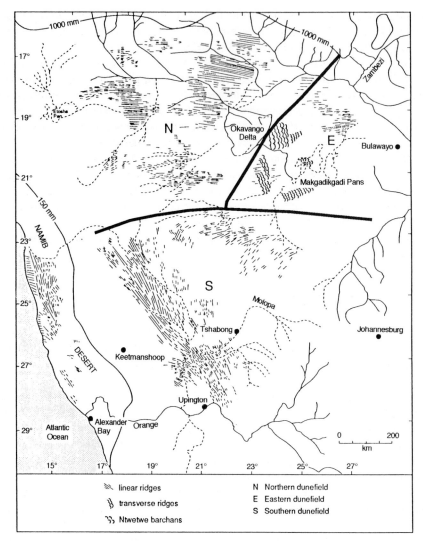

Figure 7.4   Desert dunefields of southern Africa, modified after Lancaster (1981)

parabolics). The morphometry and sedimentary characteristics of these dunefields have received detailed study (Lancaster, 1981, 1986, 1987, 1988; Thomas, 1984, 1988; Thomas & Martin, 1987).

The linear dunes have been hitherto regarded almost exclusively as relict landforms, representative of episodes of drier climate. Interpretations of the dune fields have been directed at the elucidation of the direction and strength of palaeo-wind circulations, and on identifying the extent of Quaternary aridity, based on the assumption that a maximum rainfall of 150−200 mm year^{-1} would permit dune mobility (Lancaster, 1980, 1981). Thomas (1984) pointed out that the application of this model to the erg north of the Zambezi would require an isohyet shift of 2300 km, and Lancaster (1988) has subsequently reassessed dune mobility against an index based on precipitation, potential evaporation and wind strength, concluding that the southern dunefield could be mobilised by decreases of 5°C and 30–40% precipitation, accompanied by a 117% increase in wind velocity. So far there has been no radiometrically dated evidence for shifts of this magnitude during the late Glacial.

Following the work of Ash & Wasson (1983), Wasson & Nanninga (1986) and Tsoar & Moller (1986) on the effectiveness of aeolian activity on partially vegetated linear dunes, Thomas (1989) has suggested that linear dunes are essentially sand-passing forms, capable of activity without extensive changes in morphology or movement of the dune body, and that (Thomas & Shaw, 1991) the southern dunefield is not necessarily relict, displaying evidence of dune surface activity and sand movement, under its present partial vegetation cover. Relict status is more applicable to the degraded and more heavily vegetated dunes of the northern and eastern fields, where a degree of vegetational and climatic change would be required for reactivation. We are, however, still no nearer dating these episodes of aridification, which may not necessarily be late Quaternary in age, nor in understanding when these dunefields evolved. In this context, drawing a distinction between when the different dune systems were formed and when they were last active may be an important and virtually unanswerable palaeoenvironmental problem.

QUATERNARY HISTORY

The Quaternary framework from 50 000 BP to present is fixed by a corpus of approximately 230 radiocarbon dates, of which 50% are from the Middle Kalahari, 10% from the southern and western parts, and the remainder from sites on the Kalahari periphery. With the exception of caves there are no closed sites, and archaeological and palynological evidence has contributed little so far to the palaeoenvironmental debate. Most dates refer to conditions of increased moisture availability on a local or regional scale. Palaeotemperature data is limited, and evidence for drier conditions represented by dunes and dune systems has been elusive, to the extent that aridity is assumed to have been prevalent when no evidence to the contrary exists. As the humid chronology has developed, the 'arid windows' have shrunk and shifted position (e.g. Lancaster, 1988, 1989).

A palaeoenvironmental summary is shown in Figure 7.2, and has already been discussed in relation to the lakes and caves of the Middle Kalahari. There is apparent conflict between some of the sites, which is not surprising given the fragmentary nature and poor resolution of some of the data, particularly before 20 000 BP. Well established in the sequence, however, are cold and dry conditions in the Kalahari for a relatively short period at the Last Glacial Maximum (19 000–18 000 BP), followed by a phase of greater moisture availability throughout the region at 16 000–13 000 BP, declining towards the Holocene. The Holocene record itself suggests conditions comparable to the present throughout the early millenia, with humid episodes in the Late Holocene, especially around 2000 BP. The Holocene picture is less consistent than that of the Last Glaciation.

At present the data are not sufficiently accurate to pinpoint the source of the moisture during these humid episodes. Lancaster (1979) has suggested a more southerly position of the Inter-Tropical Convergence Zone, Butzer et al. (1973) propose more effective winter cyclonic rainfall, and Rust, Schmidt & Dietz (1984) compromise on increased activity of both systems on a season-to-season basis, a likely scenario given the great latitudinal extent of the evidence. As such, the Kalahari evidence neither corroborates nor refutes palaeoclimatic models for the African continent (Deacon & Lancaster, 1988), but has implications for global climatic models that propose synchronous late glacial aridity throughout the sub-tropical arid zone (e.g. Nicholson & Flohn, 1980). Humidity in the Kalahari synchronous with glacial episodes has now been traced back to 300 000 BP (Brook, Burney & Cowart, 1990) and supports the view that the northern and southern hemispheres responded differently to changes in the earth's energy budget. Although the global glacial chronology may be revised as a result of recalibration of radiocarbon dates (e.g. Bard *et al.*, 1990), this conclusion will not be altered.

CONCLUSIONS — PROCESSES AND SENSITIVITY

The study of Kalahari geomorphology and Quaternary history has long been influenced by inferences and conclusions drawn from outside the region. Its very designation as a desert has tended to ignore its function as a depositional basin and its relatively high levels of precipitation. The undue emphasis placed on aeolian activity may not be warranted; the linear dune fields appear to be among the most stable and enduring of Kalahari landforms, for although there is evidence to suggest that sediment transport and duneform modification is taking place in southern parts of the region under a partial vegetation cover, the overall dune pattern changes little. The dune mobility often mentioned in relation to overgrazing and vegetation stripping in south-west Botswana probably relates to the local reshaping of linear dunes by blowouts once equilibrium conditions are removed. There is much to understand concerning the sensitivity of the climate—process—vegetation interactions of these forms before palaeoenvironmental certainties can be drawn from them.

The fluvio-lacustrine landforms of the Okavango—Makgadikgadi are the unique product of an unusual hydrological regime within a delicately adjusted tectonic framework. Both climatic and tectonic influences can be detected in the high degree of sensitivity revealed on Quaternary, historical and inter-annual timescales. The landforms of the swamps, and to a certain extent, of the adjoining lake basins, are the product of water volume rather than velocity. Chemical activity related to the transfer and evaporation of water has resulted in a regional suite of duricrusts and precipitates that are considerably younger than previously thought.

Influxes of groundwater and consequent chemical activity are also major long-term processes in the formation of Kalahari landforms, and in the modification of the Kalahari sediments. There has been lively debate over the past 80 years as to whether groundwater recharge of the Kalahari Group sediments has taken place *in toto* under contemporary climatic conditions. Two schools of thought exist: one (Boocock & Van Straten, 1961; Foster *et al.*, 1982) believes that recharge is limited to a capillary zone of 6 m; the other (Verhagen, Mazor & Sellschop, 1974; Mazor, 1982), based on isotope studies, suggests that recharge is uniform in neither time nor space, with concentration in zones of high infiltration (shallow sediment cover, faults, zones of vegetation penetration and bioturbation) and during years of prolonged excessive rainfall. Zones of preferential recharge (and hence a geological control) are expressed in the pan and valley landforms themselves. Experimental measurement of recharge as a function of rainfall intensity and duration has been undertaken in Botswana for the past four years and shows considerable promise (A. Gieske, 1992). Careful field mapping of duricrusts in the Kalahari should determine which are related to specific landforms, and which have formed in response to widespread groundwater inputs. The resolution of regional and localised duricrust formations would go far towards an understanding of Kalahari stratigraphy.

The palaeoclimatic pattern that has emerged for the past 50 000 years emphasises the transitional nature of the Kalahari environment in relation to atmospheric circulation systems. The climate has been drier in the past, but not as dry as previously thought. It has also been considerably wetter, or at least has had more free moisture. The current phase of palaeoclimatic research in the region seeks to differentiate between long-term climatic change, and the effects of high-intensity precipitation events, which may be difficult to differentiate from the proxy data but which have substantially different implications for palaeoenvironmental sensitivity and change.

REFERENCES

Arad, A. (1984). Relationship of salinity of groundwater to recharge in the southern Kalahari Desert. *Journal of Hydrology,* **71**, 225–238.

Ash, J. E. and Wasson, R. J. (1983). Vegetation and sand mobility in the Australian desert dunefield. *Zeitschrift für Geomorphologie*, Supplementband, **45**, 7–25.

Bard, E., Hamelin, B., Fairbanks, R. G. and Zinder, A. (1990). Calibration of the ^{14}C timescale over the past 30 000 years using mass spectrometric U–Th ages from Barbados corals. *Nature,* **345**, 405–410.

Beaumont, P. B., Van Zindren Bakker, E. B. and Vogel, J. C. (1984). Environmental changes since 32 000 BP at Kathu Pan, Northern Cape. In Vogel, J. C. (ed.), *Late Cainozoic Palaeoclimates of the Southern Hemisphere*, Balkema, Rotterdam, pp. 329–338.

Boocock, C. and Van Straten, O. J. (1961). *A Note of the Development of Potable Water Supplies at Depth in the Central Kalahari*, Bechuanaland Protectorate Geological Survey, Records 1957/58, pp. 11–14.

Brook, G. A., Burney, D. A. and Cowart, J. B. (1990). Desert palaeoenvironmental data from cave speleothems with examples from the Chihuahuan, Somali-Chalbi and Kalahari Deserts. *Palaeogeography, Palaeoclimatology, Palaeoecology,* **76**, 311–329.

Bruno, S. A. (1985). Pan genesis in the southern Kalahari. In Hutchins, D. G. and Lynam, A. P. (eds), *Proceedings of a Seminar on the Mineral Exploration of the Kalahari*. Botswana Geological Survey, *Bulletin,* **29**, 261–277.

Butterworth, J. S. (1982). *The Chemistry of Mogatse Pan, Kgalakgadi District*, Botswana Geological Survey, Report JSB/14/82.

Butzer, K. W. (1984). Late Quaternary palaeoenvironments in South Africa. In Klein, R. G. (ed.), *Southern African Palaeoenvironments and Prehistory*, Balkema, Rotterdam, pp. 1–64.

Butzer, K. W., Fock, G. J., Stuckenrath, R. and Zilch, A. (1973). Palaeohydrology of Late Pleistocene Lake Alexandersfontein, South Africa. *Nature,* **243**, 328–330.

Butzer, K. W., Stuckenrath, R., Bruzewicz, A. and Helgren, D. (1978). Late Cenozoic palaeoclimates of the Gaap Escarpment, Kalahari Margin, South Africa. *Quaternary Research,* **10**, 310–319.

Cooke, H. J. (1980). Landform evolution in the context of climatic change and neo-tectonism in the middle Kalahari of northern central Botswana. *Transactions of the Institute of British Geographers,* NS5, 80–99.

Cooke, H. J. (1984). The evidence from northern Botswana of climatic change. In Vogel, J. (ed.), *Late Cenozoic Palaeoclimates of the Southern Hemisphere*, Balkema, Rotterdam, pp. 265–278.

Cooke, H. J. and Verhagen, B.Th. (1977). The dating of cave development: an example from Botswana. *Proceedings of the 7th International Speleological Congress*, Sheffield.

Cooke, H. J. and Verstappen, H.Th. (1984). The landforms of the western Makgadikgadi basin of northern Botswana, with a consideration of the chronology of Lake Paleo-Makgadikgadi. *Zeitschrift für Geomorphologie,* **NF28**, 1–19.

Deacon, J. and Lancaster, N. (1988). *Late Quaternary Palaeoenvironments of Southern Africa*. Oxford Science Publications.

Farr, J., Peart, R., Nelisse, C. and Butterworth, J. (1982). *Two Kalahari Pans: A Study of their Morphometry and Evolution*. Botswana Geological Survey, Report GS10/10.

Foster, S., Bath, A., Farr, J. and Lewis, W. (1982). The likelihood of active groundwater recharge in the Botswana Kalahari. *Journal of Hydrology,* **55**, 113–136.

Fryberger, S. G. and Goudie, A. S. (1981). Arid geomorphology. *Progress in Physical Geography*, **5**, 420–428.

Gieske, A. (1992). *Dynamics of Groundwater Recharge: a Case Study in Semi-arid Eastern Botswana*. PhD thesis, Free University, Amsterdam.

Goudie, A. S. (1971). Calcrete as a component of semi-arid landscapes. Unpublished Ph.D. thesis, University of Cambridge.

Goudie, A. S. (1973). *Duricrusts in Tropical and Subtropical Landscapes*, Clarendon Press, Oxford.

Goudie, A. S. and Thomas, D. S. G. (1986). Lunette dunes in southern Africa. *Journal of Arid Environments*, **10**, 1–12.

Grey, D. R. C. and Cooke, H. J. (1977). Some problems in the Quaternary evolution of the landforms of northern Botswana. *Catena*, **4**, 123–133.

Gwosdz, W. and Modisi, M. P. (1983). *The Carbonate Resources of Botswana*. Botswana Geological Survey Mineral Resources Report no. 6.

Heine, K. (1978). Radiocarbon chronology of the Late Quaternary Lakes in the Kalahari. *Catena*, **5**, 145–149.

Heine, K. (1982). The main stages of late Quaternary evolution of the Kalahari region, Southern Africa. *Palaeoecology of Africa*, **15**, 53–76.

Helgren, D. M. and Brooks, A. S. (1983). Geoarchaeology at Gi, a Middle Stone Age and Later Stone Age Site in the northwest Kalahari. *Journal of Archaeological Science*, **10**, 181–197.

Lancaster, I. N. (1978). The pans of the southern Kalahari, Botswana. *Geographical Journal*, **144**, 80–98.

Lancaster, I. N. (1979). Evidence for a widespread late Pleistocene humid period in the Kalahari. *Nature*, **279**, 145–146.

Lancaster, N. (1980). Dune systems and palaeoenvironments in southern Africa. *Palaeoentology Africana*, **23**, 185–189.

Lancaster, N. (1981). Palaeoenvironmental implications of fixed dune systems in southern Africa. *Palaeogeography, Palaeoclimatology, Palaeoecology*, **33**, 327–346.

Lancaster, N. (1986). Grain-size characteristics of linear dunes in the south-western Kalahari. *Journal of Sedimentary Petrology*, **56**, 395–400.

Lancaster, N. (1987). Grain-size characteristics of linear dunes in the southwestern Kalahari — a reply. *Journal of Sedimentary Petrology*, **57**, 573–574.

Lancaster, N. (1988). Development of linear dunes in the southwestern Kalahari, southern Africa. *Journal of Arid Environments*, **14**, 233–244.

Lancaster, N. (1989). Late Quaternary Palaeoenvironments of the southwestern Kalahari. *Palaeogeography, Palaeoclimatology, Palaeoecology*, **70**, 367–376.

McCarthy, T. S. and Metcalfe, J. (1991). Chemical sedimentation in the semi-arid environment of the Okavango Delta, Botswana. *Chemical Geology*, **89**, 157–178.

Malherbe, S. J. (1984). The geology of the Kalahari Gemsbok National Park. *Koedoe*, **27**, 33–44.

Mazor, E. (1982). Rain recharge in the Kalahari — a note of some approaches to the problem. *Journal of Hydrology*, **55**, 137–144.

Mazor, E., Verhagen, B., Sellschop, J., Jones, M., Robins, N., Hutton, L. and Jennings, C. (1977). Northern Kalahari groundwaters: hydrologic, isotopic and chemical studies at Orapa, Botswana. *Journal of Hydrology*, **34**, 203–233.

Nicholson, S. E. and Flohn, H. (1980). African environmental and climatic changes and the general atmospheric circulation in the late Pleistocene and Holocene. *Climatic Change*, **2**, 313–348.

Passarge, S. (1904). *Die Kalahari*. Dietrich Riemer, Berlin.

Pickford, M. (1990). Some fossiliferous Plio-Pleistocene cave systems of Ngamiland, Botswana. *Botswana Notes and Records*, **22**, 1–17.

Rogers, A. W. (1936). The surface geology of the Kalahari. *Transactions of the Geological Society of South Africa*, **24**, 57–80.

Rust, U., Schmidt, H. and Dietz, K. (1984). palaeoenvironments of the present day arid south western Africa 30 000–5000 BP: results and problems. *Palaeoecology of Africa*, **16**, 109–148.

Shaw, P. A. (1985). Late Quaternary landforms and environmental change in northwest Botswana: the evidence of Lake Ngami and the Mababe Depression. *Transactions of the Institute of British Geographers*, **NS10**, 333–346.

Shaw, P. A. (1988). After the flood: the fluvio-lacustrine landforms of northern Botswana. *Earth Science Reviews*, **25**, 449−456.

Shaw, P. A. and Cooke, H. J. (1986). Geomorphic evidence for the late Quaternary palaeoclimates of the middle Kalahari of northern Botswana. *Catena*, **13**, 349−359.

Shaw, P. A. and De Vries, J. J. (1988). Duricrust, groundwater and valley development in the Kalahari of southeast Botswana. *Journal of Arid Environments*, **14**, 245−254.

Shaw, P. A. and Thomas, D. S. G. (1988). Lake Caprivi: a late Quaternary link between the Zambezi and Middle Kalahari drainage systems. *Zietschrift für Geomopholgie*, **32**, 329−337.

Shaw, P. A., Cooke, H. J. and Thomas, D. S. G. (1988). Recent advances in the study of Quaternary landforms in Botswana. *Palaeoecology of Africa*, **19**, 15−26.

Shaw, P. A., Cooke, H. J. and Perry, C. C. (1991). Silcrete genesis and micro-organisms in highly alkaline environments: some observations from Sua Pan, Botswana. *South African Journal of Geology*, **93**, 803−808.

Smale, D. (1973). Silcretes and associated silica diagenesis in southern Africa and Australia. *Journal of Sedimentary Petrology*, **43**, 1077−1089.

Summerfield, M. A. (1982). Distribution, nature and probable genesis of silcrete in arid and semi-arid southern Africa. In Yaalon, D. H. (ed.), *Aridic Soils and Geomorphic Processes. Catena, Supplement*, **1**, 37−65.

Summerfield, M. A. (1983). Silcrete as a palaeoclimatic indicator: evidence from southern Africa. *Paleogeography, Palaeoclimatology, Palaeoecology*, **41**, 65−79.

Thomas, D. S. G. (1984). Ancient ergs of the former arid zones of Zimbabwe, Zambia and Angola. *Transactions of the Institute of British Geographers*, **NS9**, 75−88.

Thomas, D. S. G. (1987). Discrimination of depositional environments, using sedimentary characteristics, in the Mega Kalahari, central southern Africa. In Frostick, L. E. and Reid, I. (eds), *Desert Sediments, Ancient and Modern*. Geological Society of London Special Publication 35, Blackwell, Oxford, pp. 293−306.

Thomas, D. S. G. (1988). Analysis of linear dune sediment−form relationships in the Kalahari Dune Desert. *Earth Surface Processes and Landforms*, **13**, 545−553.

Thomas, D. S. G. (1989). Aeolian sand deposits. In Thomas, D. S. G. (ed.), *Arid Zone Geomorphology*, Belhaven, London, pp. 232−261.

Thomas, D. S. G. and Martin, H. E. (1987). Grain-size characteristics of linear dunes in the southwestern Kalahari — a discussion. *Journal of Sedimentary Petrology*, **57**, 231−242.

Thomas, D. S. G. and Shaw, P. A. (1991). 'Relict' desert dune systems: interpretations and problems. *Journal of Arid Environments*, **20**, 1−14.

Tsoar, H. and Moller, J. T. (1986). The role of vegetation in the formation of linear sand dunes. In Nickling, W. G. (ed.), *Aeolian Geomorphology*, The Binghampton Symposium in Geomorphology International Series 17, Allen & Unwin, Boston, pp. 75−95.

UNDP/FAO (1977). *Investigation of the Okavango Delta as a Primary Water Resource for Botswana*, DP/Bot/71/506 Technical Report, 3 vols.

Verhagen, B., Mazor, E. and Sellschop, J. (1974). Radiocarbon and tritium evidence for direct rain recharge to groundwaters in the northern Kalahari. *Nature*, **249**, 643−644.

Wasson, R. J. and Nanninga, P. M. (1986). Estimated wind transport of sand on vegetated surfaces. *Earth Surface Processes and Landforms*, **11**, 505−516.

Wilson, B. H. (1973). Some natural and man-made changes in the channels of the Okavango Delta. *Botswana Notes and Records*, **5**, 132−153.

Wright, E. B. (1978). Geological studies in the northern Kalahari. *Geographical Journal*, **144**, 235−250.

8 Thresholds in a Sensitive Landscape: The Loess Region of Central China

EDWARD DERBYSHIRE
*Department of Geography, Royal Holloway College, University of London, UK
and Geological Hazards Research Institute, Gansu Academy of Sciences,
Lanzhou, China*

TOM A. DIJKSTRA
Faculty of Geographical Sciences, University of Utrecht, The Netherlands

ARMELLE BILLARD, TATIANA MUXART
Laboratoire de Géographie Physique, URA 141 CNRS, Meudon, France

IAN J. SMALLEY
Department of Civil Engineering, Loughborough University of Technology, UK

and

YOUNG-JIN LI
*Geological Hazards Research Institute, Gansu Academy of Sciences, Lanzhou,
China*

ABSTRACT

The western margins of the loess plateau in eastern Gansu Province are mantled by 100–300 m of aeolian silts which are quartz-rich, sometimes cemented (calcareous) and contain variable amounts of soluble salts. The conventional stratigraphic subdivision into Wucheng (Lower Pleistocene), Lishih (Middle Pleistocene) and Malan (Upper Pleistocene), to which is added a thin drape of Holocene loess, forms a useful framework for studies of loess sedimentology and geotechnical behaviour. The loess is metastable to varying degrees as a result of its fabric, which consists of subangular silt grains with bridges of fine silt and clay held together by tensional forces at prevailing low field moisture contents. Although able to maintain vertical faces, therefore, failure and liquefaction occur if specific water content threshold values are exceeded.

Thresholds of loess failure have been investigated at several scales from the field through hand specimen size to microscopic. At the geomorphological scale, an extensive mapping programme has been undertaken in an attempt to establish the nature of the relationship between selected environmental parameters and the spatial incidence of mass movements. In the finer-grained and more compact Wucheng and Lishih formations massive failure is triggered by reductions in shearing resistance with rapid rises in pore pressures in essentially undrained conditions in both wet (rain-induced) and dry (earthquake shock-induced) situations. In the Malan formation, small scale and rapid planar slides (Chinese: *tan-ta*) are common, being associated with strength reduction owing to slope undermining or a combination of rise in pore fluid pressure and slope angles of between 30° and 45°. However, the failure threshold relationship is not always straightforward: progressive small-scale failure mechanisms can contribute to weak zones, which eventually fail in response to trigger magnitudes lower than predicted.

Landscape Sensitivity. Edited by D. S. G. Thomas and R. J. Allison
© 1993 John Wiley and Sons Ltd

Hand specimens have been tested in order to determine the nature of the forces involved in peak strength conditions. Remoulded tests have provided data on the relationship between moisture content and shear strength. Vertical and horizontal permeability and infiltration capacities are relatively low, suggesting that wetting of the basal loess zone, considered a critical factor in slide initiation, is effected by groundwater movement along less permeable layers (palaeosols, bedrock) and, during heavy rain, as throughflow via the extensive piping systems.

Both undisturbed and mechanically deformed samples have been examined using scanning electron microscopy techniques. The shear systems observed macroscopically can be related to changes in fabric anisotropy, pore fluid escape fabrics, dilation, strain-hardening fabrics, and microshearing.

These studies suggest that a unique set of relationships may be involved in the mass failure of loess.

INTRODUCTION

Loess is a sensitive material: when dry and undisturbed it can sustain large stresses. Nearly vertical walls up to 20 m high are common in the loess region of China. However, when certain thresholds are exceeded, e.g. as a result of an increase in the moisture content, rapid breakdown of the loess fabric occurs. In the loess region, relative relief reaches 700 m with many slopes as steep as 30−40°. Catastrophic mass movements and loess slurries are recurrent phenomena in this area, causing loss of life and severe damage to infrastructure and farmland. Under a four-year, EC-funded research contract, the landslides and loess flows are being investigated at field sites; soil and water chemistry, mineralogy, grain characteristics and engineering properties are being tested in field and laboratory; and a major field mapping programme of the research area around Lanzhou in Gansu Province is being used to establish a landslide database, which incorporates both morphometric and causal factors (Wang & Derbyshire, 1988; Derbyshire et al., 1991).

Because collapse of the fabric of loess is the primary cause of a wide variety of mass movements, subsidence and piping phenomena, the threshold conditions of the undisturbed and remoulded strength of loess in relation to its mineralogy, chemistry and grain size distribution will be discussed first. This consideration of loess strength characteristics provides the basic framework for studies of the threshold situation on loess slopes. Examples of various types of recent mass movements will be used to describe the variables affecting threshold conditions in loess slope systems. A distinction is made between failure in thin as against thick loess mantles and also between different types of slides from deep-seated rotational failures to planar- and slab-type failures.

CHARACTERISTICS OF THE LOESS REGION

Geological Framework

The loess area of central China is situated along the north-eastern fringes of the Tibetan (Qinghai-Xizang) Plateau, its western boundary roughly coinciding with the Qilian and Riuye mountain chains along the boundary between Qinghai and Gansu Provinces (Liu et al., 1985a, Zhang & Dai, 1989; Wang & Lang, 1989; see Figure 8.1). Strong uplift since the late Cenozoic has caused extensive bedrock faulting and jointing, some of the major faults still being active as a result of the neotectonic crustal movements. Subsidence basins occur along the fault

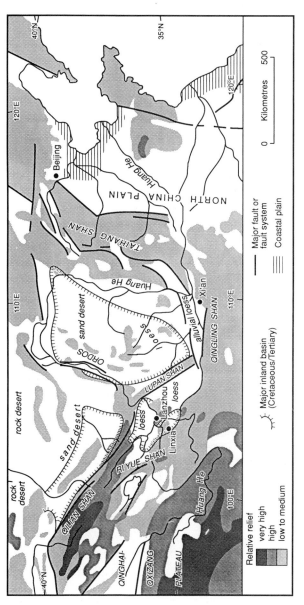

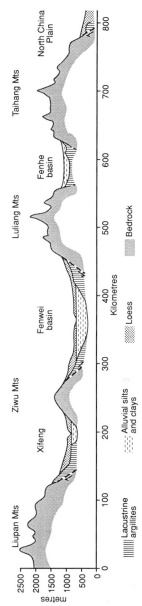

Figure 8.1 Location of the loess region of central China and schematic profile of loess deposition east of the Liupan Shan

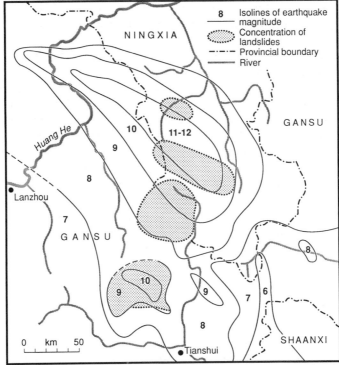

Figure 8.2 Earthquake magnitudes and areas of landslide concentration in Gansu and Ningxia provinces (after Feng & Guo, 1985)

Figure 8.3 Profile across a loess gorge, showing the effects of 'loess karst' and undercutting of the steep slopes (von Richthofen, 1877, p. 115)

River). Larger subsidence basins occur south of the loess plateau. The Weihe graben near Xi'an is the largest, its sedimentary fill of alluvial silts and clays varying in thickness from 400 m in the west to 1200 m in the east.

Neotectonic movements have been instrumental in the drying up of the large Pliocene inland lakes such as those of the Linxia basin (Figure 8.1). Such basin fills have suffered substantial erosion and incision by the stream systems. Although the magnitude and rate of neotectonic activity appears to have decreased since the Tertiary, the direction of the movements has remained essentially the same, thereby continuing to exert strong control on the regional geomorphology of the loess lands. The majority of the larger fault-controlled valleys have a north-east trend, but drainage patterns on the loess slopes show an average north-north-west trend. The Huang He is the master stream and the regional base-level for most of the central China loess plateau (Zhang & Dai, 1989; Zhang, 1989).

The neotectonic activity is clearly reflected in the earthquake history of the region. A long written record of the effects of these shocks, dating back to the period of 1129−1095 BC (Liu et al., 1979), is held on file. It is evident that many landslides in the northern China loess lands are earthquake induced, and regional landslide distribution can be related to major earthquake events (Figure 8.2).

MORPHOLOGY

Loess accumulation began in central China at the beginning of the Pleistocene (i.e. about 2.4 Ma ago: Heller & Liu, 1982; Wen, Diao & Yu, 1985; Wang & Lang, 1989; Kukla et al., 1988; Kukla, 1989; Zhang & Cheng, 1989). The predominantly windblown silty deposits accumulated in the middle reaches of the Huang He, reaching an areal extent of about 440 000 km². This essentially continuous drape covers most of the provinces of Gansu, Shaanxi, Shanxi, southern Ningxia and western Henan. Loess thicknesses of 50−100 m are common. In Gansu Province thicknesses exceed 300 m in several places, 318 m having been recorded at Jiuzhoutai Mountain, just north of Lanzhou city (cf. Derbyshire, 1984; Zhang, 1989).

The landforms of the loess plateau have been classified into three types: *yuan*, *liang* and *mao* (respectively plateau, ridge and rounded hill forms: Liu et al., 1985b; Zhang, 1986). Two hypotheses have been proposed to explain this series. Since loess was deposited as a drape on a hilly palaeolandscape with a relatively thin loess cover, old drainage patterns and relief features have been inherited and *liang* with some *mao* forms dominate. However, in areas overlain by thicker loess all three landform types exist and these have been interpreted as representing evolutionary stages in the cyclic degradation of the loess plateau (cf. Liu et al., 1985b; Zhang, 1986; Zhang & Dai, 1989).

In areas of *yuan* (loess plateau) some deeply incised gullies exist, frequently with sinkholes and pipe systems at the valley heads, commonly referred to as 'loess karst'. As a result of the high discharges generated by the throughflow in the pipes during rainfall events, vertical erosion along gully lines greatly exceeds rates of erosion of the *yuan* surfaces where infiltration and short-distance sheetwash dominate. Gullies may be 12−15 m deep with sub-vertical walls. Undercutting by stream action, often associated with liquefaction of basal gully slopes causes slab failure along the joint systems commonly found in loess. Frequently, gully incision continues down to the bedrock (Figure 8.3). As degradation of the *yuan* surface continues, gullies become integrated and the relative relief increases. Surficial erosion, including sheetwash and rill action, modifies the plateau edges resulting in rounded *liang* crests. Mass movements frequently occur on the steep slopes of such deeply incised valleys. In the *mao* areas loess slopes are typically convex and the dominant erosional process is sheetwash. The erosion rates can be extremely high, both as a result of the shape of the hill and the precipitation characteristics of the areas in which *mao* is the most common landform. In some areas erosion rates reach over 20 000 tonnes km^{-2} year^{-1} (Figure 8.4; cf. Derbyshire, 1989). The Huang He is the principal means of transporting eroded material out of the loess plateau. It is the most turbid river on earth (high suspension load: ~1.6 billion tonnes year^{-1} according to Klinkenberg, 1988; 75 million tonnes year^{-1} quoted from Kingsmill (in Smalley, 1972); >10^{8} tonnes year^{-1} as mentioned in Ferguson, 1984).

CLIMATE

The erosion rates on the loess plateau reflect the rainfall distribution (Figure 8.4). In the Lanzhou region mean annual rainfall totals are of the order of 300−400 mm year^{-1}. However, inter-annual variability is considerable, e.g. totals of 150 mm in 1980 and 540 mm in 1988. Most of the precipitation occurs in a few severe rain showers in the period July−

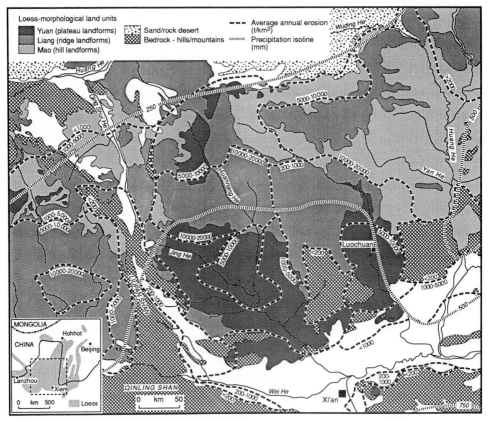

Figure 8.4 Distribution of *mao, yuan* and *liang* landforms, soil loss and rainfall distribution (after Derbyshire, 1989; Zhang, 1986)

September. Rainfall intensities may be very high with values exceeding 50 mm h^{-1} (cf. Billard *et al*., in press; Figure 8.5).

Because evaporation values are three to four times higher than the precipitation values, there is a considerable moisture deficit in the loess soils. Most of the thick loess deposits in the Lanzhou region have ambient moisture contents of the order of only 8−12%. Although

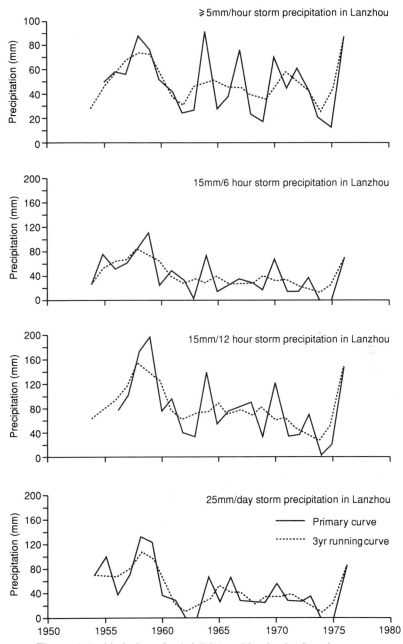

Figure 8.5 Variation of rainfall intensities in the Lanzhou area

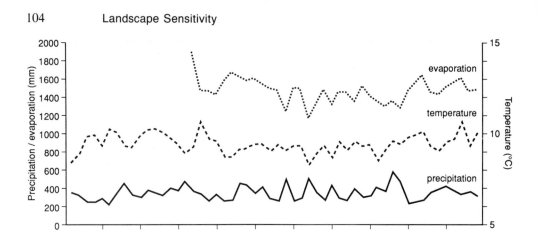

Figure 8.6   Precipitation, evaporation and temperature records for Lanzhou city

winter precipitation values are low, some snow usually falls in February and March, and during the spring melt the upper part of the loess may reach saturation. Characteristic values of precipitation, evaporation and temperature are shown in Figure 8.6.

Stratigraphy, Grain Characteristics and Mineralogy

The conventional stratigraphic division, used for several decades by Chinese workers (cf. Liu, 1985) provides a useful framework for the study of the sedimentary and geotechnical characteristics of loess. The Wucheng loess which covers most of the Lower Pleistocene, (c. 2.4−1.15 million BP) varies in thickness between 20 and 120 m and contains between 6 and 18 buried paleosols, all calcified to some extent. The Lishih loess which approximates the Middle Pleistocene (1.15−0.1 million BP) has a maximum thickness of between 90 and 200 m and, near Lanzhou, 21 paleosols are found. The Malan loess (Upper Pleistocene, Lower Holocene: 0.1 million−5000 BP) forms the most extensive cover in the loess plateau (∼60%, cf. Wen, Diao & Yu, 1987; Liu *et al.*, 1985a,b). Thicknesses commonly vary between 10 and 34 m. In addition, a Holocene loess up to 5 m thick overlying a black soil with an estimated age of 5000 years is found in some places. However, in most areas the rate of Holocene dust sedimentation was so low that it became incorporated in the soil by bioturbation and pedogenesis (Liu & Chang, 1964; Liu *et al.*, 1985a; Wen, Diao & Yu, 1985; Kukla & An, 1989).

The derivation of the loess has been deduced from studies of its sedimentology, chemical composition and grain size distribution at many sites across the loess plateau (Figure 8.7). The clear pattern of sandy loess, loess and clayey loess belts indicates a first-order source for the silt particles in the desert basins, *wadis* and *playas* to the north-west of the loess plateau (Derbyshire, 1984). The same north-west/south-east gradient is evident in the trace element content (Liu *et al.*, 1985a; Wang *et al.*, 1987; Zhao & Qu, 1981; Wen, Diao & Yu, 1987). However, doubt has been cast on whether the desert environment is capable of producing the vast amounts of silt found in north China (cf. Smalley, 1966, 1971; Smalley & Vita-Finzi, 1968). One result of this is a body of opinion in favour of a substantial silt source in the Tibetan Plateau where, it is argued, abundant energy for rock breakdown has been provided by the vigorous Cenozoic uplift of the Himalaya and the Tibetan Plateau. A third

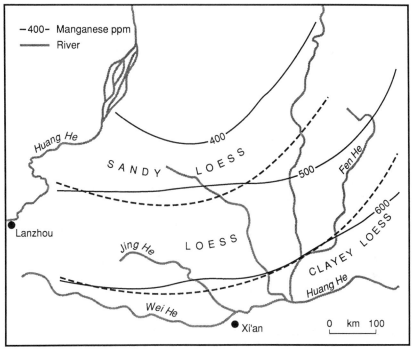

Figure 8.7   Loess grain size zones and chemical composition (manganese taken as a representative example; after Wen, Diao & Yu, 1987)

group of silt-producing mechanisms is the combined action of glacial grinding, salt weathering and to a lesser extent periglacial mechanical weathering in Tibet and the Himalayas (cf. Smalley, 1966; Lautridou, 1988; Doornkamp & Ibrahim, 1990). Following their production, silts were transported first by glaciofluvial then fluvial systems to accumulate in the arid to semi-arid areas of the highland margins, including the great sedimentation basins of the Tarim, Junggar, Qaidam and Ordos. Deflation of this fluvially concentrated silt then gave rise to the airfall loess deposits of central China (cf. Smalley, 1966, 1971; Derbyshire, 1984; Tan, 1988).

This hierarchical hypothesis of loess derivation is consistent with a number of sedimentary characteristics of the loess, including its grain surface characteristics. In any single sample of Lanzhou loess may be found quartz particles varying from well rounded to sub-angular in form and showing marked variations in their degree of weathering (cf. Smalley and Cabrera, 1970; see Figure 8.8). Such variety may stem from a number of variables including the time lag between formation and final aeolian deposition, and the relative intensities of the various weathering and erosional environments to which loess grains may be subjected (glacial, glaciofluvial, fluvial, periglacial, arid and semi-arid environments). An additional factor to be considered is variation in the source bedrock. However, further studies of the interplay between these and other potential controls is complex and there is much room for further research.

Although the clay fraction in Lanzhou loess may exceed 20%, not all consists of clay minerals. From the characteristic X-ray diffractograms shown (Figure 8.9), it can be seen that the composition of the finer than 2 μm fraction of Lanzhou loess consists of quartz,

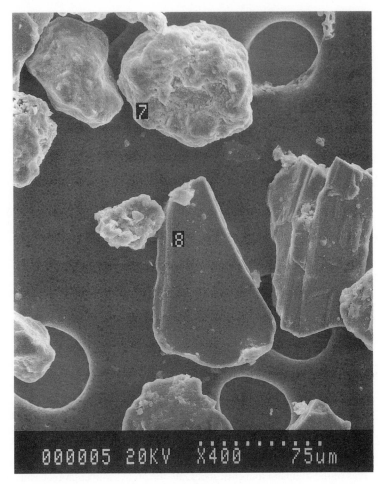

Figure 8.8 Scanning electron microscope photographs of silt-sized particles of Malan loess from Lanzhou

gypsum, feldspars, calcite, illite, kaolinite, with minor chlorite and some smectites. The clay minerals occur as coatings on the silt grains, as bridges between the silt particles or as micro-aggregates. However, owing to the low clay mineral content, these features are much less common than is the case in the Western European loess (Derbyshire *et al.*, 1988).

The fine fraction ($<10 \mu$m) strongly influences the loess fabric. These mainly platey particles frequently have a face-to-face disposition and form bridges holding the larger (20–100 μm) particles together (Krinsley & Smalley, 1973; Tan, 1988; Pye, 1987; Wang *et al.*, 1987; Derbyshire & Mellors, 1988; see Figure 8.10). Loess fabrics have been studied using the scanning electron microscope (SEM). In the undisturbed state, samples show a loosely packed fabric with typical high voids ratios and low bulk densities. However, shearing induces irreversible destruction of clay bridges and redistribution of particles and results in an anisotropic fabric with decreased voids ratios and high bulk densities (Figure 8.11 and 8.12).

In general, the loess of northern China has an average grain size of between 0.06 and 0.01 mm (phi-values varying from 5.0 to 8.0, commonly around 6.8 phi). It is poorly to

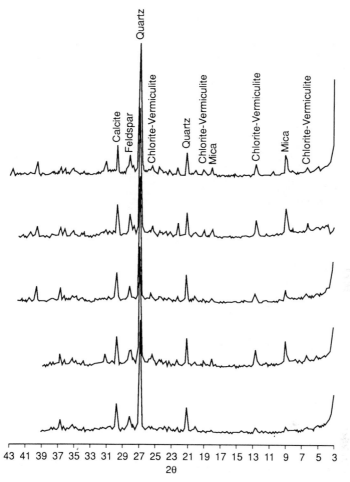

Figure 8.9   X-ray diffractograms (Cu K_α radiation) of <2 μm fraction of loess from Jiuzhoutai. Top: Malan loess; middle two: Lishih loess; lower two: Wucheng loess

very poorly sorted (Folk and Ward sorting coefficient ±1.6−3.0) and is consistently fine skewed (skewness +0.3 to +1.2: Derbyshire, 1988). The grading envelopes for Lishih and Wucheng loess indicate a relatively fine mean grain size for the Lower and Middle Pleistocene loess, Malan loess being distinctly coarser (Figure 8.13).

Loess Geochemistry, Clay Mineralogy and Engineering Properties

The characteristics of loess particles, notably the size distribution, surface texture, mineral type and chemical composition, are important influences upon the strength of the bulk material.

The application of stress, whether caused by increasing normal loads or increasing pore water pressures, results in an irreversible collapse of the cardhouse (edge-to-face) and face-to-face fabrics considered to be syngenetic in undisturbed loess (cf. Derbyshire & Mellors, 1988; Gao, 1983; Derbyshire *et al.*, 1988). Moreover, eluviation of the finer particles results in the destruction of the characteristic clay bridges: they do not reform in the illuviated

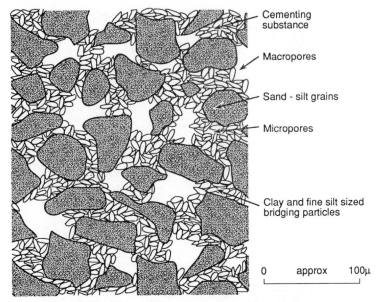

Figure 8.10   Microstructure of loess (after Tan, 1988; and others)

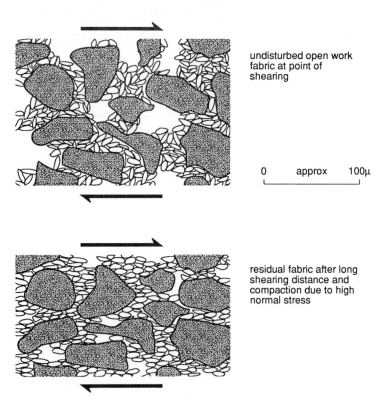

Figure 8.11   Graphical representation of the effects of shearing on the redistribution of particles and the irreversible destruction of the clay-sized bridges

Figure 8.12 Scanning electron microscope photograph of a loess sample with a dense packing after direct shear testing

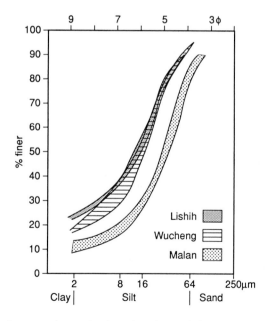

Figure 8.13 Grading envelopes for Lanzhou loess (after Derbyshire & Mellors, 1988)

horizons. The undisturbed strength of this particulate material is much greater than would be expected on the basis of the friction bonds existing between the individual particles. In order to determine the relationship between loess composition, structure and strength, geotechnical tests and geochemical analyses have been undertaken. Analyses of water samples were carried out both in the field (using portable analytical equipment — HACH apparatus) for the determination of pH, SiO_2, SO_4^{2-}, Cl^-, HCO_3^-, CA^{2+} and Mg^{2+}) and in the laboratory using atomic absorption spectophotometry of the acidified samples to determine Ca^{2+}, Mg^{2+}, Na^+ and K^+.

In the analysed samples taken from the valleys south-west of Lanzhou (Billard *et al.*, in press), the slope-foot loess left by landslides contains between 0.27 and 2.0% chlorides and sulphates (expressed as sulphur content) of between 0 and 0.27%. The *in situ* Malan and Lishih loess samples from the upper part of the slopes show that chlorides and sulphates have been eluviated, their content being minimal (generally < 0.09%). Calcium carbonate contents vary with age, being 11.2−11.6% in the Malan, 11.7−15.4 in the Lishih, and 8.7−16% in samples analysed from the Wucheng loess. The greatest observed variation in the oldest loess is related to the presence of palaeosols with their associated carbonate precipitation horizons. Laboratory geotechnical tests have shown that effective apparent cohesion (c') values in the Wucheng and Lishih loesses are higher than in the Malan and are related to differences in cementation.

X-ray diffractometry of the <2 μm fraction of both loesses and palaeosols shows the following:

- an association of illite, chlorite, kaolinite (inherited); and
- an association of illite, chlorite, kaolinite, smectite and/or interstratified chlorite−smectite. This latter association is similar to that found in the upper part of the substrate of Cretaceous sandstones.

Further, certain samples contain gypsum derived by aeolian transportation from the desert. The presence of this mineral, so vulnerable to alteration, together with the formation and preservation of smectites, indicates little or no leaching in the dry continental climate prevalent in the Gansu loess region which existed during periods of loess accumulation and during periods of moderately strong pedogenesis.

The formation of smectite is also explained by stagnating conditions either in the pores and microfissures or at certain horizons of contrasting permeability. This implies that, within parts of the loess, there is only slow water circulation or even stagnation of water lightly charged with salts (see below) among which those of magnesium and calcium are the most important. The common presence of smectite is one of the factors contributing to the structural modification and failure of the slopes of loess when moisture contents are raised by rainfall.

The chemical composition of the loess (see above) includes, albeit in small to moderate quantities, readily soluble salts (halite, gypsum) and carbonates which render the loess subject to the selective solution effects of vadose water. The chemical analyses, based on the nature and relative proportion of dissolved ions in the surface waters, have shown that water from wells, springs, and streams can be classified into four categories (Billard *et al.*, in press) as follows:

i.   Water with a significant content of chlorides (> 10 g l^{-1}) of sodium, magnesium and calcium.

ii. Water with very strong concentration of chlorides (up to 18 g l^{-1}) as well as sulphates

$(350-1000$ mg $1^{-1})$ and, to a lesser degree, bicarbonates of sodium, calcium and magnesium.

iii. Water containing more sulphates $(250-1375$ mg $1^{-1})$ than chlorides as well as bicarbonates of sodium, calcium and magnesium.

iv. Water enriched in calcium bicarbonate.

Three of the categories correspond with circulatory water in the loess (wells, springs) but also with some rivers draining the study area. They contain high to very high concentrations of soluble salts.

Spring water samples at the base of the backwall of the Sale Shan landslide in September 1987 showed the following composition; chlorides 18 200 mg 1^{-1}; sulphates 680 mg 1^{-1}; and carbonates 19 mg 1^{-1}. Spring waters sampled in the same period in the Sier valley, south-west of Lanzhou, showed lower concentrations, however; chlorides 650 mg 1^{-1}; sulphates 1375 mg 1^{-1}; and carbonates 238 mg 1^{-1}. In the neighbouring valley of Xuanjia at the foot of the ancient Tawa landslide, analyses by the Chinese team in June 1990 showed the following composition: chlorides 4987 mg 1^{-1}; sulphates 4457 mg 1^{-1}; and carbonates 167 mg 1^{-1}.

The interpretation of the water analyses data imply chemical abstraction and important removal of salt from the loess deposits, which is quite different from the results of the clay mineralogy analyses. This contradiction is only apparent, however, if account is taken of how the physico-chemical changes operate at different levels within the loess slope and how the most important changes are concentrated along preferential water circulation zones. The throughflow of water in loess is concentrated along complex channels, the velocity of this throughflow varying considerably. These factors explain the chemical solution processes, the dispersion of clays by slightly basic water and the liquefaction of the sediments which may be observed along these particular water routing channels. The interactions between loess and water that take place in these zones determine the landsliding processes which affect the slopes.

Along the water routing channels, processes related to solution and microstructural changes affect an increasingly large zone, particularly at the loess bedrock contact zone. The solution of salt particles and carbonates and the dispersion of clay mineral aggregates initiate a progressive destruction of the microfabric. As a result of the gravitational pull and rainfall events, slow plastic deformation at depth (creep) may occur. Additionally, the chemical analysis of water derived from a spring at the Sale Shan landslide site (after landsliding took place), shows very high concentrations of readily soluble salts, which are an indication of the activity of chemical solution processes at this site. During the period immediately preceding the mass movements, the interstitial water in the saturated zones was found to contain water that was highly charged with ions. This may cause suspension of the solid particles and lead to liquefaction of the material.

Determination of the $CaCO_3$ content of loess at Jiuzhoutai Mountain north of Lanzhou resulted in values of $11.19-11.59\%$ for the Malan loess, $11.67-15.37\%$ for the Lishih and $8.76-16.03\%$ for the Wucheng loess (Derbyshire & Mellors, 1988). Micromorphological and SEM analyses of this section indicate that both micro- and phanero-crystalline carbonates are present. Additionally, and more important in determining the strength of the loess, is the presence of secondary carbonate precipitates forming concretions, pore linings, encrustations and intergranular cements capable of supporting the open fabric of the loess. Solution of salts and carbonate precipitates, such as occurs during eluviation by percolating groundwater, decreases the strength of loess considerably, the process playing an important part in determining the observed high sensitivity values of loess. Thus, the subsurface loess hydrology is an important factor to consider when evaluating loess slope stability. The carbonate content, though having a tendency to decrease with depth, also shows great variability down-profile, particularly in relation to the occurrence of the palaeosols (Wen, Diao & Yu, 1985; Smalley, 1971; Derbyshire & Mellors, 1988). Parallels can be drawn between the behaviour of loess and classic 'quickclays' (Smalley & Derbyshire, 1991). The sensitivity of quickclay is related to the freefall deposition of clay-sized primary minerals in a saltwater environment, and the subsequent leaching of the salts leaving behind an openwork structure with characteristic short-range bonds. In the Chinese loess a similar openwork fabrics exists, although on a scale 10 times larger (± 50 μm versus ± 2.5 μm for quickclays). In loess the openwork fabric is the result of the freefall aeolian deposition of silt-sized primary minerals and the post-depositional leaching out of the soluble salts and carbonates. The structure is here also dependent on short-range bonds. In both cases the large differences between undisturbed and remoulded strengths are reflected in the high sensitivity values (more than 16 for most quickclays; 10 and more for loess).

Geotechnical tests have been used to determine the mechanical thresholds of loess and the differences in behaviour on either side of these thresholds. Direct shear and triaxial tests were carried out on undisturbed block samples, care being taken not to alter the original moisture content of 8–10%. Some results of these tests are shown in Figure 8.14.

In Malan loess, plastic deformation is important: no visible failure planes were observed during triaxial testing. In contrast, both the Wucheng and Lishih loesses are characterised by a pronounced brittle failure type. The shear tests show that while applying the load and increasing the stress an elasto-plastic deformation occurs until, after a sudden short strain deformation, the material fails. Since displacement after shearing occurs extremely rapidly

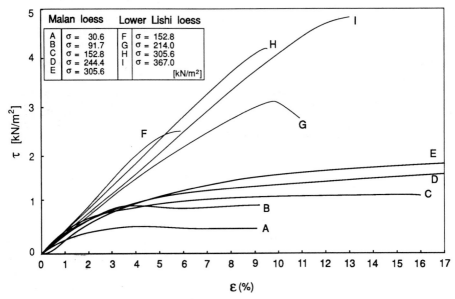

Figure 8.14 Stress–strain relationships for Lanzhou loess. Lower Lishih loess is characterised by brittle failure. Malan loess fails in a plastic mode

Figure 8.15 Scanning electron microscope photograph of vertical face of a naturally sheared sample of Malan loess (Sier landslide, 20 km south-west of Lanzhou). Microsheares and compaction of loess surrounding the macropores as a result of high pore pressures during the quick undrained shearing are clearly visible

and over large distances, it proved to be impossible to obtain realistic values for the residual strength parameters using these devices. The stress paths of the Lishih and Wucheng loess are very similar and show that cementation bonds form the basic components of the rather high cohesion values. Owing to the high initial voids ratios, consolidation of the samples during testing was considerable and is directly related to the increase of lateral pressure. The failure of bridges made up of aggregates in the loess results in a redistribution of particles and a rapid volume change at failure. However, this takes place only when the material is either unsaturated (as in the laboratory tests) or when drained conditions exist. In the field, it was found that loess failures occur very rapidly and high sliding velocities exist. Because of the relatively low permeability coefficients ($64.4-26.2 \times 10^{-5}$ cm s^{-1} in Malan loess), the high velocities and the fact that most slip planes coincide with saturated zones in the loess slopes, high positive pore water pressures can build up, resulting in liquefaction and extreme mobility of the slides. Samples were taken from a shallow planar loess failure, where the slip plane is located directly above argillaceous bedrock in a saturated groundwater zone. SEM analysis showed large pores, the surrounding loess grains having suffered from considerable pore pressures. The volume of the material in total had not changed substantially, but the redistribution of particles resulted in an enlargement of pores and a compaction of the aggregates (Figure 8.15). At other sites it is observed that, as a result of landsliding, often associated with high sliding velocities, strain hardening of the loess fabric occurs locally. This process takes place in drained situations associated with unsaturated loess, usually along the toe and lower lateral parts of the sliding bodies.

THRESHOLD CONDITIONS IN LOESS SLOPES

Based on the above discussion of the fundamental properties of loess, the threshold conditions of slopes in the loess region of central China will now be considered. When a combination of environmental factors results in exceeding a threshold, the scale of the final event will largely depend on the amount of material involved in the transition from its undisturbed and stable state to its remoulded, rearranged and less stable state. Highly variable stresses will continuously influence the upper, surficial, zone of the loess slopes. However, if these stresses exceed certain threshold values, the influence may extend deeper than the upper layer of the loess and subsequently larger amounts of material are involved in the movements. Where a combination exists between progressive failure within the loess slope associated, e.g. with groundwater activity or a less permeable layer, the amount of loess involved in the movement may reach catastrophic dimensions.

These points are illustrated below by reference to a number of examples including small-scape collapse processes in the upper reworked layer, and catastrophic, high velocity failure involving several million cubic metres of loess.

Landslide Inventory Establishment

In order to investigate the nature and relative importance of environmental parameters determining the spatial incidence and scale of loess slope instability, an extensive mapping programme has been carried out. The only maps available at the start of the research programme were topographical base maps on several scales. The best cover of an extensive area of the loess region around Lanzhou was found to be the 1:35 000 base maps. In six sheets, the area is now mapped in respect of geomorphology, geology, landslide distribution, land use, slope angle and loess flow valley systems. All landslides encountered at the time of mapping were numbered and their characteristics were stored in a landslide database. The input describes the quantitative morphometrical characteristics of the specific slide, such as altitudinal interval, slope angles before and after failure, width and length, size of landslide mass and backwall. Additionally, qualitative categorised information has been stored which includes, *inter alia*, the relative age of the slide, the probable cause and potential impact,

Figure 8.16 Upper reworked layer in Lishih loess deposits (Gaolan Shan, Lanzhou). The lower boundary of the reworked layer lies at shoulder height of Dr Meng. Below are less jointed, *in situ*, Lishih deposits

the geological composition and the present degree of stability. A total of 210 landslides are stored in this database, from which it can be seen that there exists a clear relationship between slope stability and factors such as loess thickness, slope angle and slope aspect.

Internal Drainage and Failure of Loess Slopes

Close to the land surface, breakdown of the loess fabric takes place as a result of a variety of processes. Frost penetration, wetting and drying of the soil, and solution of salts form the major processes creating a surface layer, which has a reworked appearance (Figure 8.16). The effects of frost penetration are rather limited because, for most of the time, the moisture contents required for freeze–thaw disruption are generally too low. However, effects of snowmelt and subsequent nocturnal freezing, are considered to have an effect, especially on north-facing slopes where freezing temperatures may extend to soil depths of ~ 1 m in extreme

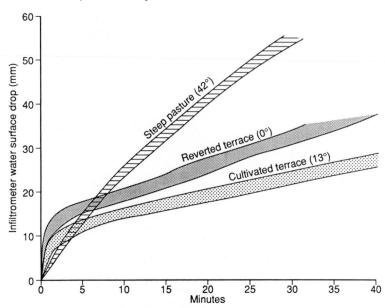

Figure 8.17 Infiltrometry variation with degree of slope and land use at West Tawa (20 km south-west of Lanzhou)

winter conditions. Infiltration of rainwater is rather limited. Some results of potential infiltration tests carried out at sites with three different types of land use showed that the strongly reworked and tilled soils of cultivated terraces have the lowest potential infiltration rates. The original aeolian loess fabric is completely disturbed at these sites, the redistribution of particles resulting in a much more compact material with a lower voids ratio. Additionally, natural joints and cracks are destroyed, with the result that ponding and overland flow occur more readily during rainfall events. On slopes underlain by uncultivated loess, meteoric water percolates into the soil by way of the joints developed in the reworked layer (Figure 8.17). Infiltration into the loess aggregates during rainfall events is usually hampered by the effect of rain beat and localised slumping and the formation of a surface crust; these processes rapidly decrease the infiltration capacity and increase overland flow. Steeper slopes enhance this trend. Maximum erosion by overland flow occurs on slopes of between 25° and 35° (Figure 8.18; Zhang & Wang, 1989).

Concentrated water flows can lead to localised liquefaction and subsequent enlargement of the joints acting as the main water routing channels. Progressive enlargement ultimately leads to the formation of nearly vertical sinkholes leading to pipe systems running sub-parallel to the surface of the slopes. These features, commonly known as loess karst, result in very high infiltration/permeability values, which exceed many times the rates measured for the undisturbed samples. The depth at which these pipe systems occur generally coincides with abrupt changes in permeability. For example, pipes occur just above palaeosols, at the Malan and Lishih/Wucheng contact zone, and directly above the bedrock. These subsurface drainage channels may reach diameters of more than 2 m: they are important in any consideration of the stability of loess slopes (Figure 8.19; cf. Pierson, 1983).

The occurrence of larger slides often coincides with a high density of sinkholes and pipes in a slope. During rainfall events, these pipe systems rapidly transmit the water through the

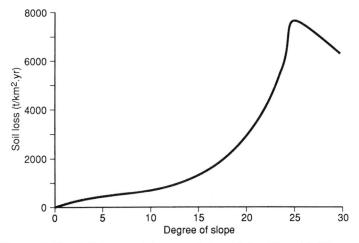

Figure 8.18   Soil loss and degree of slope (after Zhang & Wang, 1989)

Figure 8.19   Outlet of a major pipe system in the Sier Gou, 2.5 m wide by 3.5–4 m high

Figure 8.20   Salt efflorescences resulting from water percolation along joints in the loess fabric

surficial Malan deposits or even through the older Lishih and Wucheng loess so that even parts of the basal loess layer become saturated. Liquefaction in this saturated zone may lead to catastrophic failure. Where failure does not occur immediately, repetitive wetting during subsequent rainstorms progressively weakens the stable structure of the loess, while soluble salts are removed by throughflow and the (predominantly carbonate) cementation bonds are destroyed. The stability of the loess slope thus decreases progressively. Apart from salt efflorescences near spring zones at the toe of the slopes, there may be no external sign of this reduction in stability. Estimation of the threshold safety factor (Fs = 1) may, therefore, be substantially overestimated (Figure 8.20).

THRESHOLD CONDITIONS OF MASS MOVEMENTS IN LOESS

Large Complex Mass Movements

A good example of a large complex slide in loess is provided by the 1983 slide at Sale Shan. Here, a slide mass of about 50 million m³ was mobilised within seconds. Within one minute

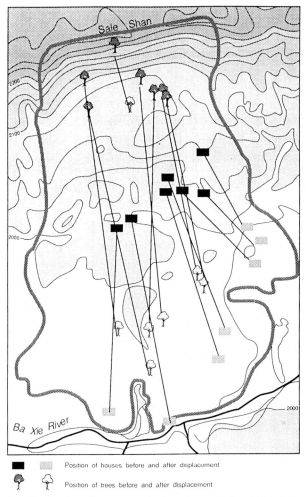

Position of houses before and after displacement

Position of trees before and after displacement

Figure 8.21 Areal extent and movement directions of the 1983 Sale Shan landslide

the catastrophic event had happened, causing the death of 220 people and destroying four villages and 223 ha of farmland (cf. Cao, 1986; Zhang, 1989; Zhu, 1989; Mahaney, Hancock & Zhang, 1990; see Figure 8.21). The mass movement consisted of a deep-seated rotational slide in the upper parts, while rapid disintegration of the lower slide blocks resulted in a highly mobile flow slide. Reconstruction of the movement shows that several thresholds were exceeded in a cascade-like system after initial triggering of the first movements. This trigger, however, has not been well defined for the Sale Shan slide. There is no evidence of either intensive rain or earthquakes at the time, so that resort must be made to other mechanisms. Two seismic shocks were recorded at the moment of sliding (~ 1.4 on the Richter scale: Cao, 1986; Zhang, 1989). It can be argued that these recordings merely picked up the event itself rather than a seismic trigger event (cf. Dijkstra *et al.*, in preparation). Moreover, Zhang (1989) has emphasised the importance of human influences on the stability of the loess slopes in the Sale area. Deforestation, building of reservoirs, irrigation works (disturbing the internal hydrological system), and the construction of roads all contributed to a possible acceleration in the reduction of the slope stability.

The Sale Shan forms part of a loess ridge (*liang*). The loess is about 150 m thick and mainly consists of the Lishih and Malan formations. Four river terraces can be distinguished with Lishih deposits only on the oldest (fourth) terrace. On all terraces a basal gravel layer, loessic alluvium and a drape of Malan loess as well as some reworked, partly colluvial, loess can be found. There is some evidence that the fourth river terrace occasionally induced landsliding, predominantly caused by failure of the argillaceous bedrock.

Liangs in this part of the Loess Plateau are strongly related to the palaeo-relief of the underlying bedrock, which consists predominantly of Pliocene lacustrine clays derived from eroded Linxia basin sediments. Locally, the bedrock surface is weathered as a result of the warm late Pliocene climate, resulting in an enrichment of swelling clay minerals along the contact zone between bedrock and loess. Analysis of surveyed cross profiles of the Sale Shan slide lead us to propose the following model.

Extensive salt efflorescences occur along spring zones at the loess—bedrock contact, reflecting the high salt content of the water. Below the thick loess the argillaceous bedrock (Pliocene red clays of the Linxia formation) is densely jointed and has characteristic cohesion and internal friction angle values of 0 kN and 4°, respectively. The weakest zones in the lower parts of the slope are thought to be the discontinuous contact zone between the lower gravels of the river terrace and the bedrock. On the basis of the reconstructed slip surface it is evident that the bulk of the strength of the Sale slope is carried by loess deposits and not by the Pliocene bedrock. The effects of the throughflow of groundwater and the resulting solution of salts and carbonates results in a progressive failure of the loess structure within the slope. Subsequently, the tensional forces of the whole slope are distributed over an increasingly smaller proportion of undisturbed loess. At this stage plastic deformation of the loess aggregates in the Lishih and the Malan begins to occur, resulting in the tension cracks to be seen along the crest of the Sale ridge in the surficial Malan loess. These cracks then act as a kind of catalyst in the failure of the whole slope. Progressive failure as a result of groundwater stagnation on top of the argillites is relatively slow. However, along the cracks accelerated subsurface erosion occurs as a result of suffosion. The main trigger factor now is time. At a certain stage the progressive failure within the slope results in a sufficient degradation of the total strength, with peak strength conditions occurring only in the crucial central part of the slope, while in the upper and lower parts of the loess structural deformation has reduced the tensile strength to residual values. At this stage only a minor trigger is necessary: this may be an increase in groundwater flow or merely some extra time. At this stage, as was demonstrated in the shearbox and triaxial tests, the brittle bonds fail and the whole mass suddenly collapses. The immediate decrease in frictional forces and the enormous release of energy potential derived from the weight of the slopes and the vertical interval over which it travels results in a rapid increase in velocity. From the toe upwards, rapid disintegration of the rotational slide blocks takes place and the whole mass spreads out over the lower river terraces as a highly mobile flow. In several outcrops and from aerial photographs taken shortly after the movement occurred, it can be concluded that only the basal layer of the flow was saturated, the upper part of which moved as an essentially dry flow. In addition, beneath the backwall dry blocks and slabs fell and broke up to produce large powder flows. This is a very common situation for slides involving large quantities of loess (cf. Derbyshire *et al.*, 1990; Dijkstra *et al.*, in preparation). This reconstruction was substantially confirmed by statements by the only survivor of this slide. He held on to a tree and travelled on top of the flow over a distance of about 800 m. The erosional capability of the lower flow part of the slide complex was very low as a result of the high degree of mobility. When sliding ceased, most of the underlying terrace edges could still be observed through the flow deposits.

Large-scale, 'Rotational' Mass Movements

Landslides have frequently been triggered by large earthquakes. Direct relationships between the effects of an earthquake and a particular slide can be established by using the written records of historical events that have had a considerable impact on daily life. One of the slides dated using these records is the Tawa slide, about 10 km south-west of the western suburbs of Lanzhou, which, on the basis of written records, could be related to the large earthquake event of 1125 AD (Figure 8.22). The slide involved part of a large *liang*, which

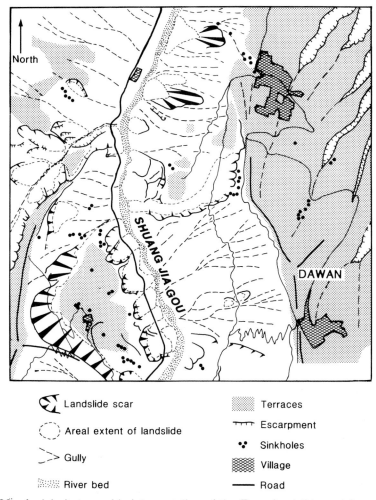

Figure 8.22 Aerial photographic interpretation of the Tawa landslide and its environs (width of figure is ~2 km)

at a maximum elevation of 2080 m, lies 380 m above the valley floor. The interpretation of aerial photographs of the slide indicates a maximum vertical displacement along the slip surface of 150 m, the shape and slope angle of the landslide scar suggesting a single major circular failure plane. However, on the basis of additional field investigations more complicated reconstructions of the mass failure are here proposed (Figure 8.23).

One of the most important factors to consider when characterising sliding in loess is the nature of the underlying bedrock topography and the relationship between pre and post failure loess thickness. In the case of the Tawa slide this is hampered by the lack of good bedrock exposures (most of which are covered by a thin drape of loessic colluvium), the apparently high bedrock relief and the existence of a large fault system outside, but very near to the mass movement, precluding the use of good bedrock exposures on adjacent slopes. In the main body of the slide the bedrock−loess boundary is exposed at the bottom of a number

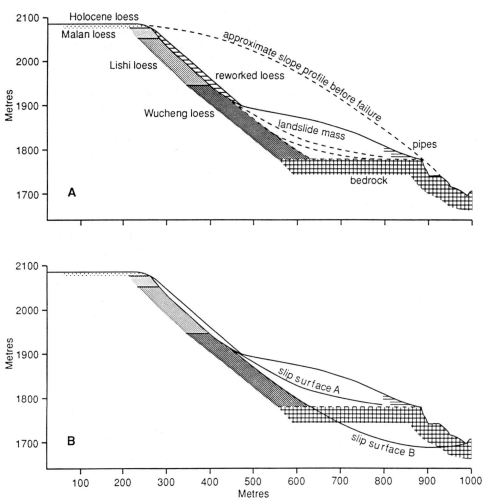

Figure 8.23 Profiles indicating the possible reconstructions of the Tawa landslide. Slip surface A: shallow slide only involving loess movement on top of the palaeo-bedrock relief. Slip surface B: deep-seated rotational slide involving bedrock

of sinkholes at depths varying between 8 and 20 m. This thickness is considerably less than expected on the basis of information derived from the back scar of the failure (assuming that no further disintegration of the slide mass has occurred). Striated boulders have been found at the lower north-western part of the slide, indicating that bedrock could have been involved in the slide process.

On the basis of the preliminary investigations at Tawa two possible reconstructions have been established. In the case of high bedrock relief, the thickest loess deposits are to be found at the highest point of the *liang* (loess ridge). Therefore, slope failure will involve only relatively small amounts of loess, thus explaining the thin drape of loess on bedrock in the central parts of the slide body. In this reconstruction it is assumed that a considerable amount of bedrock is involved in the mass movement (Figure 8.23B; slip surface B). The second reconstruction is based on an assumption of a more substantial loess cover and the coincidence of the slip surface with the bedrock–loess contact zone. In this case, mass failure results in disintegration of the loess fabric on a large scale, and a flowing movement of loess into the river valley on top of the undulating bedrock, leaving behind only a thin drape of remoulded loess (Figure 8.23B; slip surface A). The precise nature of, and the relationships between, the environmental and geotechnical parameters determining sliding processes in these large loess slides clearly need further investigation.

In both cases, as a result of the displacement, the river course has been diverted. This has led to increased undercutting of both the main sliding body and the slopes on the eastern side of the river. Continuous adjustment to new equilibrium profiles on both slopes has occurred, however, without reactivation of the complete sliding body involved in the earthquake-induced sliding process. Apparently, only at the toe of the slide can slope instability decrease sufficiently for medium-scale mass movements to occur. An additional process decreasing the stability along this zone is the throughflow of water, which is severely impeded on top of the less permeable bedrock. At the spring line, bright salt efflorescences have been observed, indicating that progressive weakening of the loess structure occurs within the slope with major consequences for the decrease in the overall stability of the old landslide mass. No evidence has been found to indicate movement of the main sliding body after the event in 1125 AD. A continuous reduction of the overall stability of the slide takes place as a result of a variety of processes, including leaching of soluble salts, removal of material (weight) along the toe of the slide and increased suffosion, and current research is investigating the specific role of salt leaching and recrystallisation in secondary failures close to rockhead.

Tan-ta

Throughout the loess region small-scale adjustments occur in response to changing slope conditions. The size of these slides is generally of the order of 10 m across, with planar slip surfaces generally at depths varying from one to several metres. After initiation of the slides rapid disintegration takes place, often associated with high sliding velocities. Chinese researchers refer to this type of mass movement as *tan-ta*. The *tan-ta* occur predominantly in the upper reworked layer of the steeper (Malan) loess slopes (Figure 8.24). Initiation of the slides is influenced by a variety of factors including infiltration of snowmelt or rainwater, undercutting of the slope and earth tremors. On many occasions *tan-ta* are initiated as a result of human activities, such as irrigation and cuts in slopes associated with roads and railways. The occurrence of a *tan-ta* may trigger additional small-scale mass movements and in several places clusters of *tan-ta* can be found. Although the size of the mass movements is relatively

Figure 8.24 Example of *tan-ta* failure affecting a road on the north-facing slopes of the Gaolan Shan near Lanzhou

limited, *tan-ta* form a substantial hazard in the loess region as a result of the high frequency of the movements and the close coincidence of this type of slope failure and human activities.

CONCLUSION

The large variety evident in the types and dimensions of mass movements in loess is dependent on the threshold conditions of a set of environmental parameters that affect loess slope stability. These parameters range from small-scale morphometrical variables to influences on engineering properties of loess on a particle to particle basis. It follows that any comprehensive slope instability assessment in the extremely sensitive loess lands demands that all aspects be investigated. Of necessity, this problem must be carried out in several substages. However, the final integration of all research results into a set of models describing the stability characteristics of loess slopes and their potential for generating a diversity of mass movements forms the major challenge of this type of research. In this paper an attempt has been made to present an overview of the major environmental parameters and their effect on the threshold conditions on loess slopes. Engineering problems in loess as a special subject, as against engineering problems in clays, sand or rock, have, however, been dealt with only occasionally. It is evident that, in order to obtain a better insight into the processes affecting stability conditions on loess slopes, further work will have to be done on the complex nature of the loess landscape at all levels described here.

ACKNOWLEDGEMENTS

This paper is contribution no. 4 of CEC contract no. CI. 1.0109.UK(H) and presents some results taken from a much larger study of landslides and mass flowage in the loess of north-central China generously supported by the Council of the European Communities and the Government of Gansu Province, PR China. Special acknowledgements are due to our Chinese research colleagues, and in particular to Professor Wang Jingtai, of the Geological Hazards Research Institute of the Gansu Academy of Sciences in Lanzhou, and to the other universities and research institutions participating in the research. Thanks are also due to Professor D. K. C. Jones (LSE, University of London, UK), who set up the original framework of the loess landslide databank, and to Dr Chris Rogers (Loughborough University of Technology, UK) and Dr Theo van Asch (University of Utrecht, The Netherlands) for their valuable discussions and for their continuing interest in the loess landsliding problem.

REFERENCES

Billard, A., Muxart, T., Derbyshire, E., Wang, J. T. and Dijkstra, T. A. (1993). Landsliding and land use in the loess of Gansu Province, China. *Zeitschrift für Geomorphologie, Supplementband*, in press.

Cao, B. (1986). The geologic characteristics of the Sale Shan type of super landslide and a model for spatial prediction. *Proceedings of the 5th International IAEG Conference*, Vol. 3, pp. 1989–1997.

Derbyshire, E. (1984). On the morphology, sediments, and origin of the Loess Plateau of Central China. In Gardner, R. and Scoging, H. (eds), *Mega-geomorphology*, Oxford University Press, pp. 172–194.

Derbyshire, E. (1988). Granulometry and fabric of Quaternary silts from Eastern Asia. In *Proceedings of the Second Conference of the Palaeoenvironment of East-Asia from the Mid Tertiary*, Vol. 1, pp. 215–245.

Derbyshire, E. (1989). Mud or dust; erosion of the Chinese loess. *Geography Review, 3*, 31–35.

Derbyshire, E., Billard, A., van Vliet-Lanoë, B., Lautridou, J.-P. and Cremaschi, M. (1988). Loess and palaeoenvironment: some results of a European joint programme of research. *Journal of Quaternary Science, 3*(2), 147–169.

Derbyshire, E. and Mellors, T. W. (1988). Geological and geotechnical characteristics of some loess and loessic soils from China and Britain; a comparison. *Engineering Geology, 25*, 135–175.

Derbyshire, E., Wang, J. T. and Smalley, I. J. (1990). Loess landslides and geotechnical classification. *Landslide News, 4*, 21–23.

Derbyshire, E., Wang, J. T., Jin, Z., Billard, A., Egels, Y., Jones, D. K. C., Kasser, M., Muxart, T. and Owen, L. (1991). Landslides in the Gansu loess of China. In Okuda, S., Rapp, A. and Zhang, L. Y. (eds), *Loess: Geomorphological Hazards and Processes, Catena Supplement, 20*, 119–145.

Dijkstra, T. A., Smalley, I. J., Derbyshire, E. and Li, Y. J. (in preparation). The Sale Shan landslide: a case study.

Doornkamp, J. C. and Ibrahim, H. A. M. (1990). Salt weathering. *Progress in Physical Geography, 14*(3), 335–348.

Feng, X. and Guo, A. (1985). Earthquake landslide in China. *Proceedings of the IVth International Conference and Field Workshop on Landslides*, Tokyo, pp. 339–346.

Ferguson, R. I. (1984). Sediment load of the Hunza River. In Miller, K. J. (ed.), *The International Karakoram Project*, Vol. 2, pp. 536–580.

Gao, G. (1983). Microstructure of loess soil in China relative to geological environment. In: Geological environment and soil properties, *Special Publication*, ASCE Geotechnology, Engineering Division, Houston, pp. 121–136.

Heller, F. and Liu, T. S. (1982). Magnetostratigraphical dating of loess deposits in China. *Nature*, **300**, 431–433.

Klinkenberg, S. (1988). Soil erosion and soil conservation on the Chinese Loess Plateau, *Vakgroep Publicatie*, **11**, Geografisch en Planologisch Instituut, Katholieke Universiteit Nijmegen.

Krinsley, D. H. and Smalley, I. J. (1973). Shape and nature of small sedimentary quartz particles. *Science*, **180**, 1277–1279.

Kukla, G. (1989). Long continental records of climate — an introduction. *Palaeogeography, Paleoclimatology, Palaeoecology*, **72**, 1–9.

Kukla, G. and An, Z. S. (1989). Loess stratigraphy in central China. *Palaeogeography, Palaeoclimatology, Palaeoecology*, **72**, 203–225.

Kukla, G., Heller, F., Liu, X. M., Xu, T. C., Liu, T. S. and An, Z. S. (1988). Pleistocene climates in China dated by magnetic susceptibility. *Geology*, **16**, 811–814.

Lautridou, J. P. (1988). Recent advances in cryogenic weathering. In Clark, M. J. (ed.), *Advances in Periglacial Geomorphology*, Wiley, Chichester, pp. 33–47.

Liu, B., Yang, T., Lai, Z., He, Y., He, N. and Wang, H. (1979). Geology and earthquakes in Gansu Province, PR China, Lanzhou Earthquake Institute Publication.

Liu, T. S. and Chang, T. H. (1964). The Huangtu (loess) of China. *Report of the 6th INQUA Congress*, Vol. 4, pp. 503–524.

Liu, T. S., An, Z. S., Yuan, B. and Han, J. (1985a). The Loess-paleosol sequence in China and climatic history. *Episodes*, **8**, 21–28.

Liu, T., Gu, X. F., An, Z. and Fan, Y. X. (1985b). *Loess and the Environment*, China Ocean Press, Beijing.

Mahaney, W. C., Hancock, R. G. V. and Zhang, L. (1990). Stratigraphy and paleosols in the Sale terrace loess section, northwestern China. *Catena*, **17**, 357–367.

Pierson, T. C. (1983). Soil pipes and slope stability. *Quarterly Journal of Engineering Geology*, **16**, 1–11.

Pye, K. (1987). *Aeolian Dust and Dust Deposits*, Academic Press, London.

Richthofen, F. von (1877). *China. Ergebnisse eigener Reisen und daran gegruendeter Studien*, Vol. 1.

Smalley, I. J. (1966). The properties of glacial loess and the formation of loess deposits. *Journal of Sedimentary Petrology*, **36**(3), 629–676.

Smalley, I. J. (1971). Nature of quickclays. *Nature*, **231**, 310.

Smalley, I. J. (1972). The interaction of great rivers and large deposits of primary loess. *Transactions of the New York Academy of Sciences*, series 2, **34**(6), 534–542.

Smalley, I. J. and Vita-Finzi, C. (1968). On the formation of fine particles in sandy deserts and the nature of desert loess. *Journal of Sedimentary Petrology*, **38**, 766–774.

Smalley, I. J. and Cabrera, J. G. (1970). The shape and surface texture of loess particles. *Geological Society of America Bulletin*, **81**, 1591–1596.

Smalley, I. J. and Derbyshire, E. (1991). Flowslide-type ground failures in the airfall loess deposits of northern China and the shallow-marine postglacial clays of eastern Canada. In Jones, M. and Cosgrove, J. (eds), *Geology of Neotectonic Environments*, Belhaven Press, London and New York, pp. 202–219.

Tan, T. K. (1988). Fundamental properties of loess from Northwestern China. *Engineering Geology*, **25**, 103–122.

Wang, J. T. and Derbyshire, E. (1988). EC launches project on landslides and debris flows in Chinese loess. *Episodes*, **11**(2), 131–132.

Wang, R. and Lang, Y. (1989). An approach to the West boundary of the Loess Plateau. In Zhang, L. and Siwei, S. (eds), *International Field Workshop on Loess Geomorphological Processes and Hazards*, pp. 140–147.

Wang, Y. Y., Wu, Z. B. and Yue, L. (1981). Constituent materials and structure of loess in Lanzhou and the date of its formation. *Northwest University Bulletin (Sci. Edn.)*, **3**, 1–27 (in Chinese).

Wang, Y. Y., Lin, Z., Lei, X. and Wang, B. (1987). Fabric and other physico-mechanical properties of loess in Shaanxi Province, China. *Catena, Supplement,* **9**, 1−10.

Wen, Q., Diao, G. and Yu, S. (1985). Geochemical characteristics of loess in the Luochuan section, Shaanxi province. In Pecsi, M. (ed.), *Loess and the Quaternary.* Akademy Kiado, Budapest, pp. 65−77.

Wen, Q., Diao, G. and Yu, S. (1987). Geochemical environment of loess in China. *Catena, Supplement,* **9**, 35−46.

Zhang, D. and Wang, D. (1989). Soil erosion features and its affecting factors in Nanxiaohe valley in Xifang, Gansu Province. In Zhang, L. and Siwei, S. (eds), *International Field Workshop on Loess Geomorphological Processes and Hazards*, pp. 44−52.

Zhang, H. and Cheng, F. (1989). The Jiuzhoutai loess profile in Lanzhou, Gansu, PR China. In Zhang, L. and Siwei, S. (eds), *International Field Workshop on Loess Geomorphological Processes and Hazards*, pp. 72−80.

Zhang, L. (1989). The landslide history and Late Cenozoic environmental factors in Sale Shan area, Dongxiang County, Gansu. In Zhang, L. and Siwei, S. (eds), *International Field Workship on Loess Geomorphological Processes and Hazards*, pp. 81−93.

Zhang, L. and Dai, X. (1989). The Loess Plateau and its formation and evolution. In Zhang, L. and Siwei S. (eds), *International Field Workshop on Loess Geomorphological Processes and Hazards,* pp. 1−17.

Zhang, Z. (1986). *Explanatory Notes to the Geomorphological Map of the Loess Plateau of China, Institute of Hydrology and Engineering Geology*, Chinese Academy of Science, Geological Publishing House.

Zhao, X. and Qu, Y. (1981). Loess of Zhaitang and Yanchi region, Beijing. *Scientifica Geologica Sinica,* **1**, 47−51.

Zhu, H. (1989). Some types of seismic landslides in loess area in China. In Zhang, L. and Siwei, S. (eds), *International Field Workshop on Loess Geomorphological Processes and Hazards*, pp. 64−71.

9 Sensitivity of Fault-generated Scarps as Indicators of Active Tectonism: Some Constraints from the Aegean Region

IAIN S. STEWART

Division of Geography and Geology, West London Institute, UK

ABSTRACT

While morphometric analysis of young normal fault scarps in Quaternary colluvium and alluvium is a powerful tool for assessing the timing of earthquake activity in a number of active tectonic terrains, normal fault scarps in the Aegean region are commonly developed in limestone bedrocks. These 'Aegean-type scarps' are underlain by a layered architecture of fault rocks of contrasting resistance to erosion that give rise to a temporally and spatially variable pattern of degradation. A more fruitful approach to the palaeoseismological dating of 'Aegean-type' fault scarps is suggested to involve the detailed analysis of the karstification of their surfaces.

In addition, in contrast to existing geomorphologically driven models of range-front evolution, the morphological characteristics of range fronts in the Aegean region are largely a function of the structural attributes of their underlying fault zones. In particular, where distributed fault splaying is promoted, a ragged, step-like scarp morphology is formed. The presence of large topographic highs or salients in the immediate hangingwall of a fault, however, may inhibit such fault splaying, producing a ramp-like scarp morphology. The recognition that these 'ramp-type' and 'step-type' morphologies, end members of a continuum of range-front forms, are independent of the rate and duration of geomorphological activity, necessitate cautious applications of purely geomorphic indices to identify contrasting degrees of tectonism along different range forms.

FAULT-SCARP STUDIES

Recognition that the instrumental and historical record of tectonic activity is insufficient to describe the long recurrence intervals of earthquakes has, in the last decade, focused attention on the identification of palaeo-earthquakes from the geological and geomorphological record. Of particular importance to this 'palaeoseismological' approach has been the recognition that the morphologies of fault-generated landforms, such as fault scarps and, to a cruder extent, range fronts, can be sensitive indicators of the spatial and temporal distribution of tectonism within a region (Figure 9.1).

The relationship between scarp morphology and age first emerged from the classic studies of G. K. Gilbert (1890, 1928) in the Basin and Range Province of the western United States, but has recently been 'rediscovered' following the study by Wallace (1977) of the

Landscape Sensitivity. Edited by D. S. G. Thomas and R. J. Allison

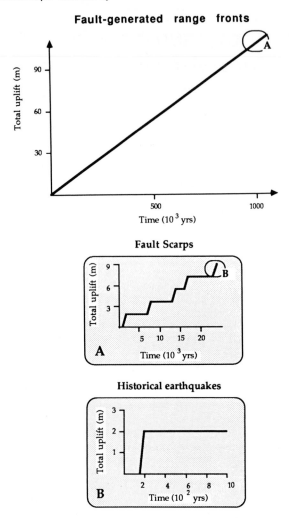

Figure 9.1 Schematic graphs showing contrasting sensitivity of fault scarps and range fronts as indicators of tectonic uplift. Range fronts, which describe the longest periods of time, preserve only a crude record of uplift, while fault scarps (A) and historical earthquake data (B) retain a more sensitive record of uplift over shorter timescales. The circled part of the range front graph denotes the time period covered in (A), while the circled part of the fault scarp graph shows the time period covered in (B). Reproduced by permission from Mayer (1986) Figure 1 in *Active Tectonics*. Copyright © 1986 by the National Academy of Sciences, Washington, DC

morphological characteristics of scarps generated by extensional (normal) faulting in the same region. Wallace's study of young fault scarps cutting colluvial and alluvial deposits in north-central Nevada showed that scarp forms appeared to degrade, over a timescale of several thousand years, via a systematic and, therefore, predictable sequence of morphologic change (Figure 9.2). In particular, Wallace (1977) identified certain key parameters of the fault-scarp profile, such as scarp height and maximum-scarp-angle (θ_m), as sensitive time-dependent attributes, as intuitive assessment later formalised by Bucknam & Anderson (1979), who demonstrated empirically that θ_m decreases with age and increases with scarp height. Numerous subsequent studies in the Basin and Range Province have employed scarp morphology to discriminate between fault scarps of contrasting ages. On the basis of this approach, piedmont scarps — single-event fault scarps cutting a piedmont surface comprising poorly cohesive sediments — can be assigned to broad age categories (e.g. Holocene, late Pleistocene) (Crone & Haller, 1991).

According to the pattern of scarp degradation proposed by Wallace (1977) to account for the morphologies of piedmont scarps in the Basin and Range Province,

> the slope of the original fault plane or scarp is replaced by one controlled by erosional processes. In the early stages of slope replacement, the dominant erosional process is gravity spalling from the free face and accompanying accumulation of debris at the scarp base. As time passes, water erosion or wash becomes the dominant process and the slope changes in angle (slope decline) as well as in configuration.   (Wallace, 1977, p. 1268)

Qualitative assessments of scarp morphologies have suggested that the initial 'slope replacement' phase could be achieved within 200—500 years, though some free faces may persist for several thousand years in particularly arid climates (Machette, 1989). Because this is within the timespan provided by instrumental and historical records of tectonic activity, and since fresh scarps possess inherently variable morphologies (Stewart & Hancock, 1990b), scarps retaining a free face are of limited use in constraining the timing of scarp formation beyond underlining the youthfulness of tectonic activity within a region. During the subsequent 'slope decline' phase, however, when the last remnants of the original tectonic landform have been buried by the debris slope, the residual scarp form increasingly becomes a function of the rate and duration of geomorphological processes acting on it. This phase lasts from many tens of thousand to several hundred thousand years. Over this time, however, the rate and degree of scarp modification progressively decreases until residual fault scarps approach a time-independent or 'characteristic' form (Brunsden & Thornes, 1979). As a consequence, the optimum period during which fault-scarp morphology can be used as an indicator of scarp age is between 2000 and 20 000 years (Machette, 1989).

Although fault-scarp morphometric analysis has been extended through quantitative modelling to provide a potentially useful numeric-age technique (Nash, 1980, 1986; Hanks et al., 1984; Machette, 1989), it is as a relative-age technique that this approach is finding increasing application beyond the Basin and Range Province. Morphometric dating, for example, has recently been applied to normal fault scarps in eastern Sinai (Gerson, Grossman & Bowman, 1985; Bowman & Gerson, 1986), China (Zhang et al., 1986) and the Andes (Cabrera, Sebrier & Mercier, 1987). This paper, however, outlines some constraints relating to its applicability to the Aegean region; an area with a comparable degree and style of active faulting to the Basin and Range Province. Although based on a five-year programme of investigating fault-generated landforms in central Greece, Crete and western Turkey, the paper will draw largely from specific examples of fault-generated scarps in central Crete.

AEGEAN-TYPE FAULT SCARPS

Similar to the Basin and Range Province, active extensional faulting in the Aegean region is manifest as a series of uplifted basement blocks, often of Mesozoic limestones, separated by subsident basins of Neogene and Quaternary sediments (Figure 9.3) (Angelier et al., 1982).

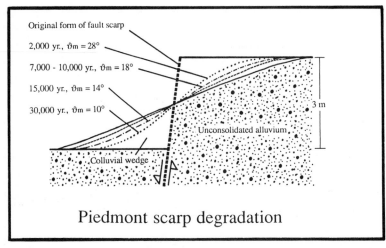

Piedmont scarp degradation

Figure 9.2 Schematic diagram showing the original form and subsequent degradation of a fault scarp formed in unconsolidated alluvium. The maximum slope angles (θ_m) for the 2000, 7000−10 000 and 15 000 year old scarps are based on empirical data for scarps of known age in the eastern Basin and Range (Bucknam & Anderson, 1979), while the θ_m for the 30 000 year old scarp is computed from a mathematical regression on the empirical data. Vertical exaggeration is ×2. Arrows show direction of motion on a fault. After Crone & Haller (1991, Figure 3)

While extension in the central Aegean has produced almost complete submergence, around the periphery of the Aegean Sea, in central Greece, Crete and western Turkey, uplifted blocks are emergent and bounded by prominent linear fault escarpments. Many of the basins within this circum-Aegean belt, however, remain partially or entirely submerged. As a consequence, the Aegean region lacks the extensive exposures of loosely consolidated colluvial and alluvial deposits that characterises the Basin and Range Province. Piedmont scarps, therefore, are a less widespread tectonic landform in the Aegean domain. Instead, bedrock fault scarps, particularly those cutting basement limestones, are a ubiquitous and well-developed form of tectonic scarp in the Aegean region. For this reason, and in order to distinguish them from scarp forms typical of the Basin and Range Province, they will be referred to here as *Aegean-type scarps*.

In addition to being a less-common expression of active tectonism, the size of fault scarps in the Aegean region is generally less than their counterparts in the Basin and Range Province. While numerous magnitude 7 or greater earthquakes in the Basin and Range Province have produced fault scarps up to 4.5 m in height (dePolo *et al.*, 1991), no historical or instrumental events exceeding magnitude 7 have been recorded in the extensional part of the Aegean domain (Ambraseys & Jackson, 1990). Instead, the more discontinuous nature of active extensional faults in the Aegean region results in moderate-sized earthquakes producing fault scarps up to a maximum height of 1.5 m (e.g. Jackson *et al.*, 1982). Morphometric analysis of such small-scale forms, which commonly vary markedly in their geomorphological expression and degree of preservation, are unable to provide useful assessments of the timing of faulting events.

While it is widely recognised that bedrock fault scarps are unsuitable for the type of morphometric analysis applied to piedmont scarps (Mayer, 1985), many workers envisage that, though characterised by slower rates of erosion, bedrock scarps evolve via a comparable sequence of scarp degradation. According to Wallace (1977, p. 1272), for example, 'after

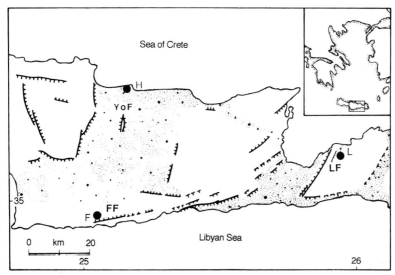

Figure 9.3 Simplified geological map of central and eastern Crete. All the faults shown are neotectonic normal faults (dashed where inferred) with barbs indicating the direction of downthrow. Faults: FF, Furnofarango fault; YoF, Youchtas faults; LF, Lastros fault. Towns: H, Heraklion; F, Furnofarango; L, Lastros. Stippled ornament denotes Neogene and Quaternary deposits, while unornamented areas represent basement rocks, predominantly carbonate bedrocks. Inset map shows position of Crete in the Aegean region

1 M.y. almost all scarps, even those in bedrock, have declined substantially in slope angle so that wash and other processes are dominant'. More recently Yuming (1989) suggested that recent fault scarps formed in gneisses and schists underwent slope decline at a rate of $0.036°$ year^{-1}.

Despite such contentions, investigations of limestone fault scarps in mainland Greece, western Turkey and, in particular, central Crete indicate that the pattern of degradation displayed by 'Aegean-type' scarps is intimately related to the 'architecture' of the fault, since faults are rarely discrete structures but, instead, are zones comprising complex networks of subparallel fault planes within broad belts of brecciation and fracturing. Although the structural attributes of normal fault zones in the Aegean region have been discussed elsewhere (Stewart & Hancock, 1988, 1990a, 1991a), such attributes have considerable implications for the sensitivity of fault scarps to subsequent geomorphic modification.

While fault zones may be characterised by tracts of highly fractured, erosionally weak fault rocks, often giving rise, for example, to 'fault-line valleys', it is important to note that erosion may be either hindered or facilitated by faulting depending on the type of fault rock generated (Tricart, 1974). Attrition and mineralisation accompanying normal faulting in limestone terrains commonly produces centimetre-thick armoured carapaces of recemented limestone fault breccia immediately underlying fault planes (Figure 9.4(a)). These *compact breccia sheets* (Stewart & Hancock, 1988) result in many fault scarps, such as that at Lastros in eastern Crete (Figure 9.3), occurring as sharply defined and laterally persistent features (Figure 9.4). Underlying these sheets are metre-wide belts of intensely fractured and disaggregated *incohesive breccias* (Figure 9.4(a)), formed by localised fracturing as individual fault planes propagated towards the surface (Stewart & Hancock, 1990). By contrast, the near-surface flexing of bedrock

Figure 9.4   (a) A limestone normal fault scarp in central Greece. The 0.5 m thick, pedogenically stained strip along the scarp base was exposed during the 1981 Corinth earthquakes (Jackson *et al.*, 1982). The 0.5 cm thick compact breccia carapace (C) is breached by solutional pipes (S) and stress-release fractures (F) to expose the underlying incohesive breccia belt (I). Note how the appearance of the scarp varies markedly over a distance of a few metres. Tape measures 2 m. (b) A prominent limestone normal fault scarp cutting across the hillside to the west of Lastros, Crete. The scarp is 4–5 m in height and possesses a well-developed compact breccia carapace. Offset glacial deposits nearby suggest the scarp is Wurm in age. View towards the south-west. (c) The Lion's Gate at Mycenae, central Greece, which was constructed around 3300 years ago and which is cut by, but not offset by, a limestone fault scarp. The scarp retains a centimetre-thick compact breccia carapace scarred by centimetre-sized karstic grooves

and pits, indicating that such carapaces may be preserved for many thousands of years.
(d) A corrugated normal fault plane emergent along the base of the Furnofarango range
front in south-central Crete. The resulting 30–50 m high, moderately inclined fault scarp
gives the range front a smooth, ramp-like appearance. Breaches of the thick compact
breccia carapace along this part of the range front are largely confined to corrugation
troughs. View towards the south-west

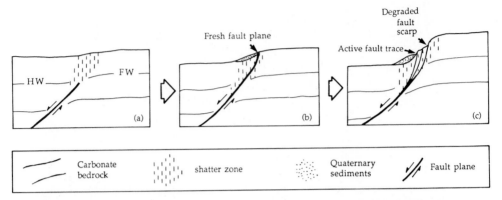

Figure 9.5 Sequential evolution of a normal fault zone in the Aegean region. Emergence of the fault at the surface is preceded by the development of a wide shatter zone in the hangingwall (HW) (a). Later increments of fault motion propagating a fault through this zone result in further fracturing and brecciation (b). The preferential migration of the active fault plane from the footwall (FW) towards the hangingwall (HW) (intrafault-zone hangingwall collapse) of the main fault results in a stepped fault scarp and a layered arrangement of fault rocks (c). Modified from Hancock & Barka (1987, Figure 8)

limestones prior to the emergence of the main fault zone produces a *shatter zone* (Vita-Finzi & King, 1985; Hancock & Barka, 1987; Stewart & Hancock, 1990a), a several hundred metre-wide tract of regularly spaced, subvertical fractures that strike parallel with and perpendicular to the buried fault (Figure 9.5).

With the progressive evolution of a fault zone, overprinting by different deformation mechanisms may give rise to a complex assemblage of fault rocks of contrasting resistance to erosion. Overprinting is often encouraged by the tendency for some active normal faults to splay preferentially basinward (i.e. into the hangingwall of earlier fault traces — a process called *intrafault-zone hangingwall collapse* [Stewart & Hancock, 1988]) (Figure 9.5(b)). This tendency results in many Aegean-type fault zones developing a layered architecture of alternating compact and incohesive breccias set in a broad shatter zone (Figure 9.5(c)). As a consequence of this, degradation rates vary with the type of fault rock encountered; slowest through the compact breccia carapaces, fastest through incohesive breccia belts and proceeding at more moderate rates through the wide shatter zone (Figure 9.6).

Although it is difficult to assign erosion rates to the contrasting fault rocks, some qualitative assessments can be made. Degradation of the uppermost compact breccia occurs predominantly by its progressive karstification. A limestone fault scarp, for example, which cuts but does not offset the Lion's Gate at Mycenae, central Peloponnesus, constructed around 3300 years ago, retains a 5–10 cm thick compact breccia carapace that is scarred by dissolution pits and karstic grooves several centimetres in length (Figure 9.4(c)). The well-defined scarp at Lastros possesses a compact breccia carapace that, though breached in places, remains several tens of centimetres thick. Although the age of the last increment of faulting along this scarp is not known, Angelier (1979) estimates it to be Wurm in age, based on offset glacial deposits nearby. It is clear, therefore, that compact breccia carapaces may be preserved for several thousand and probably several tens of thousands of years. Where breaching of a carapace has been promoted, however, degradation can, over the same timespan, result in the almost complete removal of the compact breccia sheet (Figure 9.4(a)).

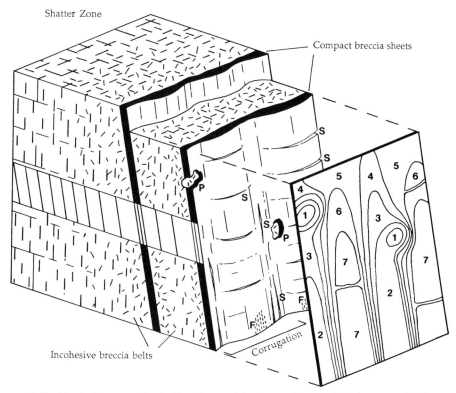

Shatter Zone

Compact breccia sheets

Incohesive breccia belts

Corrugation

Figure 9.6 Block diagram illustrating the architecture of normal fault scarps in the Aegean region. Alternating layers of compact breccias and incohesive breccias immediately underlying the scarp give way to a less intensely fractured shatter zone. The unornamented section through the fault scarp highlights the varying susceptibility of the different fault rock to erosion. Each 'rhombic box' represents a volume of rock imagined to be removed in a standard period of time. Highest degradation rates characterise incohesive breccia belts, moderate degradation rates characterise shatter zone and lowest degradation rates characterise compact breccia sheets. The fault scarp surface is chacterised by a variety of structures which breach the armoured compact breccia carapace. Contoured surface shows idealised variations in degradation potential arising from the uneven scarp surface (highest rates correspond to lowest numbers). S, stress-release fractures; P, pluck holes, F, frictional-water striae

In addition to temporal variations in scarp degradation, spatial variations occur in the breaching of the erosionally resistant compact breccia carapace (Figure 9.6). Breaching, for example, may preferentially occur at tectonically plucked cavities (*pluck holes* [Hancock & Barka, 1987]), subsurface solutional flow conduits or at fault-plane fractures initiated by stress-release when the fault plane became emergent at the surface (Figure 9.4(a)). More commonly, however, breaches are coincident with along-strike anomalies in the fault architecture since fault scarps rarely possess uniformly planar surfaces but are, instead, undulatory along strike. These undulations or *corrugations* (Hancock & Barka, 1987) are commonly up to several metres in wavelength and several tens of centimetres in amplitude (Figure 9.4(d)). The troughs of corrugations are commonly prime sites for localised breaching since they are underlain by thinner compact breccia sheets, cut by closely spaced stress-release fractures and encourage the preferential flow of surface wash. In contrast, corrugation crests are areas of comparatively little geomorphic modification (Figure 9.6). As a consequence of this uneven pattern of scarp degradation, in which degradation is initiated at different parts of the scarps and progresses at different rates through it, many scarps develop highly cavitated and ragged morphologies.

FAULT-GENERATED RANGE FRONTS

The geomorphological expression of several hundred metre high fault-generated mountain fronts possess a longer but considerably cruder record of tectonic activity (Figure 9.1) (Mayer, 1986). Existing models of range-front evolution have focused primarily on the role of geomophological processes in modifying range front morphology, and have attempted to relate particular morphometric characteristics to the degree of tectonism. In a study of range fronts in north-central Nevada, Wallace (1978) proposed that range fronts were subjected to a pattern of slope modification broadly comparable to that characterising piedmont scarps, but augmented by the effects of lateral dissection of the front by consequent drainage. According to this scheme, fresh steep fault scarps (Figure 9.7(a)) become dissected by minor rills and gullies and eroded back by slope replacement (Figure 9.7(b)). With time, progressive decline in slope angle concomitant with broadening of channel courses transforms the range front into

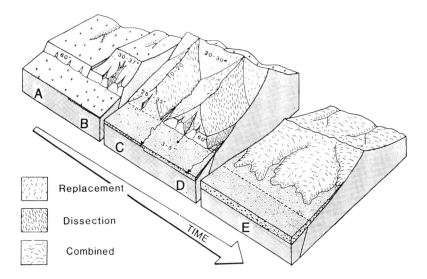

Figure 9.7   Block diagram illustrating the sequential development of a range front. Initial faulting creates a steep, linear scarp front (A) (scale of each increment of fault displacement greatly exaggerated) which begins to retreat by scarp replacement and to be dissected by minor gullies and ravines while fresh scarps are created by further faulting at the base (B). With increasing geomorphic modification the range front is characterised by moderately inclined triangular facets and deep, V-shaped valleys (C), with geomorphological activity being continuously rejuvenated by repeated faulting at the base of the range front (D). With the cessation of tectonic activity, uplift is terminated and erosional processes are unhindered, downwearing the range front to produce a deeply embayed, low relief landform (E). Reproduced by permission from Stewart & Hancock (1991b) based on Wallace (1978, Figure 3)

a series of moderately inclined triangular facets or faceted spurs (Figure 9.7(c)), which may be preserved for thousands of years or even a few million years. Some range fronts, such as the Wasatch Front in the eastern Basin and Range Province, possess flights of faceted spurs that give rise to a stepped profile, a morphology generally interpreted as the product of episodic uplift whereby phases of tectonic uplift and scarp formation alternate with phases of tectonic quiescence and scarp retreat (Hamblin, 1976). Although while uplift progresses fresh scarps continue to form at the base of the range front (Figure 9.7(d)), with the cessation of tectonic activity, the range front becomes worn down by denudation into a highly embayed, low-relief form (Figure 9.7(e)).

Recognition of large-scale denudation chronologies such as this has permitted comparative assessments to be made of the degree of tectonism exhibited by contrasting range fronts based on simple morphometric indices (Bull & McFadden, 1977; Wallace, 1978; Mayer, 1986). According to Bull (1987, p. 193), for example,

> rapid uplift . . . results in straight mountain−piedmont junctions, narrow V-shaped valleys with steep longitudinal profiles, and truncated spur ridges with well-defined triangular facets. Slow uplift . . . results in embayed sinuous mountain-piedmont junctions, broad U-shaped valleys with gentle longitudinal profiles and highly degraded spur ridges.

While it is recognised that climatic and lithological heterogeneity along a range-bounding fault zone may serve to complicate its geomorphic expression, range-front morphology remains predominantly interpreted as a function of geomorphic process.

Range-front Morphology and Structure

Investigations in the Aegean region indicate that marked variations in the morphology of individual range fronts may also reflect the heterogeneous structural architecture of many fault zones (Stewart & Hancock, 1991a, 1991b). A segment of the Furnofarango range front, for example, part of an east−west trending normal fault zone in south-central Crete (Figure 9.3), exhibits marked morphological contrasts over distances of a few hundred metres which cannot be attributed to different geomorphological regimes or uplift histories (Figures 9.8 and 9.9(a)). The range front, for example, which declines progressively in elevation eastwards, possesses a ragged, stepped appearance at its western end (profile C in Figure 9.9(b)) and a smooth, ramp-like appearance at its eastern end (profiles A and B in Figure 9.9(b)). The appearance of the youngest fault scarp along the base of the range front similarly varies along strike (Figure 9.8). In the east it is 30−50 m high and laterally continuous, only being breached where deeply entrenched streams emerge from the front (Figure 9.4(d)). Towards the west, however, the scarp becomes less prominent until, at its western limit, it is 4 m in height and laterally discontinuous.

Morphological change along the Furnofarango range front expresses along-strike changes in the underlying architecture of the range-bounding fault zone, from a series of high-angle (60−90°) normal faults capped by thin, compact breccia carapaces in the west, to a several metre-wide layered carapace of 0.5−1.0 m thick compact breccia sheets bounded by moderate-angle (35−50°) fault planes (Figure 9.9(a)). The thinner carapaces in the west have permitted the fault scarp to become extensively breached and the underlying incohesive breccia belt exposed, producing a highly cavitated scarp form and allowing a thick talus wedge to build up at the expense of the fault scarp. In contrast, the sheeted nature of the fault scarp in the east has resulted in a form of toppling failure, whereby slabs of compact breccia, bounded

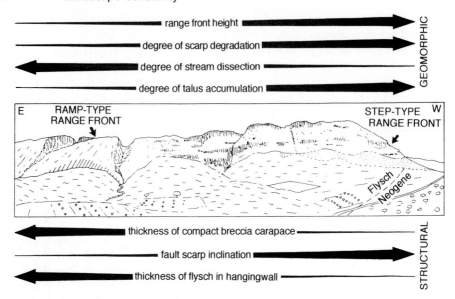

Figure 9.8 Panoramic line drawing showing the contrasting structural and geomorphic characteristics along the Furnofarango fault zone. See text for details

by stress-release fractures, become detached (possibly during seismic shaking) from the underlying fault plane (Figure 9.10). Initial failure occurs near or at the scarp base and progresses up scarp, with removal of one block promoting instability in the overhanging one. The relative absence of talus build-up in the eastern part of the range front indicates that such toppling failure constitutes a slower mechanism for scarp retreat than slope replacement processes operating further east (Figure 9.8).

In addition to preserving the smooth appearance of the fault scarp, this form of scarp degradation may result in higher parts of a fault scarp being less degraded and, therefore, apparently younger than its basal parts. This is an important corollary to the interpretation of scarp age based on the degree of surface karstification.

A Model of Range-front Evolution in the Aegean Region

Evidence from studies of active fault zones in the Aegean region suggest that the geometry of an evolving fault zone, and, as a result, its subsequent topographic expression, is highly sensitive to the shape of the pre-faulting land surface (i.e. the 'free' surface) towards which, and through which, it propagates. While minor topographic features, such as hills and depressions, are known to induce deflections in the traces of active strike-slip faults (Berril, 1988) and normal faults cutting through surficial Quaternary deposits (Mercier *et al.*, 1983; Wallace, 1984), Stewart & Hancock (1991a) suggested that the presence of isolated topographic highs in the hangingwall of a fault (*hangingwall salients*) may serve to increase locally overburden pressure and, therefore, inhibit fault splaying. Instead, deformation is concentrated along a relatively narrow, moderately inclined zone to produce a thick, sheeted, compact breccia complex, such as that characterising the eastern end of the Furnofarango fault segment. Here, the hangingwall of the faults is overlain by a thick wedge of Mesozoic flysch and indeed the flysch completely buries the fault zone further a few hundred metres east of profile A

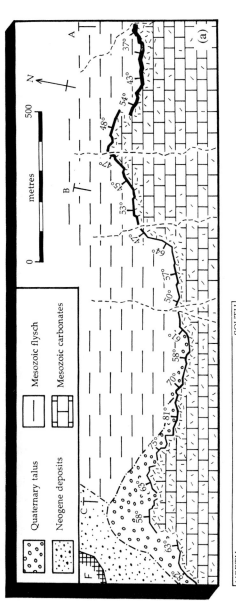

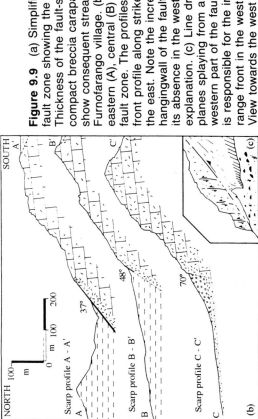

Figure 9.9 (a) Simplified geological map of the Furnofarango fault zone showing the eastward decline in fault-scarp inclination. Thickness of the fault-scarp trace relates to the thickness of the compact breccia carapace along the zone. Thin dashed lines show consequent streams cutting across the fault zone. F, Furnofarango village. (b) Series of slope profiles taken across the eastern (A), central (B) and western (C) parts of the Furnofarango fault zone. The profiles show the contrasting form of the range-front profile along strike, from step-like in the west to ramp-like in the east. Note the increased thickness of Mesozoic flysch in the hangingwall of the fault zone in profile A relative to profile B, and its absence in the westernmost profile (C). See text for explanation. (c) Line drawing showing minor, high-angle fault planes splaying from a moderate-angle fault plane in the south-western part of the fault zone. It is envisaged that this geometry is responsible for the irregular and step-like morphology of the range front in the west. Arrows denote sense of fault movement. View towards the west

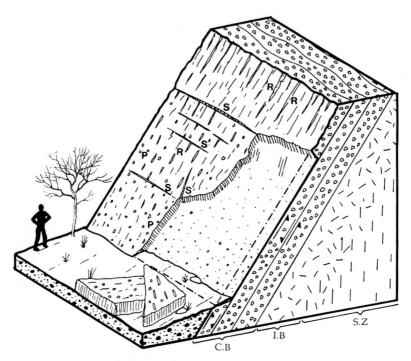

Figure 9.10 Geomorphic and structural characteristics of the uppermost levels of the Furnofarango fault zone at its eastern end. A moderately dipping fault scarp, coincident with the uppermost fault plane, is underlain by metre-thick sheets of compact breccias (CB) overlying more disaggregated incohesive breccias (IB) and shatter zone (SZ). Stress-release fractures (S) segment the compact breccia carapace into blocks that detach along the underlying fault plane and accumulate in boulders fields along the scarp base. This process is initiated at the scarp base and proceeds upscarp, with the most recently removed slab exposing a smooth and often striated fault plane. By contrast, the progressive karstification of the fault scarp surface, through the etching out of dissolution pits (P) and rillenkarren (R), results in a surface that is increasingly degraded upscarp. Pecked line shows removed portion of the fault carapace. Arrows denotes the sense of fault movement. Figure shows approximate scale

in Figure 9.9(b). This flysch cover dies out westwards, however, to be replaced by a thick cover of Quaternary colluvial and alluvial deposits that postdate the main phase of fault movement. The relative lack of pre-faulting overburden in the hangingwall in this western part of the fault zone encouraged high-angle minor faults to splay from the main, moderately dipping fault into the immediate hangingwall of earlier fault traces, thereby giving rise to a stepped morphology (Figure 9.9(c)).

The geomorphic expression of the West Youchtas fault zone, in central Crete (Figure 9.3), records a comparable evolution, with the central ramp-like part of the emergent fault zone coincident with an upstanding block of Miocene sediments in its immediate hangingwall (B in Figure 9.11). Here the underlying main fault plane is relatively moderately dipping. In contrast, the form of the range front less than a hundred metres north and south of this section (A and C in Figure 9.11), where the Miocene sediments have been substantially exhumed by fluvial and human action, is markedly stepped and cavitated, and underlain by a network of parallel, high-angle normal faults.

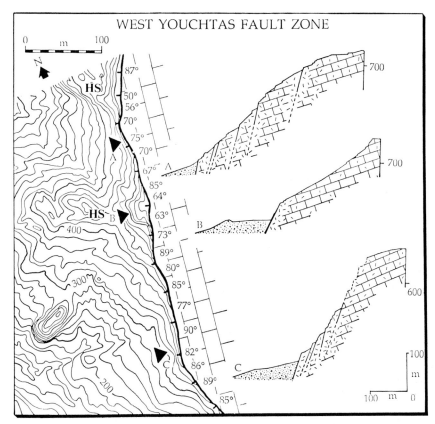

Figure 9.11 Series of range-front profiles across northern (A), central (B) and southern (C) parts of the West Youchtas fault zone, central Crete. Thick solid line denotes the range-bounding fault with barbs indicating downthrow to the west. Angles (e.g. 79°) relate to the inclination of the range-bounding fault plane. The contrasting appearance of the profiles is attributed to marked along-strike variations in surface topography in the hangingwall of the fault zone. In particular, areas of upstanding topography (hangingwall salients: HS) are coincident with more moderately inclined fault scarps and a less irregular, step-like range-front profile (B). Contours in metres

While much of the West Youchtas range front displays roughly 300 m of topographic relief, this declines sharply in that part of the range front coincident with the moderately dipping fault planes (Stewart & Hancock, 1991a, Figure 11). A similar relationship is found along the Furnofarango range front, which becomes increasingly prominent towards the west. Stewart & Hancock (1991b) have suggested that moderate-angle faulting may give rise to lower relief fronts than high-angle faulting.

SUMMARY AND CONCLUSIONS

The contention by many previous studies that the form of the fault scarp profile preserves a sensitive indicator of the age of the landform finds little application to limestone faults scarps in the Aegean region. Instead of exhibiting a time-dependent evolution, the rate and style

(a)

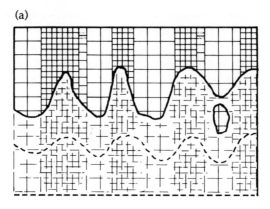

(b)

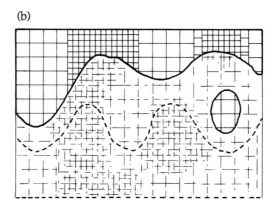

Figure 9.12   Development of scarp forms in massive sandstones which display contrasting fracture spacing. Areas of closely spaced fractures (denoted by smallest boxes) are associated with rapid scarp retreat whereas areas of widely spaced fractures (denoted by larger boxes) coincide with much slower scarp retreat. The laterally variable pattern of scarp retreat results in the irregular development of embayments and headlands along the cliff line. Over time a uniform (a) or varied (b) pattern of scarp retreat can occur depending upon whether fracture spacing is held constant or changes in the direction of scarp retreat. Redrawn from Nicholas & Dixon (1986, Figure 8)

of degradation of Aegean-type scarps is intimately related to the arrangement of contrasting fault rocks immediately underlying the fault scarp. The resulting uneven style of morphologic change bears a closer affinity to sequences of scarp retreat described from sandstone cliffs in south-eastern Utah than to previous models of fault-scarp degradation (e.g. Wallace, 1977, 1978). Oberlander (1977), for example, attributed uneven scarp retreat through time to the discontinuous nature of planes of weakness in otherwise uniform sandstones, while Nicholas & Dixon (1986) stressed the role of variable fracture spacing in determining whether the recession of sandstone cliffs was 'uniform' or 'varied' (Figure 9.12). Similarly the degradation of fault scarps may become enhanced, sustained or dampened over time as contrasting fault rocks are encountered and the effectiveness of geomorphic processes varies.

Given that it is not possible to relate the changes in the profile of limestone fault scarps to the ageing or degradation progress due to the erosional resistance of compact breccia carapaces, an alternative assessment of the ages of these scarps may be determined by considering the degree of karstification of the limestone fault surfaces. Limestone fault surfaces immediately above those parts exposed during the 1981 Corinth earthquakes and likely to be of the order of a few hundred to a thousand years in age, display millimetre-sized solution pits (Figure 9.4(a)), while scarps of the order of several thousand years in age exhibit more deeply etched karstic features. In the Basin and Range Province, for example, Wallace (1984) estimated centimetre-wide dissolution pits on the pre-1915 limestone fault surface of the Pleasant Valley earthquake scarp to be more than 1200 years old but considerably less than 12 000 years old, and karstic phenomena of similar scale on Aegean fault scarps appear to be a similar age range. Stewart & Hancock (1990b) have suggested that an additional time-related attribute of limestone fault scarps may be the range of fault-plane phenomena preserved on the fault plane. It is known, for example, that while a fine pedogenic staining occurs on many recent limestone fault scarps (Figure 9.4(a)), this veneer is absent on scarps older than 80 years (Wallace, 1984; Blumetti, Michetti & Serva, 1988). On these historical limestone scarps, however, tectonic phenomena such as frictional-wear striae are preserved, while comparable features are often denuded on older scarps, such as along the Furnofarango range front, where, instead, larger-scale lineations such as corrugations have survived. Although clearly requiring a considerably more detailed inventory, qualitative lines of inquiry such as these may prove to be the most fruitful for palaeoseismological investigations of Aegean-type fault scarps.

In addition, the premise, underpinning many regional tectonic analyses, that differences in the geomorphic expression of fault-generated range fronts are diagnostic of contrasting degrees of tectonic activity is questioned in the light of more recent studies of range-front structure. Individual range fronts, such as along the Furnofarango and West Youchtas fault zones, for example, vary markedly in their morphology along strike, reflecting underlying changes in fault-zone architecture. In particular, contrasting degrees of splaying into the hangingwall of the fault zone in response to along-strike variations on overburden, may result in a continuum of range-front morphologies varying from step-type to ramp-type forms.

Some recent investigations of mountain fronts in the Basin and Range Province (Menges, 1987, 1990) have supported the contention that fault structure is an important control of range-front morphology. A study of the Sangre de Cristo range front in the western United States, for example, demonstrated that 'many geomorphic patterns of the range front may reflect large-scale variations in the net displacement and structural geometry of the range-front fault zone' (Menges, 1987, p. 219). Indeed, certain specific morphological characteristics may be reinterpreted in the light of structural studies. The stepped appearance of range fronts such as the Wasatch Front, for example, can be accounted for by a preferential hangingwall-directed migration of the active fault trace similar to that described above from Cretan range fronts (Stewart & Hancock, 1988), or in terms of along-strike discontinuities within a range-bounding fault (Menges, 1988). An additional complication, beyond the scope of this study, is the recognition that abrupt changes in range-front morphology may be surface manifestations of the deeper-level segmentation of the fault zone, a characteristic believed to control earthquake rupturing (Wheeler, 1987). Such changes are larger-scale equivalents of the topographic contrasts between areas of high-angle and moderate-angle faulting described from the Cretan range fronts. Although more detailed studies of fault-generated range fronts are clearly required, recent studies from the Aegean region and the Basin and Range Province warn against the over zealous use of morphological attributes as indicators of the timing of tectonism along range fronts.

REFERENCES

Ambraseys, N. N. and Jackson, J. A. (1990). Seismicity and strain in central Greece between 1890 and 1988. *Geophys. J. 15.*, **101**, 663−708.

Angelier, J. (1979). Recent Quaternary tectonics in the Hellenic Arc: some examples of geological observations on land. *Tectonophysics*, **52**, 267−275.

Angelier, J., Lyberis, N., Le Pichon, X., Barrier, E. and Huchon, P. (1982). The tectonic development of the Hellenic Arc and the Sea of Crete: a synthesis. *Tectonophysics*, **86**, 159−196.

Berril, J. B. (1988). Diversion of faulting by hills. *Quarterly Journal of Engineering Geology*, **21**, 371−374.

Blumetti, A. M., Michetti, A. M. and Serva, L. (1988). The ground effects of the Fucino earthquake of January 13th, 1985; an attempt for the understanding of the recent geological evolution of some tectonic structures. In Margottini, C. and Serva, L. (eds), *Workshop on Historical Seismicity of the Central Mediterranean Region*, ENEA Rome.

Bowman, D. and Gerson, R. (1986). Morphology of the latest Quaternary surface faulting in the Gulf of Elat region, eastern Sinai. *Tectonophysics*, **101**, 97−119.

Brunsden, D. and Thornes, J. B. (1979). Landscape sensitivity and change. *Transactions of the Institute of British Geographers*, **NS4**, 463−484.

Bucknam, R. C. and Anderson, R. E. (1979). Estimation of fault-scarp ages from a scarp-height slope-angle relationship. *Geology*, **7**, 11−14.

Bull, W. B. (1987). Relative rates of long-term uplift of mountain fronts. In Crone, A. J. and Omdahl, E. M. (eds) *Proceedings of Workshop XXVIII — Directions in Paleoseismology*, US Geological Survey, Open-file Report 87−673, pp. 192−202.

Bull, W. B. and McFadden, L. D. (1977). Tectonic geomorphology north and south of the Garlock fault, California. In Doehring, D. O. (ed.), *Geomorphology in Arid Regions, Binghampton Symposia in Geomorphology International Series*, Vol. 8, pp. 115−137.

Cabrera, J., Sebrier, M. and Mercier, J. L. (1987). Active normal faulting in High Plateaus of Central Andes, the Cuzco region (Peru). *Annales Tectonicae*, **1**, 116−138.

Crone, A. J. and Haller, K. M. (1991). Segmentation and coseismic behaviour of Basin and Range normal faults: examples from east-central Idaho and south-western Montana. *Journal of Structural Geology*, **13**, 151−164.

dePolo, C. M., Clark, D. G., Slemmons, D. B. and Ramelli, A. R. (1991). Historical surface faulting in the Basin and Range province, western North America: implications for fault segmentation. *Journal of Structural Geology*, **13**, 123−136.

Gerson, R., Grossman, S. and Bowman, D. (1985). Stages in the creation of a rift valley — geomorphic evolution along the southern Dead Sea Rift. In Morisawa, M. and Hack, J. T. (eds), *Tectonic Geomorphology, Binghampton Symposia in Geomorphology, International Series*, Vol. 15, pp. 53−74.

Gilbert, G. K. (1890). Lake Bonneville. *US Geological Survey Monograph*, Vol. 1, pp. 000−000.

Gilbert, G. K. (1928). Studies on basin range structure. *Professional Paper US Geological Survey*, **153**, 1−92.

Hamblin, W. K. (1976). Patterns of displacement along the Wasatch fault. *Geology*, **4**, 619−622.

Hancock, P. L. and Barka, A. A. (1987). Kinematic indicators on active normal faults in western Turkey. *Journal of Structural Geology*, **9**, 573−584.

Hanks, T. C., Bucknam, R. C., Lajoie, K. R. and Wallace, R. E. (1984). Modification of wave-cut and fault-controlled landforms. *Journal of Geophysical Research*, **89**, 5771−5790.

Jackson, J. A., Gagnepain, J., Houseman, G., King, G. C. P., Papadimitriou, P., Soufleris, C. and Virieux, J. (1982). Seismicity, normal faulting, and the geomorphological development of the Gulf of Corinth (Greece): the Corinth earthquakes of February and March 1981. *Earth and Planetary Science Letters*, **57**, 377−397.

Machette, M. (1989). Slope-morphometric dating. In Forman, S. L. (ed.), *Dating Methods Applicable to Quaternary Geologic Studies in the Western United States, Utah Geological and Mineral Survey*, 89-7, pp. 30−42.

Mayer, L. (1985). Tectonic geomorphology of the basin and range — Colorado Plateau boundary in Arizona. In Morisawa, M. and Hack, J. T. (eds), *Tectonic Geomorphology, Binghampton Symposia in Geomorphology, International Series*, Vol. 15, pp. 53−74.

Mayer, L. (1986). Tectonic geomorphology of escarpments and mountain fronts. In *Active Tectonics, Studies in Geophysics*, National Academy Press, Washington DC, pp. 181–194.

Menges, C. (1987). Temporal and spatial segmentation of Pliocene-Quaternary fault rupture along the western Sangre de Cristo mountain front, northern New Mexico. In Crone, A. J. and Omdahl, E. M. (eds), *Proceedings of Workshop XXVIII — Directions in Paleoseismology*, US Geological Survey Open-file Report 87-673, pp. 203–222.

Menges, C. (1988). Tectonic origin of facet benches on a normal fault-bounded mountain front. an alternative model. *Geological Society of America Abstracts with Programs*, **20**, 215.

Menges, C. M. (1990). Late Quaternary fault scarps, mountain-front landforms, and Pliocene–Quaternary segmentation of the range-bounding fault zone, Sangre de Cristo Mountains, New Mexico. *GSA Reviews in Engineering Geology*, **8**, 131–156.

Mercier, J. L., Carey-Gailhardis, E., Mouyaris, N., Simeakis, K., Roundoyannis, T. and Anghelidhis, C. (1983). Structural analysis of recent and active faults and regional state of stress in the epicentral area of the 1978 Thessaloniki earthquakes (northern Greece). *Tectonics*, **2**, 577–600.

Nash, D. B. (1980). Morphologic dating of degraded normal fault scarps. *Journal of Geology*, **88**, 353–360.

Nash, D. B. (1984). Morphologic dating of fluvial terrace scarps and fault scarps near West Yellowstone, Montana. *Bulletin of the Geological Society of America*, **95**, 1413–1424.

Nash, D. B. (1986). Morphologic dating and modelling degradation of fault scarps. In *Active Tectonics, Studies in Geophysics*, National Academy Press, Washington DC, pp. 181–194.

Nicholas, R. M. and Dixon, J. C. (1986). Sandstone scarp form and retreat in the Land of Standing Rocks, Canyonlands National Park. *Zeitschrift für Geomorphologie*, **30**, 167–187.

Oberlander, T. M. (1977). Origin of segmented cliffs in massive sandstones in southeastern Utah. In Doehring, D. O. (ed.), *Geomorphology in Arid Regions, Binghampton Symposia in Geomorphology International Series*, Vol. 8, pp. 79–114.

Stewart, I. S. and Hancock, P. L. (1988). Fault zone evolution and fault scarp degradation in the Aegean region. *Basin Research*, **1**, 139–153.

Stewart, I. S. and Hancock, P. L. (1990a). Fracturing and brecciation within neotectonic normal fault zones in the Aegean region. In Knipe, R. J. and Rutter, E. H. (eds), *Deformation Mechanisms, Rheology and Tectonics*, Geological Society of London Special Publication, Vol. 54, pp. 105–110.

Stewart, I. S. and Hancock, P. L. (1990b). What is a fault scarp? *Episodes*, **13**, 256–263.

Stewart, I. S. and Hancock, P. L. (1991a). Scales of structural heterogeneity within neotectonic fault zones in the Aegean region. *Journal of Structural Geology*, **13**, 191–204.

Stewart, I. S. and Hancock, P. L. (1991b). Neotectonic range-front fault scarps in the Aegean region. In Cosgrove, J. and Jones, M. E. (eds), *Neotectonics and Resources*, Belhaven Press, London, pp. 93–107.

Tricart, J. (1974). *Structural Geomorphology*, Longman, London.

Vita-Finzi, C. and King, G. C. P. (1985). The seismicity, geomorphology and structural evolution of the Corinth area of Greece. *Philosophical Transactions of the Royal Society*, **A314**, 379–407.

Wallace, R. E. (1977). Profiles and ages of young fault scarps, north-central Nevada. *Bulletin of the Geological Society of America*, **88**, 1267–1281.

Wallace, R. E. (1978). Geometry and rates of change of fault-generated range fronts, north-central Nevada. *Journal of Research of the US Geological Survey*, **6**, 637–650.

Wallace, R. E. (1984). Fault scarps formed during the earthquakes of October 2, 1915, Pleasant Valley, Nevada and some tectonic implications. *Professional Papers of the US Geological Survey*, **1274-A**.

Wheeler, R. L. (1987). Boundaries between segments of normal faults: criteria for recognition and interpretation. In Crone, A. J. and Omdahl, E. M. (eds), *Proceedings of Workshop XXVIII — Directions in Paleoseismology*, US Geological Survey Open-file Report 87-673, pp. 385–398.

Yuming, Z. (1989). Slope variation and ages of fault scarps and recurrence intervals of great earthquakes on the Koktokay–Ertai fault. *Earthquake Research in China*, **1**, 377–388.

Zhang, B., Liao, Y., Guo, S., Wallace, R. E., Bucknam, R. E. and Hanks, T. C. (1986). Fault scarps related to the 1739 earthquake and seismicity of the Yinchuan graben, Ningxia Huizu Zizhiqu, China. *Bulletin of the Seismological Society of America*, **76**, 1253–1287.

10 Shallow Failure Mechanisms during the Holocene: Utilisation of a Coupled Slope Hydrology – Slope Stability Model

SUE M. BROOKS
Department of Geography, University of Bristol, UK

KEITH S. RICHARDS
Department of Geography, University of Cambridge, UK

and

MALCOLM G. ANDERSON
Department of Geography, University of Bristol, UK

INTRODUCTION

The study of long-term slope development due to mass movement has focused on the role of temporal changes in the regolith angle of internal friction by direct measurement of its value for different soils (Carson & Petley, 1970; Carson, 1975; Rouse & Farhan, 1976). Although such studies address the role of changes in pore water pressure, they do so in a simplistic way, assuming that the water table rises to the ground surface and the limiting slope angle is then reduced to the semi-frictional value. In some soils, especially those that are freely draining, unsaturated conditions may control the slope stability. For example, shallow failures on slope angles of between 28° and 38° occurred in Glen Livet, Scotland after a high-magnitude storm in February 1989 (Figure 10.1). These slopes had not failed in previous storms though the slope angles are in excess of the frictional angle of the regolith. The shallow failures were in freely draining podsol soil profiles, which possess three hydrologically distinct horizons, each having a different response to rainfall, and experiencing complex variations in moisture content, suction and stability during and after rain. These issues suggest that alterations in the probability of shallow failures may occur over Holocene timescales as the regolith develops. To begin with, the hydrological control of failure is more complex than considered in previous studies, since such failures take place during rainstorms that do not necessarily lead to complete saturation of the regolith. Thus dynamic fluctuations occur in soil moisture status during a storm, and these vary depending on the magnitude – frequency characteristics of the rainstorms. Secondly, the hydrological response alters over time as the regolith develops and becomes increasingly differentiated. Thus the probability of shallow failure increases until rainstorm and pedogenic conditions combine to initiate failure.

Landscape Sensitivity. Edited by D. S. G. Thomas and R. J. Allison
© 1993 John Wiley and Sons Ltd

The changing sensitivity of slopes to shallow failure is governed by the concurrent development of the regolith and the secular variations in the climate. Elucidation of this dual role of pedogenesis and climatic change can be approached via the application of a physically based coupled hydrology—stability model, and it is the aim of this chapter to show how this can be achieved. The coupling of a hydrological model to a stability analysis enables more detailed assessment of the hydrological control of slope stability, and its physical basis enables application to time periods that are unavailable for direct observation. Complexities in the failure mechanisms can be evaluated by comparing behaviour in the hydrology—stability model with observed failures. This chapter therefore provides an initial assessment of the combined role of pedogenesis and climatic change in accounting for alterations in the probability of shallow planar failures over the Holocene, and introduces a physically based modelling approach as a potentially valuable means of assessing such geomorphic problems of landscape sensitivity.

PEDOGENIC AND CLIMATIC FACTORS THAT GOVERN THE PROBABILITY OF MASS MOVEMENT

Climatic variation during the Holocene had widespread and varied consequences for the intensity of geomorphic processes, including mass movement on slopes (Starkel, 1966; Grove, 1972). Synchronous phases of enhanced activity have been identified over wide regions, in association with regional climatic deterioration. However, some debate has focused on the relative importance of anthropogenic factors and climate in inducing mass movement, though with severe limitations imposed by data availability. Most studies aim to date the occurrence of slope failures, and to correlate these with contemporaneous changes in the likely forcing factors (Harvey *et al.*, 1981; Innes, 1983; Brazier & Ballantyne, 1989). This approach provides suggestive evidence for the effect of Holocene climatic changes on slope stability, but involves association rather than causal connection, since it provides no insight into the mechanisms that control the processes, or into the nature and extent of changes required for failure to occur. Underlying mechanisms must be understood if we are to be able to establish causes of failure, and physically based modelling can be used to achieve this.

Climate and climatic change have two roles to play in influencing slope stability (Innes, 1983). The first influence is the triggering action of individual storms of different intensity and duration, and varying interarrival times which determine antecedent moisture levels. The second is the climatic control of the rate of soil development, which determines the interaction between the rainfall event and moisture redistribution within the soil profile; generally, better developed soils will respond to rainfall in a different way from young soils.

Studies of climatic control of slope stability particularly invoke a correlation between high-magnitude rainfall events and landslide occurrence (Common, 1954, 1956; Baird & Lewis, 1957; Rapp, 1960; Rice, Corbett & Bailey, 1969; Caine, 1980; Cooke, 1984; Iida & Okunishi, 1983). However, this ignores the multifaceted role of climate, and its spatial and temporal variability. The effects of variation in evaporation rates, interstorm periods, rainfall intensities, rainstorm profiles and storm frequencies are required for a complete assessment of slope stability, but accurate assessment is restricted by data limitations especially in the steep terrain where failure is most dominant. The spatial pattern of slope failure also reflects the inherent variability of storm intensity profiles and storm cell tracks across the landscape. These factors all serve to confuse interpretation of the effects of climate on slope stability over Holocene or longer timescales.

The significance of climatic properties for slope stability is further complicated by variation in the soil profiles that form the slope mantle (Freeze, 1987). Variations in texture and bulk density are important controls on soil hydraulic behaviour (Masch & Denny, 1966; Arya & Paris, 1981; Mishra, Parka & Singhal, 1989). These properties determine soil profile response to a given rainstorm, and hence the subsequent pore water pressure or suction distribution, and susceptibility to failure. Variability in soil hydraulic properties therefore complicates the assessment of the role of climate. Since soil profiles vary greatly within a catchment, across individual hillslopes and at any given location, such assessment must incorporate the unique combinations of soil hydraulic behaviour and rainfall intensity distributions that occur for any location and time period, and which are crucial controls of slope failure.

The soil property that is particularly significant for slope stability is its vertical differentiation into horizons, producing hydrological discontinuities that can lead to zones of reduced suction or increased pore water pressure. The hydraulic properties of all horizons within a soil profile need to be considered, so that the importance of their interaction can be assessed. In earlier studies of slope failure the angle of internal friction was normally measured for the horizon which 'was thought to possess the least shear strength' (Carson & Petley, 1970, pp. 78−79). Thus the emphasis was on an incomplete and simplified set of controls. Both the total soil depth (Iida & Okunishi, 1983) and the vertical variation in soil physical properties control failure probabilities through their control of soil hydrological and strength properties. The present study explicitly considers hydrological and strength variation between horizons within a soil profile. In considering changing climatic influences on slope stability, the inherent spatial and vertical variability of soils is not the only significant pedological issue. Pedogenesis causes an increase in soil thickness and greater horizon differentiation. Hence *changing* failure probabilities can be linked to *altering* soil hydraulic behaviour, and the different moisture distributions within the profile that result from a *given* rainstorm. Soil chronosequences (Jenny, 1941) provide a means of assessing the magnitude of changes in hydrology and regolith shear strength for the different horizons within the soil profile.

In order to evaluate empirically the value of physical modelling in providing insight into shallow failure mechanisms, the changing stability of evolving podsolic soils on slopes in the Cairngorms is studied in this chapter. Figure 10.1 shows that such failures occur, but it is of interest to consider the changing sensitivity of slope soils to failure as they evolve. This is calibrated by using the Glen Feshie chronosequence, described by Robertson-Rintoul (1986), which includes a series of freely draining podsols developed on river terraces during the Holocene, with dates of 1000, 3600, 10 000 and 13 000 BP marking the start of soil formation on each terrace. These soils have a similar morphology and hydrological behaviour to those observed to have failed on slopes in this region of eastern Scotland. This sequence of podsol development can therefore be used to provide appropriate parameters for the coupled hydrology−stability model, in order that the effect of Holocene pedogenesis on slope stability can be assessed. The effect of Holocene climatic change is also included by separate specification of appropriate values for rainstorm character, evaporation rates and storm interarrival times; this chapter uses information on soils of differing age, but its emphasis is on the changing sensitivity of slopes to failure at different stages of the Holocene as a result of the interaction of changing storm and soil properties.

Figure 10.1   Shallow failures in well-developed podsols, Glen Livet Estate, eastern Scotland

A PHYSICALLY BASED COUPLED HYDROLOGY–STABILITY MODEL

Physically based hydrological models are increasingly applicable to a variety of geomorphological issues. The problem of evaluating and understanding the mechanisms that lead to slope failure is a particular area in which they are becoming valuable research tools, given the complexity which underlies moisture redistribution within the soil, and given the fundamental importance of this to failure initiation. For analysis of mechanisms such models must have a physical basis, incorporating relationships between measurable properties of the soil that have physical meaning, and whose changing values can be related to the underlying processes in order to understand the behaviour of the system. The current model application is a two-stage process, first simulating the moisture redistribution taking place during rainstorms, and then coupling this with a stability analysis to discover the precise role of hydrology in slope failure. The model is described in one dimension in Anderson & Howes (1985), where it was used to assess slope stability under a variety of rainstorms and slope

NB: Slope angle and form is governed by cell geometry

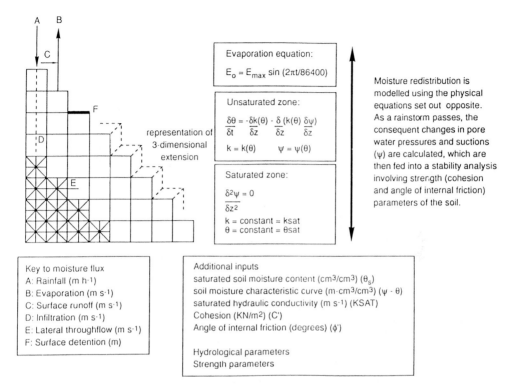

Figure 10.2   Representation of the coupled hydrology−stability model

angles in Hong Kong, and a more recent extension to two-dimensional analysis is described by Lloyd (1990).

Representing any continuous system as a numerical model on a computer requires discretisation, and one method available is that of finite differences, where the spatial dimensions of the system are represented as a set of contiguous grid squares (Figure 10.2). Also shown in Figure 10.2 are the equations that govern soil moisture redistribution in the unsaturated and unsaturated zones, along with the input parameters required by the model. For a single soil profile a vertical stack of cells can represent the arrangement of the horizons. Podsols at different stages of development are considered in the following analysis since recently observed shallow failures have occurred in well-developed podsols in Scotland (Figure 10.1). Younger soils can be represented by altering the cell geometry and the hydrological properties appropriately (Figure 10.3), using data from the Glen Feshie chronosequence. For each horizon a set of properties that define the soil hydraulic behaviour is required, including the saturated soil moisture content, the saturated hydraulic conductivity and the soil moisture characteristic curve (Figure 10.4). Unsaturated rates of flow are calculated using the Millington−Quirk method (Millington & Quirk, 1959), which is based upon the saturated hydraulic conductivity and the soil moisture characteristic curve, and has been shown to be accurate over a range of suctions (Jackson, Reginato & van Baval, 1965; Kunze, Nehava

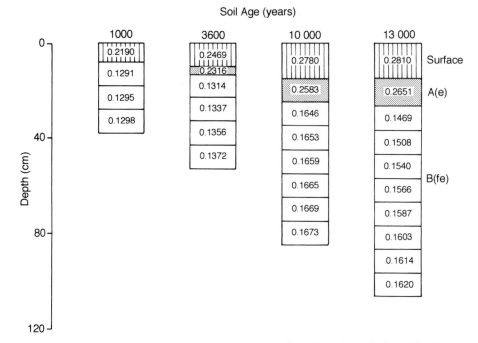

Figure 10.3 Arrangement of cells to represent the Glen Feshie podsol profiles in one dimension; numbers refer to initial moisture content

& Graham, 1968; Nielsen, Biggar & Erh, 1973; Cameron, 1978). These form the hydrological requirements for solving the equations of flow that govern moisture redistribution within the profile (Figure 10.2). Coupled to this is a stability analysis, which simply requires the angle of internal friction in these cohesionless podsols, as well as the constantly altering suctions or pore water pressures within the soil, which are obtained from the hydrological submodel. An infinite planar stability analysis is appropriate given the shallow translational slides typical of the sandy podsols of the region.

In order to assess the changing regolith stability, the factor of safety is calculated for each cell after every hour of a simulated storm. The model outputs therefore include the variation in the factor of safety with depth throughout and following the storm, which is important in determining both the timing of the failure, and the position of the slip plane within the soil profile. Both will vary depending on the rainfall intensity, antecedent moisture conditions, and pedogenic properties. Thus a complete analysis of the changing sensitivity of slopes to failure during the Holocene requires detailed consideration of these aspects of the variation in factors of safety.

The fact that the stability analysis is carried out for moisture contents below the saturated value introduces complications related to the strength enhancement due to suction. Research has shown that although the effect of suction is to increase the shear resistance, utilisation of suction in the pore water pressure term of the Coulomb equation is not strictly accurate, though it may be regarded as satisfactory up to suctions of 100 cm (Fredlund, 1987). The infinite planar slide analysis used here employs this direct substitution, which is reasonable given that the moisture contents approach saturation and certainly involve suction values of well below 100 cm. However, where such an approach involves stability assessment under

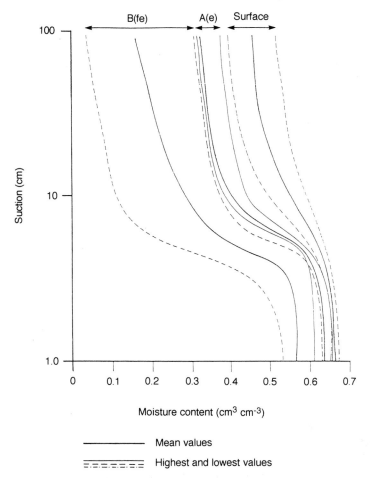

Figure 10.4 Soil moisture characteristic curves for three horizons of a well-developed podsol (13 000 years old) from the Glen Feshie chronosequence

higher suctions, such as in some of the applications involving the design of artificial slopes in the tropics (Anderson, Kemp & Lloyd, 1988), the effect on the angle of internal friction also needs to be taken into account.

The hydrological element of the model provides the key to enhanced understanding of failure mechanisms, involving continuous moisture redistribution during and after a rainstorm, which results in complex variation in soil moisture status within the profile. Once the maximum moisture content is reached, failure may take place if the slope angle, overburden pressure and angle of internal friction are appropriate. This maximum moisture content governs the minimum suction, which is the crucial control on stability. The rate at which failure conditions are approached depends on moisture redistribution within the profile determined by changes in the soil hydraulic conductivity as the profile becomes wetter, and these are calculated by the model after each specified iteration. The timing and depth of failure are determined by the hydrological properties of the soil, and also depend on the prevailing rainstorm characteristics. The model requires rainfall totals over specified periods, usually of 1 hour.

By altering this rainfall incremental period and the total amount falling within each, the effects of variation in storm intensity and duration can be assessed. During daytime interstorm periods, when no rainfall is occurring, evaporation takes place from the top of the profile at a rate that is calculated from the positive part of a 24 h sinusoidal function peaking at a specified maximum midday evaporation rate. Through this, the effect of variability in the interstorm period and evaporation rates can be investigated and different initial moisture contents can be specified accordingly. By applying a variety of rainstorm characteristics and interstorm periods the effects of climatic variation can be considered at a variety of scales.

The physically based coupled hydrology–stability model adopted here can incorporate the effects of variability both deterministically and stochastically. Deterministic variability involves the specification of different mean values for each of the properties for each soil horizon, as well as specification of the climatic characteristics (Freeze, 1987). Stochastic variability is included through incorporation of random generation of each parameter from a specific distribution, defined by the mean and standard deviation. Thus distributions of factors of safety are generated from several simulations which can be used to assess the probability of failure. Physically based models have been used for prediction (Abbott *et al.*, 1986a, 1986b; Bathurst, 1986; Beven, 1989), usually of catchment response to rainfall, since they are thought to provide an improvement on earlier conceptual models. Stochastic variation makes precise prediction of specific landslides impossible, and the key role of physically based models lies in their ability to enhance understanding of the mechanisms that govern the occurrence of observed failures. These mechanisms involve the rate of moisture delivery and its continuous redistribution within the soil, which are determined by the association between climatic and pedogenic conditions.

The utilisation of a physically based model to explore the complex system behaviour that leads to shallow failure in freely draining regoliths, developing during the Holocene in the Cairngorms, eastern Scotland, represents a major development in the use of physically based models. This extends beyond predicting contemporary hydrological response of catchments of hillslopes to understanding the mechanisms that govern the behaviour of geomorphological systems over longer timescales. In order to achieve this, some preliminary model results based around initial simulations under set climatic conditions need to be considered, to comprehend the hydrological mechanisms that produce failure within shallow, freely draining regolith. Following this, the results will be discussed in the light of current climatic data from Aviemore and probable past rainstorm characteristics for different periods of the Holocene, evaluated using a variety of sources.

HYDROLOGICAL PROPERTIES OF PODSOLS SIGNIFICANT IN CONTROLLING SHALLOW MASS MOVEMENT

Seasonal increase in groundwater level — a major control of slope failure (Campbell, 1975; Brunsden & Jones, 1976) — is determined by storm frequency, evaporation rate and soil drainage. The last is dependent on the soil moisture characteristic curve and unsaturated flow rates. As the soil dries during the interstorm period, the rate of moisture loss slows, and after a certain period little further drainage takes place. The hydrological model used here can simulate the effects of variable drainage properties of the soil, and the results presented in Figure 10.5 show changes in the volumetric moisture content for the horizons of three of the Glen Feshie podsols. Initially the soils drain rapidly, and after a period of approximately

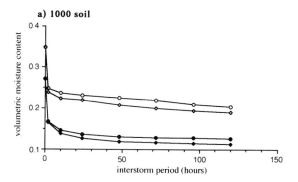

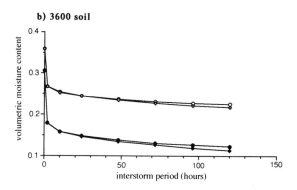

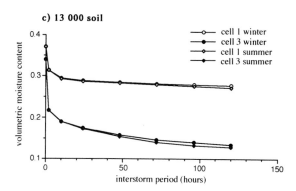

Figure 10.5 Effect of variation in the interstorm period on initial moisture contents for different soils of the Glen Feshie chronosequence for the surface (cell 1) and top of B(fe) horizon (cell 3)

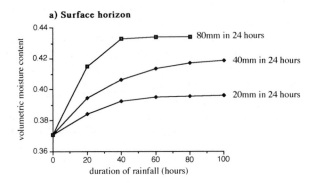

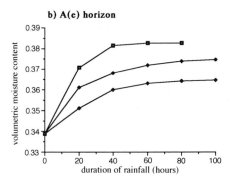

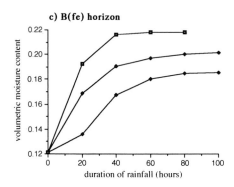

Figure 10.6 Time to attain the maximum moisture content under different rainfall intensities for the three horizons of the 13 000 year old podsol

two days any increase in the interstorm period does not have a significant effect on values for the moisture content. However, soon after the end of a storm when the soil has a volumetric moisture content close to saturation, a subsequent event could have severe consequences, making the interval between storms instrumental in causing failure. The period after which the storm interval ceases to be significant depends on the evaporation rate, with higher rates associated with shorter periods. For the podsol profile considered here, there are clear differences between the horizons reflecting their differing hydraulic properties. The model can simulate interstorm drainage and provide suitable values for initial moisture contents for a variety of interstorm periods and evaporation rates.

The hydraulic parameters are also important controls on the rate of increase in profile moisture content following the onset of rainfall. There exists a potential maximum moisture content, and hence a minimum suction, which can be reached in each horizon of freely draining profiles, and this is determined by the rainfall intensity and the relationship between soil hydraulic conductivity and moisture content. The value of the maximum moisture content for a given rainfall intensity is different for each horizon, and substantial differences are also apparent between high and low intensity storms (Figure 10.6). The significance of this for failure lies in the association between moisture content and suction, with the maximum moisture content determining the minimum suction. For a higher rainfall intensity, lower suctions are apparent, producing a greater probability of failure. Since this relationship differs between horizons, a different probability of failure occurs in each, and failure conditions are further complicated by the increasing overburden pressure deeper in the profile (Figure 10.7). In the sandy soil profile modelled in Figure 10.7, the downslope shear stress increases with depth at a greater rate than the additional strength, making the deeper cells more prone to failure. That the B(fe) horizon also has a potentially lower minimum suction compounds this effect, making it the one most likely to fail. However, the precise position of the failure plane

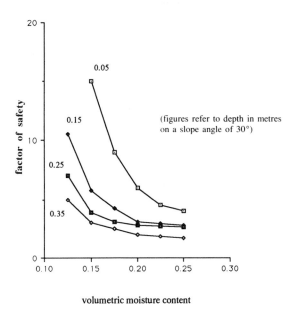

Figure 10.7 Combined effect of volumetric moisture content and depth on the factor of safety of the B(fe) horizon of the 13 000 year old podsol (depths are cell mid-points)

160

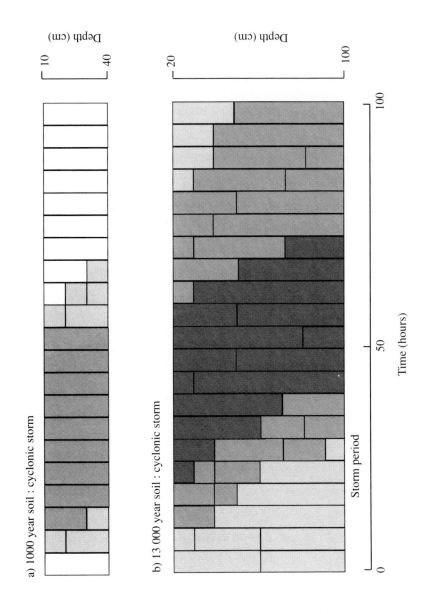

a) 1000 year soil : cyclonic storm

b) 13 000 year soil : cyclonic storm

Depth (cm)

Depth (cm)

Time (hours)

Storm period

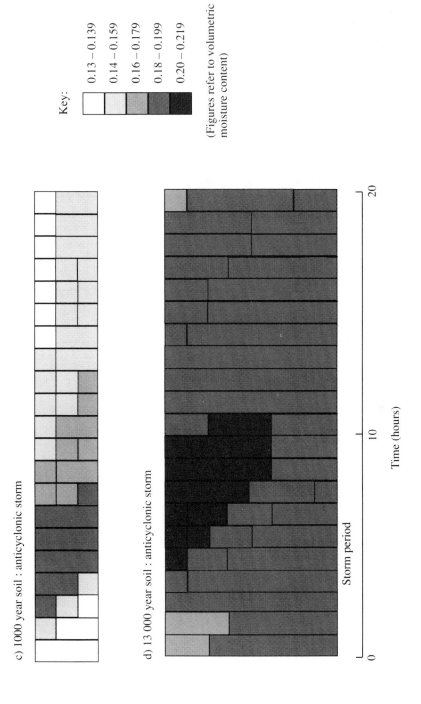

Figure 10.8 Changes in moisture content with depth and time during and after rain, for different rainfall intensities and soil ages

depends on the rainfall intensity since cells closer to the surface become wetter earlier in the storm and may therefore attain the failure conditions first. Since the rainfall intensity dictates the rate of increase in profile moisture content and unsaturated hydraulic conductivity, it will control the differential change in each cell and hence the likely position of the slip surface. Model simulations for well-developed podsols show that under high intensity rainfall, the surface cells become wetter and do so more rapidly (Figure 10.8), thus making surface failure more likely. However, the rainfall intensity is coupled with the storm duration to govern failure, especially at low intensities when sufficient time needs to be available to enable the minimum suctions to develop.

The lag between rainfall and the rise in moisture content means that the intensity has to be sustained for some time before the maximum moisture contents defined above can be established. At greater depths the lag increases and the storm duration needed for failure is greater. These relationships can be explored using the model. For high rainfall intensities and surface horizons the response is quick, and maximum moisture contents are readily attained. Lower intensities require longer durations to promote failure, and this is increasingly the case with depth. In reality high intensities tend to occur over short durations, whereas the opposite is true for low intensities. Thus high intensity rainfall is more likely to bring about shallow failure, while low intensity rainfall may be associated with deeper failure. Furthermore, the increasing lag in response from the top to the base of the soil profile means that the upper horizons may experience a decrease in their factor of safety first, with lower horizons possibly only responding after rainfall has ceased (Figure 10.9). This is an important consideration, since an early drop in the factor of safety below unity in upper horizons would

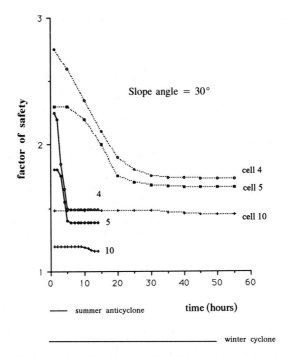

Figure 10.9 Changes in the factor of safety under high and low intensity storms at different depths in the B(fe) horizon of a 13 000 year old podsol

cause them to fail before lower horizons, possibly precluding failure at greater depths. This effect is intimately associated with the precise distribution of rainfall intensity within a storm, and with the varying soil hydraulic properties down-profile, in a finely balanced situation which makes forecasting of failure occurrence and slip-plane position unreliable. Inclusion of stochastic variability compounds the difficulty of interpretation. However, it is possible to investigate the combined effect of all these factors in an assessment of slope stability, through sensitivity analysis using the coupled hydrology−stability model. The results can be used to show how changing pedogenic and climatic factors can alter the probability and nature of failure, as the following example from eastern Scotland will demonstrate.

AN EMPIRICAL EXAMPLE: HOLOCENE FAILURES IN SCOTLAND

Having described the model and shown the basic simulation results, it can now be applied to the issue of climatic variability during the Holocene, and its implications for slope instability. Given the structure of the above assessment, the following analysis will explore the effect of variation in interstorm period and evaporation rates in controlling initial moisture contents, the values of potential maximum moisture contents, and the durations of storm require to attain this for a series of soil profiles representing different stages of pedogenesis. Thus by combining different climatic and pedogenic conditions to represent various phases during the Holocene, it is possible to assess the changing senstivity of slopes to shallow mass movement.

Climatic Data for Aviemore Appropriate for Evaluating Climatic Control of Failure

The most readily available rainfall data for recent periods relate to daily totals, with records for stations close to the observed slope failures (shown in Figure 10.1) back to 1947. However, data at this level of resolution are insufficiently detailed to capture the worst groundwater conditions during a rainstorm, since they mask the rainfall intensity changes that take place during storms. The effect of changing the level of resolution of the data can easily be demonstrated by using the model (Figure 10.10), and it is clear that a finer resolution than 24 h is required. The meteorological station at Aviemore provides a record of hourly totals from 1983, which is invaluable in assessing likely rainstorm characteristics in the region at the level of detail required for landslip analysis. Aviemore is situated about 40 km to the south-west of the field area in which the landslips were observed. The altitudes and mean annual rainfall totals are similar for both locations. The altitudes are 229 and 215 m, and the mean annual rainfall totals are 930 and 950 mm for Aviemore and the field area, respectively.

In order to identify significant rainstorms from the hourly record, these were defined as having a minimum total rainfall of 5 mm, and if rain ceased to fall for over six consecutive hours then the storm was considered to have ended. In this way, hourly totals were used to define durations and rainfall totals for each storm in the months of December, January, February, June, July and August since 1983. The storms were then grouped according to season of occurrence (summer or winter) and synoptic origin (cyclonic or anticyclonic), the latter from consulting the daily synoptic charts. This revealed considerable differences between the storm characteristics (Figure 10.11) associated with different synoptic situations, which

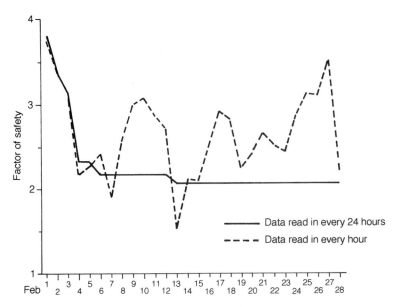

Figure 10.10 Comparison of minimum factors of safety obtained from different levels of subdivision of daily rainfall totals for February, 1988, applied to the 13 000 year old podsol on a slope angle of 30°

provides a potentially valuable means of considering the effect of longer-term climatic changes. The Holocene has involved several shifts in the prevailing synoptic situation over eastern Scotland, and hence the rainstorm characteristics experienced there. Thus if failure is more likely under particular rainfall characteristics, certain periods might have had conditions more conducive to failure. Futhermore, since the rainfall intensity also determines the position of the slip surface, then different types of failure might have occurred at different times.

Scottish Evidence for Holocene Climatic Change and Slope Failure

During the Holocene four major climatic phases have been identified, linked to post-glacial rise in sea level, changes in the position of the westerly air stream and changes in the strength of the general circulation (Lamb, 1977). The Boreal was a period of increased continentality with lower annual rainfall totals and greater seasonal variability in temperatures. From 7500 BP the Atlantic period experienced conditions that were warmer and wetter, being more dominated by depressions. A return to more continental conditions occurred after 5000 BP in the sub-Boreal, and the current sub-Atlantic period is again dominated by low-pressure systems and cyclonic activity. The rainfall characteristics during each of these periods are likely to have been different, as is illustrated by the current climatic data presented in the previous section. More continental conditions probably involved high-intensity, short-duration storms, with the above data suggesting 24 h totals of around 40 mm. However, individual studies of 'extreme' rainfall events in Scotland suggest that such a total can be distributed over shorter periods, and that over a single hour rainfall totals of up to 60 mm can be received (Green, 1958, 1971; Jenkins et al., 1988). During the more maritime phases, longer-duration storms are likely to have delivered greater rainfall totals, but at a lower rate. The climatic

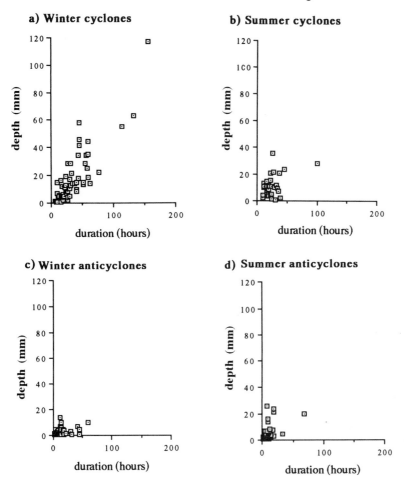

Figure 10.11 Depth–duration relationships for storms of different synoptic origin from Aviemore

data suggest that 24 h totals of 40 mm have a low recurrence interval, with high totals being likely. Thus as soil developed it would have been subjected to different types of storm at various times, possibly leading to phases of enhanced mass movement as particular rainfall events combined with soil profiles at particular stages of development.

Evidence of phases of enhanced erosion in Scotland is provided by cores of lake sediment which reflect both erosion within the catchment and the dominant vegetation assemblage (Birks, 1970; Birks & Mathewes, 1978). It has been suggested that post-glacial mass movement in Scotland is confined to the immediate period following glaciation, and to very recent times due to anthropogenic activity (Ballantyne & Eckford, 1984), with no occurrence at other times. However, phases of heightened mass movement have been identified in Scotland before about 2000 BP, and after 1500 AD (400 BP) (Brazier & Ballantyne, 1989), when debris cones accumulated. Although river undercutting has been suggested as a major factor behind debris cone instability, accumulation of cone material takes place due to instability in the valley-side slopes at higher altitudes, which can be explained climatically. Innes (1983) has identified dates for organic material accumulation on talus slopes during stable periods when talus accumulation was limited, namely between 2300 and 1800 BP and also between 1000 and 600 BP. The hiatuses between these stable periods were times of slope instability, around 2500 and 1200 BP. The earlier date for mass movement activity has been linked to a wet, cool phase during the late sub-Boreal around 2500 BP, and appears to be consistent between these two studies. Earlier synchronous phases of climatic deterioration and mass movement activity have been identified for other regions (Starkel, 1966; Hutchinson & Gostelow, 1976), occurring at the beginning of the Atlantic, and supporting a climatic explanation for mass movement activity. During the Little Ice Age, a period of climatic deterioration during the sub-Atlantic, enhanced mass movement has also been identified (Grove, 1972), providing a similar date to the later one obtained by Brazier & Ballantyne and, again, supporting a climatic explanation. However, the later dates for mass movement occurrence in Scotland also suggest that anthropogenic activity provided a control of enhanced slope failure, and this has been suggested for other regions (Harvey et al., 1981). Despite the faith put in both explanations, there has been little consideration of the precise causes related to either factor, in terms of the impact of changed rates and pathways of moisture redistribution within the soil either following vegetation clearance or climatic change. Although the model has not been used to address the issue of the effect of vegetation change, this is a potential development. For now, the causal links between climatic deterioration and slope activity can be considered using ideas inherent in the model simulations described above.

Results of Model Simulations and Discussion of Mass Movement Occurrence During the Holocene

The climatic control of mass movement is clearly linked to changing interarrival times and intensity—duration—frequency characteristics of rainstorms, along with varying evaporation rates during the interstorm period. However, there are concurrent alterations in the soil profile character that also need to be considered as the soils have developed through the post-glacial period. For the demonstrative modelling described above, only the well-developed podsol was used (i.e. that with a horizon structure reflecting 13 000 years of pedogenesis), but in the following analysis the simulations involve soil profiles at different stages of development. Each soil profile was subjected to a variety of simulated rainstorms of varying intensity and duration to examine temporal change in stability (Figure 10.12). It is clear that increasing soil development reduces slope stability, but this is most marked under low intensity storms. Younger profiles are highly stable under long-duration storms, and require high intensities (>5 mm) for failure. As the soil develops, anticyclonic conditions do become more likely to initiate failure, but because the rainfall intensity that can cause failure is reduced, cyclonic storms could also be effective. The cause of this can be established through consideration of the soil hydraulic properties at different stages of development, and the ways these control the potential maximum moisture content which can be attained under different rainfall intensities (Figure 10.13). The oldest soil has the potential to develop higher moisture contents within the B(fe) horizon under a variety of rainfall intensities, provided these are sustained for a sufficient period of time. The results of a variety of storms show the temporal decline of stability, and the greater sensitivity of younger soils to storms of different intensity (Figure 10.14). For the storm characteristics considered, slopes of 30° appear to be stable for all

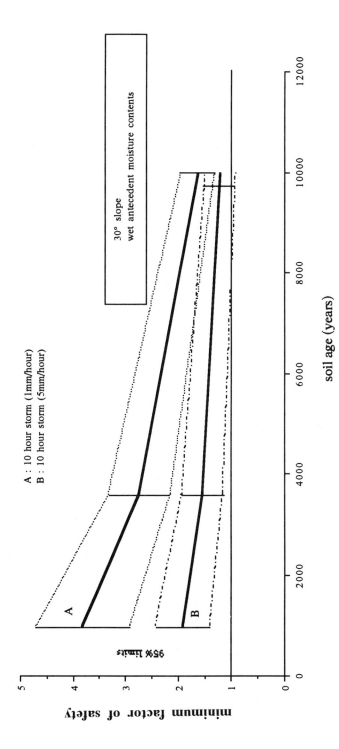

Figure 10.12 Changes in the stability for soils of different stages of development under 10 h rainstorms of 1 and 5 mm h^{-1}

volumetric moisture content

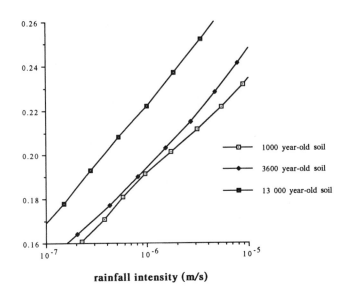

rainfall intensity (m/s)

Figure 10.13   Changes in the potential maximum moisture content attained under different rainfall intensities for soils in the Glen Feshie chronosequence

soil profiles and rainfall intensities, but slopes of 40° just become unstable under high intensity rainfall applied to soils of 3600 years of age. With soils of greater development, instability can occur at a wider range of rainfall intensities.

Interpreting these results in the light of Holocene slope stability can be more readily achieved by considering the range of slope angles which might fail under different rainfall conditions. Plotting the minimum factor of safety against slope angle for a variety of rainfall conditions and soil profiles shows the range of vulnerable angles (Figure 10.15). The soil of 1000 years of age is stable under cyclonic storms, but 80 mm of rainfall in 4 h can cause instability on slopes in excess of 40°. Since it is unlikely that soil will develop on such high slope angles, and given that 80 mm represents an 'extreme' high-intensity rainfall total, it seems that slopes having soil which has only reached 1000 years of development are unlikely to fail. After 3600 years of development, instability can occur on slopes down to 37° under similar anticyclonic storms, though such a soil remains stable under longer-duration storms even with high rainfall totals; 13 000 years of soil development takes instability down to angles of 30° under both cyclonic and anticyclonic storms. Thus as the soil develops there are changing domains of slope instability under different rainstorms; periods of enhanced mass movement are related to these conditions, and to whether recent previous failure has taken place, removing the soil and precluding the development of subsequent instability.

The current analysis considers the interactive effect on slope stability of climatic change and the development of freely draining soils. The pedological constraint simplifies discussion and enables improved understanding of the hydrological changes which reduce slope stability. However, it is important to note that not all soils in this region of Scotland are freely draining, and that many profiles outside the limits of the Loch Lomond Readvance possess indurated horizons that are thought to have formed at the close of the last ice age under periglacial

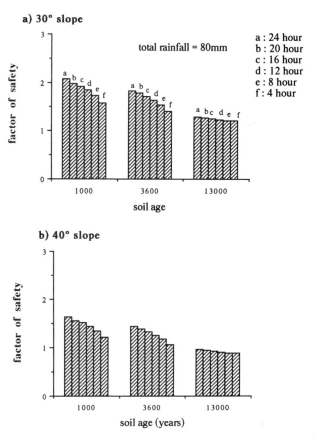

Figure 10.14   Changes in the factor of safety for a variety of rainfall intensities (3.33–20 mm h^{-1}) for soils in the Glen Feshie chronosequence

influences (FitzPatrick, 1956). Such horizons are widely reported to have a high bulk density as well as a high percentage of silt-plus-clay (Yassaglou & Whiteside, 1960; Habecker, McSweeney & Madison, 1990), which will change the hydrological response of such soils. Initial simulations indicate that this 'fragipan' layer behaves differently under different prevailing moisture contents. For example, the idea that this horizon is likely to lead to the development of perched water tables and a reduction in slope stability appears to be borne out for conditions close to saturation, and rainfall intensities that enable these to be sustained. However, under storms of lower intensity there appears to be a different response; higher moisture retention in the fragipan horizon gives it a higher hydraulic conductivity than the surrounding soil. A later paper will consider more fully the role of the fragipan layer under a range of rainfall intensities. For the present purposes, the combined role of climatic change and progressive development of freely draining profiles during the Holocene will form the basis of discussion.

In the Boreal the degree of soil development would have been related to the amount of post-glacial exposure to which the parent materials had been subjected. It is therefore crucial whether a given slope had been covered with ice during the Loch Lomond Readvance of 12 500 BP (Synge, 1956; Sissons, 1979). Within the Readvance limits soils would have been

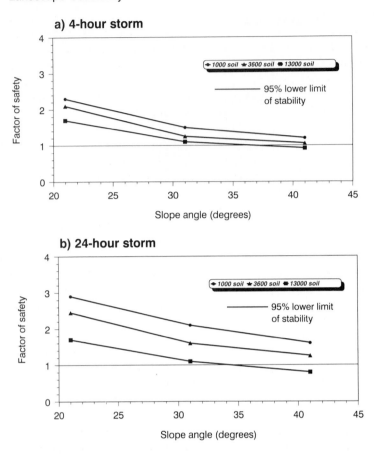

Figure 10.15   Relationship between factor of safety and slope angle for different soils and storms

young during the Boreal, reflecting about 1000 years of pedogenesis. Outside the readvance limits soils would have attained a higher degree of profile differentiation, possibly attaining similar characteristics to the 3600 year old soil, though the climate of the time was unlikely to have been conducive to chemical leaching (Birks, 1970). Thus the anticyclonic conditions of the Boreal would probably not have initiated failure on any slopes within the readvance limits, since they would have had to be in excess of 41°, but outside these limits better-developed soils, being unstable on slopes of about 37°, might have undergone failure. Following this, the warmer, wetter conditions of the Atlantic would have been superimposed on the soils within the ice limits, which, by then, would have been similar to those of 3600 years in age in the chronosequence. Outside these limits soils developing on slopes below 37° would have had about 6000 years of development, whereas those on slopes above 37° would have had younger soils. Thus the most vulnerable slopes would have been those developing on angles about 37° within the limits of the Loch Lomond Readvance, which had not undergone previous failure, and under 37° outside these limits. The lower limiting angle of instability in this latter case is difficult to establish since there is no soil of around 6000 years of development available for study. However, it is likely to be around 34° since this is intermediate between the angles appropriate to the 3600 year old and 10 000 year old soils. During the Atlantic, the shift in climatic conditions added a complication since under cyclonic conditions the younger soils are more stable. Thus failure is more likely on slopes supporting mature soils, namely those within the limits of the Readvance that did not fail during the earlier period.

The subsequent sub-Boreal saw a return to more anticyclonic conditions with the possibility of reactivation of mass movement on steeper slopes that had failed during the earlier Boreal but not during the Atlantic. Soils on the steepest slopes would most probably have reflected 3600 years of development, the approximate interval since the last failure. Thus slopes down to 37° would have undergone shallow mass movement outside the Loch Lomond Readvance limits, but within them slopes would have been stable. The wetter phase at around 2300 BP, referred to above, would have coincided with soils of slightly greater maturity, with the result that failure might have occurred on somewhat lower slope angles. Thus at the start of the sub-Atlantic, the region is likely to have consisted of a mosaic of differentially developed soils on a range of slope angles, each having a different probability of failure under any given rainfall event. The steepest slope angles would have been the least likely to be stable in earlier periods, so would have had younger soils which only fail under anticyclonic conditions. Conversely, slopes of 30−35° could have had uninterrupted soil development over 10 000 or 13 000 years, making them vulnerable under both cyclonic and anticyclonic rainstorms. Thus there is a complex set of phases of stable and unstable periods, which relates to the rate of soil development and the independently changing climate. The existence of such cycles, as revealed by the model simulations, supports earlier ideas relating to K-cycles which are responsible for the complex soil−slope associations in the landscape (Butler, 1959).

At the start of the sub-Atlantic wetter conditions could well have initiated failure on a range of lower slope angles that had previously been stable. Thus the dates identified by Innes (1983) and by Brazier & Ballantyne (1989) can be linked to enhanced mass movement during the early sub-Atlantic. Also, it is suggested that during this period both short-lived, high-intensity anticyclonic rainstorms and longer-duration, lower-intensity cyclonic storms could initiate failure on slopes that had well-developed soil profiles. Innes has associated the later date of instability with anthropogenic causes, because this period is commonly thought of as rather drier and more continental. However, under such conditions high-intensity anticyclonic rainstorms could well initiate failure, without the need to invoke an anthropogenic cause. It is also likely that increased storminess during the Little Ice Age explains the enhanced mass movement at that time. The altering probabilities of failure appear to be related to progressive pedogenesis creating the conditions within the soil which allow higher moisture contents and reduced suctions to develop. Climatic control of the hydrological behaviour of the soil has, however, formed the main emphasis of this paper, with two main intentions. First, it has been shown that phases of climatic deterioration during the Holocene in eastern Scotland could easily be responsible for enhanced mass movement, but the relationship with slope angles and soil development requires careful interpretation. Secondly, the hydrological mechanisms that govern failure have been shown to be complex, and physically based models can be used to evaluate their operation in a way that is not possible by direct observation.

SUMMARY AND CONCLUSIONS

A physically based model has been used to demonstrate how well-developed podsols become unstable down to angles of 30°, because of climatic conditions likely to have prevailed during the late Holocene in eastern Scotland. It has also shown that earlier phases of enhanced mass movement are likely, but that these probably occurred on steeper slopes because of the young soils that would have existed. The hydrological mechanisms leading to failure have been discussed, and these have been shown to relate to changes in the hydrological properties of the soil following pedogenesis, and to the varying climatic conditions. Thus variations in both soil and climate have been shown to produce a changing sensitivity of slopes to failure during the Holocene.

Focusing on the mechanisms controlling failure is a more fruitful approach than trying to make regional correlations between periods of increased slope activity and widespread changes in other variables which possibly control the process (Grove, 1972; Harvey *et al.*, 1981; Innes, 1983; Brazier & Ballantyne, 1989). Such correlations are often based on the most readily available evidence, without consideration of possible influences provided by other variables for which evidence is not readily available. These studies provide useful dates for the hiatuses in the stratigraphic sequence marked by buried soils, which can be used to aid explanations. Such research generates ideas and knowledge of environmental changes with likely consequences, but is reliant on the accuracy of the dating, and on the linking of dates for events that have been ascertained by different methods. The mechanisms whereby environmental changes can influence slope processes are not considered explicitly. It is likely that vegetation removal or climatic deterioration can cause failure, but it is important to know more about how and under what circumstances each factor becomes effective. Simple correspondence between events which are thought to be causally linked, though helpful, is incomplete until the mechanisms behind the changes are investigated. Such ideas have been expressed previously in relation to the study of phases of fluvial erosion and sedimentation (Richards *et al.*, 1987). The establishment of causal mechanisms and the use of physically based models also allow investigation of the extent of change required to initiate failure on slopes of different angle with varying regolith character. For example, vegetation clearance might initiate failure if it is carried out on certain slopes with a particular type of soil, but slopes could remain stable if appropriate conditions of soil type and slope angle are not met. Instability, however, also depends on the nature of the storms in a region, and this affects whether vegetation clearance will initiate failure.

It has been shown how physically based modelling can explore and explain such possibilities, thereby providing a means of testing possible scenarios under which failure can take place. It is difficult to consider the role of climatic change through direct field observation, and modelling can therefore provide an important complementary research tool for such analysis. It can reveal which climatic parameters, or combination of parameters, are significant to failure, and which slopes are vulnerable under the various combinations. It also enables consideration of the consequences of previous failures. As an exploratory tool for ascertaining the likely consequences of future changes, modelling provides invaluable insight. However, it cannot overcome all of the problems that result from the inherent complexity of natural systems, especially related to the interactions which exist from time to time and place to place. For example, the course of slope development often involves failure and redeposition of material at the slope base at a lower angle. This material overlies the soil and becomes the parent material for the next phase of pedogenesis. Thus the timing of subsequent failure is complicated by the rate of pedogenesis upslope on the old slip surface, and at the slope base on the newly deposited material, with neither the rates nor the slope angles being the same for each section. Thus if the basal deposits undergo failure before the upper slope sections, this could initiate progressive failure, but if the upslope sections fail first then the nature of the deposits at the slope base becomes highly complex. Other such limitations need to be discussed, since they reveal areas in which modelling must be attempted with care. It is clear that their interpretation requires insight which can only be verified through field observation, and that the strength of the outputs is dictated by the quality of the data which are available as inputs. However, it is hoped that this research has revealed how physically based models can be valuable in elucidating the operation of complex geomorphological processes in both a specific and a general sense in relation to the changing sensitivity of slopes to mass movement over Holocene timescales.

ACKNOWLEDGEMENTS

The research for this paper was funded by an NERC studentship awarded to S. M. Brooks tenable in the Department of Geography of the University of Cambridge. Thanks are also due to the estate managers at Glen Feshie and Glen Livet for permission to carry out fieldwork.

REFERENCES

Abbott, M. B., Bathurst, J. C., Cunge, J. A., O'Connell, P. E. and Rasmussen, J. (1986a). An introduction to the European Hydrological System — Système Hydrologique Européen, 'SHE', 1. History and philosophy of a physically based, distributed modelling system. *Journal of Hydrology*, **87**, 45–59.

Abbott, M. B., Bathurst, J. C., Cunge, J. A., O'Connell, P. E. and Rasmussen, J. (1986b). An introduction to the European Hydrological System — Système Hydrologique Européen, 'SHE', 2. Structure of a physically based, distributed modelling system. *Journal of Hydrology*, **87**, 61–77.

Anderson, M. G. and Howes, S. (1985). Development and application of a combined soil water-slope stability model. *Quarterly Journal of Engineering Geology*, **18**, 225–236.

Anderson, M. G., Kemp, M. J. and Lloyd, D. M. (1988). Applications of soil water finite-difference models to slope stability problems. *5th International Landslide Symposium*, Lausanne, pp. 525–530.

Arya, L. M. and Paris, J. F. (1981). A physicoempirical model to predict the soil moisture characteristic from particle size distribution and bulk density data. *Soil Science Society of America Journal*, **45**, 1023–1030.

Baird, P. D. and Lewis, W. V. (1957). The Cairngorm Floods, 1956: summer solifluction and distributary formation. *Scottish Geographical Magazine*, **73**, 91–100.

Ballantyne, C. K. and Eckford, J. D. (1984). Characteristics and evolution of two relict talus slopes in Scotland. *Scottish Geographical Magazine*, **100**, 20–33.

Bathurst, J. C. (1986). Physically based, distributed modelling of an upland catchment using the Système Hydrologique Européen. *Journal of Hydrology*, **87**, 79–102.

Beven, K. J. (1989). Changing ideas in hydrology — the case of physically based models. *Journal of Hydrology*, **105**, 157–172.

Birks, H. H. (1970). Studies in the vegetational history of Scotland. I A pollen diagram from Abernethy Forest, Inverness-shire. *Journal of Ecology*, **58**, 827–846.

Birks, H. H. and Mathewes, R. W. (1978). Studies in the vegetational history of Scotland. V. Late Devensian and early Flandrian pollen and macrofossil stratigraphy at Abernethy Forest, Inverness-shire. *New Phytology*, **80**, 445–484.

Brazier, V. and Balantyne, C. K. (1989). Late Holocene debris cone evolution in Glen Feshie, western Cairngorm Mountains, Scotland. *Transactions of the Royal Society of Edinburgh: Earth Sciences*, **80**, 17–24.

Brunsden, D. and Jones, D. K. C. (1976). The evolution of landslide slopes in Dorset. *Philosophical Transactions of the Royal Society of London*, **A283**, 605–631.

Butler, B. E. (1959). Periodic phenomena in landscapes as a basis for soil studies. CSIRO Australian Soil Publication 14.

Caine, N. (1980). The rainfall intensity–duration control of shallow landslides and debris flows. *Geografiska Annaler*, **62A**, 23–27.

Cameron, D. R. (1978). Variability of soil water retention curves and predicted hydraulic conductivities on a small plot. *Soil Science*, **126**, 364–371.

Campbell, R. H. (1975). Soil slips, debris flows and rainstorms in the Santa Monica Mountains and vicinity, Southern California. *United States Geological Survey, Professional Paper* 851.

Carson, M. A. (1975). Threshold and characteristic angles of straight slopes. *Proceedings of the 4th Guelph Symposium on Geomorphology*, pp. 19–34.

Carson, M. A. and Petley, D. J. (1970). The existence of threshold slopes in the denudation of the landscape. *Transactions of the Institute of British Geographers*, **49**, 71–95.

Common, R. (1954). A report on the Lochaber, Appin and Benderloch floods, May, 1953. *Scottish Geographic Magazine*, **70**, 6–20.

Common, R. (1956). The Border floods, August, 1956: observations and comments. *Scottish Geographical Magazine*, **72**, 160–162.

Cooke, R. U. (1984). *Geomorphological Hazards in Los Angeles. A Study of Slope and Sediment Problems in a Metropolitan County*, Allen and Unwin, London.

FitzPatrick, E. A. (1956). An indurated soil horizon formed by permafrost. *Journal of Soil Science*, **7**, 248–254.

Fredlund, D. G. (1987) Slope stability analysis incorporating the effects of soil suction. In Anderson, M. G. and Richards, K. S. (eds), *Slope Stability*, Wiley, Chichester, chap. 4, pp. 113–144.

Freeze, R. A. (1987) Modelling interrelationships between climate, hydrology and hydrogeology and the development of slopes. In Anderson, M. G. and Richards, K. S. (eds), *Slope Stability*, Wiley, Chichester, chap. 12, pp. 381–403.

Green, F. H. W. (1958). The Moray Floods of July and August, 1956. *Scottish Geographical Magazine*, **74**, 48–50.

Green, F. H. W. (1971). History repeats itself — flooding in Moray in August, 1970. *Scottish Geographical Magazine*, **87**, 150–152.

Grove, J. M. (1972). The incidence of landslides, avalanches and floods in western Norway during the Little Ice Age. *Arctic and Alpine Research*, **4**, 131–138.

Habecker, M. A., McSweeney, K. and Madison, F. W. (1990). Identification and genesis of fragipans in Ochrepts of north central Wisconsin. *Soil Science Society of America Journal*, **54**, 139–146.

Harvey, A. M., Oldfield, F., Baron, A. F. and Pearson, G. W. (1981). Dating of postglacial landforms in the central Howgills. *Earth Surface Processes and Landforms*, **6**, 401–412.

Hutchinson, J. N. and Gostelow, T. P. (1976). The development of an abandoned cliff in London Clay at Hadleigh, Essex. *Philosophical Transactions of the Royal Society of London*, **A283**, 557–604.

Iida, T. and Okunishi, K. (1983). Development of hillslopes due to landslides. *Zeitschrift für Geomorphologie*, **46**, 67–77.

Innes, J. L. (1983). Lichenometric dating of debris flow deposits in the Scottish Highlands. *Earth Surface Processes and Landforms*, **8**, 579–588.

Jackson, R. D., Reginato, R. J. and van Baval, C. H. M. (1965). Comparison of measured and calculated hydraulic conductivities of unsaturated soils. *Water Resources Research*, **1**, 375–380.

Jenkins, A., Ashworth, P. J., Ferguson, R. I., Grieve, I. C., Rowling, P. and Stott, T. A. (1988). Slope failure in the Ochil Hills, Scotland, November, 1984. *Earth Surface Processes and Landforms*, **13**, 69–76.

Jenny, H. (1941). *Factors of Soil Formation*, McGraw-Hill, New York.

Kunze, R. J., Nehava, G. and Graham, K. (1968). Factors important in the calculation of hydraulic conductivity. *Soil Science Society of America Proceedings*, **32**, 760–765.

Lamb, H. H. (1977). *Climate: Past, Present and Future*, Vol. 2, Methuen, London.

Lloyd, D. M. (1990). Modelling the hydrology and stability of tropical cut slopes. Unpublished Ph.D. thesis, University of Bristol.

Masch, F. D. and Denny, K. J. (1966). Grain size distribution and its effect on the permeability of unconsolidated sands. *Water Resources Research*, **2**, 655–677.

Millington, R. J. and Quirk, J. P. (1959). Permeability of porous media. *Nature*, **183**, 387–388.

Mishra, S., Parker, J. C. and Singhal, N. (1989). Estimation of soil hydraulic properties and their uncertainty from particle size distribution data. *Journal of Hydrology*, **108**, 1–18.

Nielsen, D. R., Biggar, J. W. and Erh, K. T. (1973). Spatial variability of field-measured soil–water properties. *Hilgardia*, **42**, 215–259.

Rapp, A. (1960). Recent development of mountain slopes in Karkevagge and surroundings, northern Scandinavia. *Geografiska Annaler*, **42**, 73–200.

Rice, R. M., Corbett, E. S. and Bailey, R. G. (1969). Soil slips related to vegetation, topography and soil in southern California. *Water Resources Research*, **5**, 647–659.

Richards, K. S., Peters, N. R., Robertson-Rintoul, M. S. E. and Switsur, V. R. (1987). Recent valley sediments in the North York Moors: evidence and interpretation. In Gardiner, V. (ed.), *International Geomorphology 1986, Part I*, Wiley, Chichester, pp. 869–883.

Robertson-Rintoul, M. S. E. (1986). A quantitative soil stratigraphic approach to the correlation and dating of postglacial river terraces in Glen Feshie, south-west Cairngorms. *Earth Surface Processes and Landforms*, **11**, 605–617.

Rouse, W. C. and Farhan, Y. I. (1976). Threshold slopes in south Wales. *Quarterly Journal of Engineering Geology*, **9**, 327–338.

Sissons, J. B. (1979). The Loch Lomond Stadial in the British Isles. *Nature*, **278**, 199–203.

Starkel, L. (1966). Postglacial climate and the moulding of European relief. *Proceedings of the International Symposium of World Climate 8000 to 0 BC*, Royal Meteorological Society of London, pp. 15–32.

Synge, F. M. (1956). The glaciation of north-east Scotland. *Scottish Geographical Magazine*, **72**, 129–143.

Yassaglou, N. J. and Whiteside, E. P. (1960). Morphology and genesis of some soils containing fragipans in northern Michigan. *Soil Science Society of America Proceedings*, **24**, 396–407.

11 Numerical Experiments on the Orthogonal Profile with Particular Reference to Changing Environmental Parameters

JACK HARDISTY

School of Geography & Earth Resources, University of Hull, UK

ABSTRACT

Beach and seabed response to changing wave and tidal conditions and to eustatic and isostatic sea-level fluctuations can be analysed through the use of mobile bed numerical models. In particular, the geomorphological relaxation time can be estimated and set against the rate of change of the input parameters in order to gauge the sensitivity of the system to environmental stress. An orthogonal profile model is described which utilises first- and second-order wave theory, a slope-inclusive, excess-stress bedload function and a suspended local concentration profile referenced to peak bedload rates. Nett transport rates are integrated over the wave cycle and over the bottom 100 cm of the water column and utilised to compute orthogonal profile response. The results demonstrate the highly non-linear behaviour of the system but confirm that, typically, the profile approaches a state of equilibrium over a period of $10^1 - 10^3$ h, though the relaxation time varies with distance from the shoreline. The results, therefore, suggest that the orthogonal profile is sensitive to the longer-term environmental changes and that it will respond by constantly and rapidly re-establishing an equilibrium form providing that the required sediment is locally available.

INTRODUCTION

The orthogonal profile, for the purposes of the present chapter, will be taken to be a line lying normal to incident wave crests from the water's edge to wave base along which there is an accumulation of non-cohesive sediment having a form and texture controlled by wave dominated processes. The upper sections therefore represent coastal beach deposits. A number of adjacent profiles together constitute the three dimensional form of the nearshore zone. Neglecting local longshore variations in tidal or residual currents in deeper water or in radiation stress within the surf zone, the two-dimensional, orthogonal transport controls nearshore bathymetry and the two-dimensional profile offers a suitable and accessible representation of nearshore form.

Environmental changes that influence the orthogonal profile can be divided into *hydrodynamic input controls* and *sediment input controls*. The hydrodynamic inputs are the incident wave characteristics and fluctuating sea-levels due to short-term storm surges or to atmospheric

Landscape Sensitivity. Edited by D. S. G. Thomas and R. J. Allison
© 1993 John Wiley and Sons Ltd

warming or to any consequential shifts in local amphidromic systems. The rate of supply of sediment through fluvial inputs to the coastal zone is dealt with elsewhere in this volume and will not be considered in this chapter. Here, we firstly consider the magnitudes of variations in the hydrodynamic inputs and techniques for specifying geomorphological relaxation time, before continuing to describe the construction and operation of a mobile bed numerical model for the orthogonal profile.

ENVIRONMENTAL CHANGE

Latitudinal shifts in weather systems, which may be consequential upon climatic changes, can result in local increases in 'storminess' and therefore in incident wave parameters. For example, the 50 year maximum wave height in the central North Sea is presently estimated to be some 20−25 m (Hardisty, 1990a, chapter 4), though frictional loss, wave breaking and wave refraction considerably reduce this value for coastal sites. There is some evidence of a 10% wave height increase in this region over the last quarter of a century but it is presently difficult to separate short-term oscillations from long-term trends. Coastal wave heights may also increase (or decrease) due to changing refraction patterns following a sea-level rise, but little work has yet been carried out to quantify these effects. The fluctuation in sea-level that results from climatic change is a subject of some debate at the present time. Warrick & Oerlemans (1990) describe the results of the Intergovernmental Panel on Climatic Change (IPCC) which suggest a best estimate global mean temperature rise of 0.3 °C per decade if no steps are taken to limit greenhouse gas emissions. They suggest that this will lead to ocean thermal expansion, to low-latitude glacier melt and, later, to peripheral high-latitude ice sheet melt. The best estimate of the resulting sea-level rise suggests a figure of some 18 cm by 2030 and 66 cm by 2100. Consequent upon such sea-level rises will be changes in the local amphidromic systems, which may lead, therefore, to increased tidal ranges and to increases in high water levels that are in excess of these figures. This chapter is concerned with an attempt to place these predictions within a geomorphological context through estimates of the relaxation time of the orthogonal profile.

LANDFORM SENSITIVITY AND RELAXATION TIME

An appreciation of the stability, or otherwise, of an environmental system is achieved through the use of sensitivity analyses. This approach generally recognises that the mathematical representation of the system will respond to either changes in the input parameters (parameter sensitivity) or to changes in the representation of the internal transfer functions (process sensitivity). The use of process sensitivity techniques in the analysis of orthogonal profile models was described by Hardisty (1990b) and showed that the orthogonal system was most sensitive to the inclusion of breaking waves, the form of the pick-up function used to define suspended load concentration profiles and the inclusion of bedslope effects in the transport functions. In effect, the shorewards transport of sediment due to flow asymmetry is counteracted by offshore, downslope transport as equilibrium is approached. The present paper concentrates on parameter sensitivity and utilises the relaxation time, T_R, which is generally taken to be the time required for a system to re-establish equilibrium following changes in the input parameters. For geomorphological purposes (e.g. Thornes & Brunsden,

1977; Hardisty, 1990b, p. 195) the relaxation time can be defined as:

$$T_R = \frac{\text{change in elevation}}{\text{rate of elevation change}} = \frac{(z - z_e)}{\partial z / \partial t} \tag{1}$$

where z is the elevation of the landsurface, z_e is the equilibrium elevation and the response function, $\partial z / \partial t$, is the rate of surface erosion or accretion. Since the response function is likely to vary both spatially (i.e. along the orthogonal profile) and temporally (i.e. as the system approaches an equilibrium state) then equation (1) must be integrated numerically in order to estimate the relaxation time. The objective of the present paper is therefore to estimate the relaxation time and the sensitivity of the orthogonal profile in terms of some scenarios for environmental change.

ORTHOGONAL PROFILE MODELS

Early attempts are predicting the two- and three-dimensional form of wave-dominated nearshore bathymetries concentrated on statistical correlations between the beach gradient and some measure of the incident waves and were later developed to include the full wave-controlled orthogonal profile (cf. King, 1972, for a review of the empirical models). Models based upon a better understanding of the processes began with Bagnold (1963), and have evolved into either numerical models based upon computer simulations (e.g. Davidson-Arnott, 1981; Horikawa, 1988), which 'run' the process equations over a given initial bathymetry, or analytical models, which attempt to solve the process equations directly for an equilibrium seabed gradient (e.g. Bowen, 1980; Bailard & Inman, 1981; Hardisty, 1986, 1990c). Although the analytical approach offers an understanding of the beach response through the use of state-space representations of system stability, the equations can presently be solved only for small subsections of the system and the predictive capability is therefore rather limited. Alternatively, the numerical approach promises a powerful predictive capability, but the computer programs are becoming so complex that it is difficult to understand the response of the beach system to changing inputs. Anderson & Sambles (1988) discuss this problem in more detail for general, geomorphological modelling and the present paper utilises relaxation time as a state-space variable in order to interpret a mobile bed, numerical model and hence to quantify orthogonal profile responses.

GOVERNING EQUATIONS

A mobile bed, numerical model has been written for the orthogonal profile system (SLOPES: the shoreline and orthogonal process emulation system) and is described in detail in Hardisty (1990b). The models shoals a monochromatic description of the deep water wave conditions using first- and second-order theory, a description of wave breaking and of seabed frictional losses:

$$L = \frac{gT^2}{2\pi} \tanh\left(\frac{2\pi h}{L}\right) \tag{2}$$

$$n = \frac{1}{2}\left(1 + \frac{2kh}{\sinh kh}\right) \tag{3}$$

$$H_2 = H_\infty \sqrt{\frac{1}{2n}\frac{C_\infty}{C}} \tag{4}$$

$$\alpha = \sin^{-1}\left(\frac{\sin \alpha_\infty \, C}{C_\infty}\right) \tag{5}$$

$$\Delta H_r = H_\infty \sqrt{\frac{C_\infty}{2nC}\frac{\cos \alpha_\infty}{\cos \alpha}} \tag{6}$$

$$H_b = 0.72\, h(1 + 6.4 \tan \beta) \tag{7}$$

The local wave amplitudes are used to compute the nearbed oscillatory and drift currents using second-order hydrodynamic equations:

$$u(t) = A_a \frac{\pi H}{T \sin kh} \cos \omega t \tag{8}$$

$$u(t) = A_a \left(\frac{\pi H}{T \sinh kh} \cos \omega t + \frac{3}{4}\left(\frac{\pi H}{L}\right)^2 C \frac{1}{\sinh^4 kh} \cos 2\omega t\right) \tag{9}$$

$$U = \frac{5}{4}\left(\frac{\pi H}{L}\right)^2 C \frac{1}{\sinh^2 kh} \tag{10}$$

The currents drive an excess stress bedload equation that includes the effect of threshold and bedslope on the transport rate, and is integrated over the wave cycle to determine the nett bedload mass transport for each wave:

$$i_b(t) = A_b(t)\, k_b\, [u(t)^2 - B_b(t)^2\, u_{cr}^2]\, u(t) \tag{11}$$

$$A_b = \frac{\tan \phi}{\cos \beta \left(\tan \phi + \dfrac{u(t) \tan \beta}{|u(t)|}\right)} \tag{12}$$

$$B_b = \sqrt{\frac{\tan \phi + \dfrac{u(t) \tan \beta}{|u(t)|}}{\tan \phi}\cos \beta} \tag{13}$$

The maximum bedload concentration is used as a reference, C_a, to determine the vertical profile of the suspended load concentration using a diffusion equation. The concentration profile is scaled with the vertical profile of the non-oscillatory currents and integrated for the lowest 1 m of the water column and over the wave cycle to determine the nett suspended load mass transport for each wave:

$$C_z = C_a \left(1 - A_c \ln \frac{z}{z_a} \right) \tag{14}$$

$$U_z = \frac{1}{4} \left(\frac{\pi H}{T} \right) \left(\frac{\pi H}{L} \right) \frac{1}{\sinh^2 kh}$$

$$\times \left\{ 2 \cosh \left[2kh \left(-\frac{z}{h} - 1 \right) \right] + 3 + 2kh \left(3 \frac{z^2}{h^2} + 4 \frac{z}{h} + 1 \right) \sinh 2kh \right.$$

$$\left. + 3 \left(\frac{\sinh 2kh}{kh} + \frac{3}{2} \right) \left(\frac{z^2}{h^2} - 1 \right) \right\} \tag{15}$$

$$\bar{i_s} = \frac{1}{T} \int_{t=0}^{t=T} \int_{z=0}^{z=1} U_z C_z \, dz \, dt \tag{16}$$

Local hydrodynamic and sediment transport parameters values and time series are calculated in 37 cells from deep water to the shoreline and the nett transports are transferred between the cells to model the evolution of the profile for a single wave. The procedure is repeated so that geomorphological changes provide feedbacks into the shoaling transformations, into the hydrodynamic equations and into the sediment transport functions through the bedslope effects.

RESULTS

The sensitivity of the model to the various process functions is detailed in Hardisty (1990a), and only the longer-term response of the system is examined here. Figure 11.1 shows the results of running monochromatic waves of period 5 s and deep water wave height 1.5 m over an initial profile given by:

$$h = 0.075x^{2/3} \tag{17}$$

where h is the water depth at a distance x from the shoreline.

Figure 11.1(a) shows the shoaling wave height (solid line) and peak oscillatory current speeds; Figure 11.1(b) shows the bedload (broken line), suspended load (dotted line) and total (solid line) mass transport rates; and Figure 11.1(c) shows the profile response, $\partial z / \partial t$. The values computed by the model are not unreasonable. For example, the wave height increases shorewards to some 1.8 m before breaking at a distance of about 80 m from the shoreline, while the flows increase from about 50 cm s^{-1} some 0.5 km offshore to a maximum in excess of 3 m s^{-1} at the breakpoint. The corresponding transport rates are very low in deeper water, reaching a total load maximum of about 21 gm cm^{-1} s^{-1} at the breakpoint, of which more than 75% is bedload transport in the present example. The profile responds to increasing shorewards transport by slight erosion (some 10 mm day^{-1}) in the offshore zone balanced by rapid accretion at the breakpoint (0.6 mm day^{-1} at the beginning of the simulation). These values are in general agreement with field results reported for prototype sites.

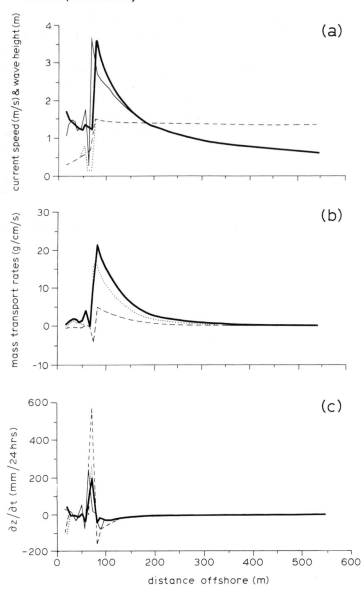

Figure 11.1 Results of the profile simulation described in the test showing (a) wave height and oscillatory currents, (b) bedload, suspended load and total load mass transport rates and (c) profile response, $\partial z/\partial t$. Symbols are defined in the text

The resulting changes in the profile are shown in Figure 11.2(a) where it can be seen that a bar develops beneath the breakpoint some 80 m from the shoreline. The sensitivity of the profile to these conditions is shown as a graph of $\partial z/\partial t$ through time in Figure 11.2(b) for two contrasting sites on the profile. First, the upper plot represents the bar crest ($x = 72$ m and open squares) where the rapid initial accretion continues at a lower rate on the evolving, barred profile. Conversely, the lower plot in Figure 11.2(b) represents a site offshore from the bar ($x = 98$ m) at which the initial profile shows erosion which continues but at a decreasing rate as the profile evolves. These two examples demonstrate the inherent stability of the system with the rate of change of the response function, $\partial^2 z/\partial t^2$, being negative when $\partial z/\partial t$ is positive, i.e. where accretion is occurring, and $\partial^2 z/\partial t^2$ being positive when $\partial z/\partial t$ is negative, i.e. where erosion is occurring. In both cases the shorewards-dominated flow asymmetries are becoming balanced by offshore-directed gravitational transport enhancement. That is, where erosion or accretion occurs it does so at a decreasing rate through time until profile stability is achieved. These are typical characteristics of a dynamically stable geomorphology, and the profile therefore represents an attractor in state space.

It is apparent, however, that stability is achieved at a rate that varies along the profile. Typically, the outer, eroding section of the profile achieves equilibrium (i.e. $\partial z/\partial t$ returns to zero) some 100 h after new wave conditions have been imposed, while the bar crest appears to require a longer interval to adjust. In general, however, it appears that a relaxation time of tens to thousands of hours would be required to achieve stability in this particular example.

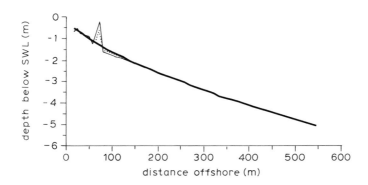

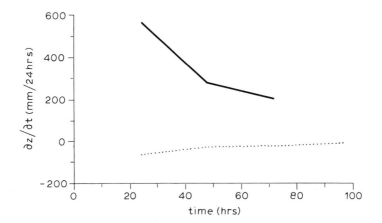

Figure 11.2   Results of the profile simulation described in the test showing (a) profile evolution and (b) the differences between $\partial^2 z/\partial t^2$ at the bar crest and at a position offshore from the bar. Symbols are defined in the text

Similar experiments with small changes in sea-level also produced relaxation time; intervals of 10^2-10^3 h were typical for water depth changes of up to 10 cm. These results appear to suggest that the orthogonal profile exhibits a relaxation time that is very many orders of magnitude shorter than the timescales involved in the types of changing environmental parameters which were outlined earlier.

FUTURE DEVELOPMENTS

Experiments with the model using random rather than monochromatic wave fields showed that, because of the geomorphological feedbacks, different sequences of waves from the same, random distribution generate different profile forms. Additionally recent work is beginning to show that beaches are not controlled by energy at the frequency of the incident waves in the gravity band but rather by nearshore resonances in the long-wave and infra-gravity bands (e.g. Huntley & Hanes, 1987; Beach & Sternberg, 1988) and therefore the orthogonal profile model described here is being developed to analyse the processes in the frequency rather than the time domain in order to include long wave energy (Hardisty, Hoad and Davidson, in preparation).

CONCLUSIONS

This chapter has demonstrated a numerical model for the orthogonal system and shown that the profile operates as a stable geomorphological attractor and exhibits a relaxation time of the order of 10^1-10^3 h. Since this is very many orders of magnitude faster than the rates of sea-level rise or long-term wave climate response to environmental change, these results can be interpreted as suggesting that the profile will continue to re-establish a long-term equilibrium in response to changing environmental parameters.

SYMBOLS

A_a attenuation coefficient for seabed currents
A_b slope coefficient for bedload rate parameter
A_c scaling parameter for concentration profiles
B_b slope coefficient for threshold velocity
C_a sediment concentration at z_a above bed (kg m^{-3})
C_z sediment concentration at z above bed (kg m^{-3})
C_∞ deep water wave celerity (m s^{-1})
ΔH_r change in wave height due to refraction (m)
H local wave height (m)
H_b breaking wave height (m)
H_s shoaled wave height (m)
H_∞ deep water wave height (m)
$i_b(t)$ instantaneous bedload rate (kg m^{-1} s^{-1})
$\bar{i}_s$ mean suspended load transport rate (kg m^{-1} s^{-1})
L local wave length (m)

L_∞ deep water wave length (m)
h water depth (m)
k wave number $(=2\pi/L)$
k_b bedload rate parameter (kg m^{-4} s^{-2})
n shoaling abreviation
t elapsed time (s)
T wave period (s)
T_R relaxation time (s, h, etc.)
u_{cr} threshold velocity (m s^{-1})
$u(t)$ seabed current (m s^{-1})
U_z velocity of second-order drift current (m s^{-1})
x distance offshore (m)
z depth below sea surface (positive upwards) (m)
α local wave orthogonal direction
α_∞ deep water wave orthogonal direction
β seabed gradient
ϕ angle of internal friction of sediment
ω wave radian frequency $(=2\pi/T)$

REFERENCES

Anderson, M. G. and Sambles, K. M. (1988). A review of the bases of geomorphological modelling. In Anderson, M. G. (ed.), *Modelling Geomorphological Systems*, Wiley, Chichester, pp. 1–32.

Bagnold, R. A. (1963). Mechanics of marine sedimentation. In Hill, M. N. (ed.), *The Sea*, Vol. 3: *Ideas and Observations*, Interscience, New York, pp. 507–526.

Bailard, J. A. and Inman, D. L. (1981). An energetics model for a plane sloping beach: local transport. *Journal of Geophysical Research*, **86**, 2045–2053.

Beach, R. A. and Sternberg, R. W. (1988). Suspended sediment transport in the surf zone: response to cross-shore infragravity motion. *Marine Geology*, **80**, 61–79.

Bowen, A. J. (1980). Simple models of nearshore sedimentation, beach profiles and longshore bars. In McCann, S. B. (ed.), *The Coastline of Canada*, Geological Survey of Canada, pp. 1–11.

Davidson-Arnott, R. G. D. (1981). Computer simulation of nearshore bar formation. *Earth Surface Processes and Landforms*, **6**, 23–34.

Hardisty, J. (1986). A morphodynamic model for beach gradients. *Earth Surface Processes and Landforms*, **11**, 327–333.

Hardisty, J. (1990a). *British Seas: An Introduction to the Oceanography and Resources of the North-west European Continental Shelf*, Routledge, London.

Hardisty, J. (1990b). *Beaches: Form & Process*. Unwin–Hyman, London.

Hardisty, J. (1990c). A note on suspension transport in the beach gradient model. *Earth Surface Processes and Landforms*, **15**, 91–96.

Hardisty, J., Hoad, J. P. and Davidson, M. (in preparation). Numerical experiments with gravity and infragravity waves on a macro-tidal beach profile. In Short, A. D. (ed.), *Beach and Surf Zone Morphodynamics. Journal of Coastal Research*, Special Issue.

Horikawa, K. (ed.) (1988). *Nearshore Dynamics and Coastal Processes*, University of Tokyo Press.

Huntley, D. A. and Hanes, D. M. (1987). Direct measurement of suspended sediment transport, *Proceedings of Coastal Sediments '87*, ASCE, pp. 723–737.

King, C. A. M. (1972). *Beaches and Coasts*, 2nd edn, Arnold, London.

Thornes, J. B. and Brunsden, D. (1977). *Geomorphology and Time*, Methuen, London.

Warrick, R. A. and Oerlemans, H. (1990). Sea level rise. In Houghton, J .T., Jenkins, G. J. and Ephraums, J. J. (eds), *Climate Change: The IPCC Scientific Assessment*, Cambridge University Press, Cambridge, pp. 257–282.

Part 2

LAND USE AND LANDSCAPE SENSITIVITY

12 Sensitivity of the British Landscape to Erosion

ROBERT EVANS
Department of Geography, University of Cambridge, UK

ABSTRACT

The sensitivity of land to erosion depends on the level of the geomorphological thresholds that have to be crossed before erosion can take place. High levels of resistance make the land insensitive to erosion. The factors that interact to cause erosion, such as the weather, slope factors like rock and soil strength, the presence or absence of vegetation cover and the actions of people, are assessed for different scenarios. The scenarios range from when woodland was at its maximum extent in Britain, to landscapes covered by heath, grass, moor or arable. Land use is the factor that dominates the sensitivity of the landscape to erosion: 'wildwood' is the least sensitive, arable land the most. The extent of arable land is probably greater now than ever before. Sensitivity increases as thresholds are lowered by people's actions; mankind's ever-improving technology, used on the farm and for draining the land and planting forests, further lowers erosional thresholds. For the above reasons more of Britain is presently more sensitive to erosion that it has been at any time since woodland clearance began. However, much of the country remains insensitive to erosion. In the lowlands the surface is being lowered by soil stripping, whereas in the uplands channels and gullies are slowly extending. Identifying those areas most sensitive to erosion, and the processes acting on them, will allow us to combat erosion and reduce its impacts.

INTRODUCTION

The sensitivity of the landscape to erosion depends on the level of the threshold at which erosional forces triggered by the weather or, rarely in Britain, earthquake shocks, in association with gravity, overcome the resistances of rock, soil and vegetation. The perception of the sensitivity of the landscape to erosion is closely bound up with the perception of the magnitude, frequency and impact of an erosional event. Sensitivity is high when the thresholds that have to be crossed to trigger erosion are low so that erosion takes place frequently, though the impact of an individual event on the landscape is not great. A landscape of low sensitivity suffers erosion only when subject to a large stress which occurs rarely, but the impact on the landscape is great and evidence of the erosional event persists over a long period of time.

The thresholds that have to be overcome before erosion takes place may change over time. Thus, weathering and leaching of the soil profile under a vegetation cover may make it more or less susceptible to erosion by mass movement; a change in vegetation cover as climate changes may also inhibit or encourage erosion. An increase in rainfall amount and intensity may be accompanied by more erosion as a geomorphological threshold is subsequently crossed

---

Landscape Sensitivity. Edited by D. S. G. Thomas and R. J. Allison

more frequently. But changes in threshold levels as soils and vegetation alter over time as climate changes are often slow in comparison with the lowering of geomorphological thresholds brought about by people's actions.

How geomorphological thresholds have changed over time, primarily in response to changes in land use brought about by the farmer, and how these have affected the sensitivity of the land to erosion are described here. It is difficult to be precise about the sensitivity of the British landscape to erosion in terms of how often erosion events are likely to take place, and their magnitude, frequency and extent. This is a preliminary assessment of sensitivity, taking a broad approach. The sensitivity of coastal zones to changes in sea-levels and to storminess are not dealt with.

NATURAL WOODLAND AND SENSITIVITY TO EROSION

As the climate ameliorated after the last ice age woodland gradually spread and Britain was forested to its maximum extent 7000–8000 years ago (Birks, 1988). Only northern and western Scotland, the highest mountains, screes and mires had few or no trees, and even here heath similar to that now found in parts of Norway would protect the ground. Under such conditions most of the landscape was insensitive to erosion.

In England and Wales there is little evidence of eroded soil being deposited as alluvium or colluvium in valley floors at this time (Evans, 1990a). Where the landscape was wooded sediments in lakes were not derived from eroded slopes or stream banks (e.g. Pennington, 1970; Pennington et al., 1972; Tinsley & Derbyshire, 1976).

Within the woodlands it is likely, if these are analogous with Norway or New Zealand for example, that mass movements of one kind or another and destabilisation of stream channels and banks would not occur unless storms were of large magnitude (Selby, 1976). Such an event was described by White (1788). January, February and early March 1774 were remarkable for great melting snows and 'vast gluts' of rain, leading on the night of 8 and 9 March to the collapse of a wooded slope near Selborne, Hampshire, leaving a high freestone cliff naked and bare.

Rapid mass movements would be concentrated in those localities where slopes were steep and lithologies vulnerable to sliding (Doornkamp, 1990), e.g. where alternating shales and sandstones occurred dipping valleyward, as in Longdendale (Johnson, 1980) in the Peak District. But Tallis & Johnson (1980) hypothesised that slides may not have happened when the slopes were most wooded between 7000 and 5000 years ago. In the present-day, debris slides that formed in rare large storms in the southern Uplands (Learmonth, 1950) and western Highlands (Common, 1954) of Scotland were less common in mature woods than elsewhere.

Soil creep — the cyclic daily and seasonal wetting and drying of the soil — was probably as ineffective a process under wooded slopes as it is now under grass or moor (e.g. Evans, 1974; Anderson & Cox, 1984), and can be discounted as an agent for shaping the surface of the land. Soils probably moved downslope fastest where they were wettest and contained most organic matter (Anderson & Cox, 1984), as in minor valley floors or flushes now (Evans, 1974), and not where slopes were steepest. The rate of creep does not relate well to slope angle (Anderson & Cox, 1984; Evans, 1974; Finlayson, 1981; Slaymaker, 1972), probably because soils on steeper slopes are often coarser textured and more stable.

Within clearings in woods and on heathlands some overgrazing by animals could have occurred, leading to erosion, but this was probably rare as herds were free to roam and graze.

Above the tree-line where the vegetation cover was thin or absent, the slopes may have been more sensitive to the weather and erosional processes. Frost action could have caused rock shattering, sorting and mass movements, similar to those acting now or in the recent past at high altitudes (Ballantyne, 1986a; Caine, 1963), especially where the scree was of well shattered rock. However, Ballantyne & Eckford (1984) considered that most screes in Scotland are relict ones dating from the Loch Lomand Stadial 10 000 years ago.

About 7500 years ago, as the climate became moister than now and soils more leached, peat began to grow on gentle slopes in the uplands above about 365 m OD, and this was associated with the decline of woodland (e.g. Conway, 1954; Cundill, 1977; Simmons, 1963; Pennington et al., 1972). Where sites were poorly drained peat formation may have started even earlier (Turner et al., 1973). As in northern Norway now, thin peat could also have accumulated on stable rock and scree slopes.

WOODLAND CLEARANCE AND SENSITIVITY TO EROSION

Woodland clearance by farmers could have contributed to early peat growth (Evans, 1975; Moore, 1968; Jacobi, Tallis & Mellars, 1976). Clearance began on a large scale about 5000 years ago and continued intermittently (e.g. Birks, 1988; Evans, 1975; Cundill, 1976, 1977; Moore & Chater, 1969; Pennington, 1970; Pennington et al., 1972; Simmons & Cundill, 1974; Smith & Taylor, 1969; Tallis, 1964; Tallis & Switsur, 1983; Turner, 1964; Turner et al., 1973) until the cutting down of the last (pine) woods in Scotland in the 18th and 19th centuries (Steven, 1951; Werrity, 1984). There were two periods when woodland recovered somewhat during population declines, after about 450 AD and following the Black Death in the 14th century. In the late 17th century only about 8% of England and Wales was wooded (King, 1698, in Hoskins & Stamp, 1963), a figure that had probably changed little over many centuries. By the turn of this century about 6% of Britain was wooded (Best & Coppock, 1962) and now after much coniferous afforestation, this figure has risen to about 10% (Forestry Commission, 1989).

Clearance of the woods led not only to peat formation in the uplands but also lowered the sensitivity of the landscape to erosion. It is since clearance that alluvium and colluvium have been accumulating in valley floors (Evans, 1990a).

The evidence of the impacts of woodland clearance are most clearly seen in the uplands. In the lowlands it is likely that much of the evidence has been removed by later erosion of arable land. Tree clearance was by axe and fire, and charcoal is commonly present in soil and peat layers.

Removal of woodland would increase runoff, because less rainfall would be intercepted and evaporation from grassland would be lower. This extra runoff would destabilise stream channels, lead to bank erosion and headward extension of channels. In many valleys in the central Pennines (Figures 12.1 and 12.2), the North York Moors (Curtis, 1965) and the Bowland Fells, streams have exposed layers of minerogenic sediment and peat over head or till, indicative of periods of erosion, deposition and peat formation. In places in the southern Pennines, stratified minerogenic and peat sediments are exposed in scars on slopes, presumably gully infills (Figure 12.3) and fans. Slump deposits, triggered by stream undercutting or weakening of soil strength as tree roots decomposed (Rogers & Selby, 1980), can be seen in upland valleys in the northern Pennines, the Howgill (Harvey, Oldfield & Baron, 1981) and Bowland Fells (Harvey & Renwick, 1987) and the North York Moors (Richards, 1981;

Figure 12.1   Section exposed in Greave Clough, Boulsworth Hill, West Yorkshire, showing a poorly sorted head/terrace deposit below a laminated sandy and reworked peat deposit separated by peat. The sequence suggests two periods of erosion and deposition, probably related to woodland clearance, separated by a period of landscape stability

Richards *et al.*, 1986), burying former soils or peat and the trees growing within them. Stripping of a thin (organic) soil cover, where there was one, and reworking of screes in the Highlands of Scotland was probably associated with clearance and subsequent grazing by sheep (Innes, 1983).

In the uplands, erosion took place when the woodland was cleared; after that slopes and channels probably stabilised and have altered little since. This was probably not so in the lowlands, e.g. topsoil in valley floors in the southern chalk Downlands was removed down to an indurated (Bt) layer, which was later buried by colluvium (Smith, personal communication).

Initially, after woodland clearance, apart from small arable plots, much of the land would be grazed by sheep, cattle and, possibly, goats. Grazing would inhibit tree regeneration not only in clearings but also in the adjacent woodlands, so creating lowland heaths (Dimbleby, 1962) and grasslands, and in the uplands, acidic grasslands or moors.

Figure 12.2 Section exposed in Broad Head Clough, Boulsworth Hill, Lancashire, showing at depth a head/till deposit, overlain by a laminated, mostly fine-textured water-lain deposit containing wood in its upper part which, in turn is overlain by a dominantly sandy deposit with occasional bands of reworked peat

Figure 12.3 Section exposed in sheep scar, Hey Clough, Peak District, South Yorkshire, showing laminated deposits laid down in gullies cut in head

SENSITIVITY OF GRASSLAND, HEATH AND MOOR TO EROSION

Once grassland, heath or moor was well established after woodland clearance (undisturbed landscapes), it would not be very sensitive to erosion, until the actions of the land user led to the disruption of the vegetation cover (disturbed landscapes).

Undisturbed Landscapes

A distinction needs to be made here between semi-natural rough grasslands which cover(ed) the uplands and steep slopes of escarpments and flanking streams in the lowlands, and improved grasslands. The present-day limit between the two is the upper limit of enclosed fields.

Within enclosed fields, constant grazing and fertilising of the land has improved the pasture, leading to a dense sward of better quality grasses. Often, since the agricultural revolution of the 18th century, the more gently sloping fields have been ploughed and put down to rye grass and clover leys. Once established, these are very resistant to erosion except by rapid mass movements on clayey soils after very heavy rainfalls, as on the Carboniferous strata of the Culm measures in Devon or the Coal Measures in the north of England, or on Jurassic clays or clayey tills. Often these features are small in extent and are not a significant feature of the landscape. Rarely, where cattle have concentrated in gateways, sufficient bare soil has been exposed for rilling to be initiated.

In rough grassland, slips, slides and flows have occurred on slopes generally steeper than about 15° (with a range of 5−37°) (Ballantyne, 1986b; Beven, Lawson & McDonald, 1978; Common, 1954; Duckworth, 1969; Gifford, 1953; Learmonth, 1950; Newson, 1980a; Statham, 1976). They could be widespread within a locality after a heavy rainfall, usually more than 75 or 100 mm, but it could be as low as about 50 mm (Arkell, 1955; Baird & Lewis, 1957; Beven, Lawson & McDonald, 1978; Bleasdale & Douglas, 1952; Common, 1954; Duckworth, 1969; Jenkins et al., 1988; Harvey, 1986; Learmonth, 1950; Newson, 1980a; Statham, 1976), falling on ground that had already become saturated. Occasionally, drier weather, which had caused soil cracking, preceded the wetter period. Selby (1976) suggested that in New Zealand a lower amount of rainfall was needed to trigger landslides on grassy slopes than to initiate slides in woodland.

Often the narrow (< 10 m), shallow (< 1−2 m) slips and slides, not more than a few hundred metres in length, were within first-order valley heads, which had no channel but a saturated, frequently rushy and often peaty central zone. Subsoil pipes could be present at the junction of more and less permeable layers. Debris flows too were small features, often < 20 m^3 (Innes, 1985). Larger mass movements could be triggered on steep slopes (often > 30°) flanking streams by undercutting during flooding. Similar features on lower slopes of the last incision phase of the stream could be due purely to slope failure under saturated conditions. These features could be hundreds of metres in length and 50 m or more wide, but formed a small, though often spectacular, part of the landscape (e.g. Baird & Lewis, 1957; Fairbairn, 1967; Harvey 1986). Slips and slides become overgrown by vegetation but can still be identified decades later by their morphology (e.g. Anderson & Calver, 1977).

The replacement of birch woodland by heather moor could lead to podsolisation of the soil and formation of an iron-pan (Dimbleby, 1962; Miles, 1985), so increasing a slope's sensitivity to erosion. On steep slopes saturation of the upper soil layers over the impermeable pan during rare rainfall events could trigger mass movements, as in the Howgill Fells in 1982.

On gentler slopes, less than about 10°, as peat under moorland continued to grow, its sensitivity to erosion increased. Early in the 19th century Vancouver (1808) noted that peat could become unstable and 'prodigious slips' were observed at different times on Dartmoor. Where peat became sufficiently deep (> 1 m) so that when it was saturated and a storm — probably exceeding 65 mm (Carling, 1986; Hudleston, 1930) — fell over a few hours, it failed and slid over the impermeable indurated layer beneath. The slides were often associated with slope convexities (Bowes, 1960; Carling, 1986; Werrity & Ingram, 1985) and could be in valley heads or seepage hollows (Crisp, Rawes & Welch, 1964; Bowes, 1960; Hudleston, 1930) and associated with piping. Slipping and tearing of the turf could have started at the convexities where the peat was thinnest and where cracking had occurred in the dry weather which often preceeded the storm. Water seeping down these cracks could have increased the pore-water pressure and lubricated the base of the peat.

The peat slides were often much larger than the slips, slides and flows described above in mineral soils or flushes. They were often tens of metres across and a hundred or more metres in length. In the north Pennines their extent varied from 0.3 to 2.7 ha (Carling, 1986); however, they covered a tiny fraction, about 0.4%, of the drainage basin in which they occurred. Nearby features similar in form to those described, but colonised by vegetation, were clearly visible for many years (Crisp, Rawes & Welch, 1964; Carling, 1986; Hudleston, 1930; Werrity & Ingram, 1985), one of the 1963 slides described by Crisp, Rawes & Welch (1964) being covered by only 31% vegetation 14 years later (Evans, unpublished), and still clearly visible from the air in the mid-1980s (Carling, 1986). Crisp, Rawes & Welch (1964) mapped the sites of three older peat slides, presumably still clearly visible on the ground, one dating from about 1870, the others 1689.

Hemingway & Sledge (1945) described a 'bog-burst' in peat in the North York Moors, still clearly visible six years later; it was larger in extent than the peat slides described above, about 4.7 ha. It was in a gently sloping large depression near the head of a stream and the 'burst' removed a section of the stream channel. A 'porridgy peat' had flowed down the valley from the burst, carrying with it some of the more coherent upper peat. The flow resulted in the ground surface falling about 1.5 m, which caused echelon cracking and the creation of a chaotic jumble of peat blocks. A heavy thunderstorm of about 46 mm rain was the trigger for the 'burst'.

A similar feature, also in a gently sloping depression (1° steepening to 4°) within a valley head, formed during a very localised intensive storm falling over 2 h in the central Pennines, May 19 1989. A rainfall gauge near the slide registered a questionable 193 mm (Acreman, 1989). The rainfall triggered widespread stream channel widening and deepening, as well as a rotational slide in head (Acreman, 1989; Evans, unpublished). The stream now heads in a 1 m high 'knick point' in indurated head, a few metres upslope of where the downvalley cracks delineating the margins of the burst meet the stream (Figure 12.4). The peat was about 4 m deep, and the burst was about 300 m long and 250 m wide. What appeared to have happened was that peat 'slurry' (Figure 12.5) flowed out from the base, so undercutting it, and this was followed by collapse of the peat into a chaotic jumble of arcuate slips. There is a similar, though vegetated, feature close by.

Although the slides and 'burst' described above are impressive when seen on the ground, they cover only a tiny fraction of the landscape within which they occur. And even though similar older features occur close by, they are still uncommon with regard to the total area of adjacent peat moor. In contrast, land users can increase the sensitivity to erosion of large areas of grassland, heath and moor. They do this not so much by triggering mass movements but by exposing soils to the weather and animals.

Disturbed Landscapes

Fire has been used for many centuries as a tool by lowland farmers to improve heaths for grazing sheep (Webb, 1986), as part of a cycle that removed fertility (as food) from the heaths to the arable (as dung). Often the exposed exhausted soil bore no traces of vegetation for

Figure 12.4   (A) Looking down into the 'bog-burst', on the Lancashire/West Yorkshire boundary, Boulsworth Hill, showing the stream headcut in the drained bog and chaotic mass of grassed blocks of peat which collapsed into the bog when the underlying saturated peat flowed down valley. (B) as (A) but showing original ground level of peat and echelon cracking of peat around the northern and eastern margins of the 'burst'

Figure 12.5   Peat slurry deposited on a stony overbank deposit, Walshaw Dean, about 1.5 km below the 'bog-burst'

years, as on Waltham Chase, Hampshire in the 18th century (White, 1788). The bare peaty or sandy soils were extremely prone to windblow, so creating a desert-like landscape, e.g. Bagshot Heath in the 18th century, estimated to cover about 40 000 ha (Defoe, 1724—26).

To improve the uplands for grazing by sheep and to increase the number of grouse reared and shot by increasing the area of moor under heather of variable age, the farmer and gamekeeper burned and drained the moors, and still does so.

Fire has been used deliberately in the uplands since the late 18th century and especially between the Victorian era, when it became fashionable to shoot grouse, and the Second World War. Exposing the mineral or peat soil surface to the weather led to widespread erosion in Scotland (McVean & Lockie, 1969). Sediment layers deposited in a valley head in the North York Moors contained Victorian pottery, and the layers were attributed by Curtis (1965) to erosion of burnt heather moors. Imeson (1971) demonstrated that more recent burning of the North York Moors had initiated erosion and higher sediment loads in streams.

Fires caused accidentally, often during times of drought, are known to have destroyed large areas of vegetation in the Peak District over the last 100 years (Tallis, 1981), parts of the central Pennines, the North York Moors (Alam & Harris, 1987) and South Wales. It is likely that many other still eroding patches of mineral and peat soil can be attributed to past fires.

Ditches have been dug to drain wet moorlands probably only since the mid-19th century. It is unlikely that they were/are efficient (Stewart & Lance, 1983). Where the gradient steepened, incision into the ditch floor occurred, followed by headward retreat. Widening of the channel took place by frost heave and wind erosion undercutting the banks and sheep, probably as now, broke down the overhanging turf and so the sides of the ditch retreated. There are many eroded ditches in the Pennines and Bowland Fells, for example, now recolonised by vegetation. But drains continue to be dug and erode (Stewart & Lance, 1983).

The uplands have been grazed by cattle and sheep for many centuries, mostly during the summer months only (Roberts, 1959). Until the late 18th century cattle were the main animals grazing the uplands of Wales (Hughes *et al.*, 1973), the south-west, the Pennines and Scotland

(Trow Smith, 1957). After this time, sheep became the main grazing animal, and the uplands are now generally stocked all year round. Formerly, sheep were shepherded, now they are generally allowed freely and selectively to graze the fells. Although grazing has induced vegetation changes it seems unlikely that erosion was initiated in earlier times. For instance, Vancouver (1808) noted on Dartmoor that even though the moorland sheep were the best he had even seen, the grass was still only half consumed at the beginning of November. The number of sheep in the uplands, except probably in northern Scotland (Coppock, 1971), was generally greater in 1965 than in 1875 when agricultural censuses started, and they were probably higher in 1875 than ever before. Numbers have continued to increase since then.

Erosion ascribed to overgrazing by sheep has been described from the Highlands (McVean & Lockie, 1969) and islands (Birnie & Hulme, 1990) of Scotland, the southern Uplands (Tivy, 1957), central (Thomas, 1965) and north Wales (Baker et al., 1979), and proved in the Peak District (Evans, 1977) and Lake District (Carr & Evans, unpublished). Grazing densities year round of about 0.5 ha per sheep was sufficient firstly to cause a decline in heather moorland, and then to initiate erosion in the form of sheep scars (Carr and Evans, unpublished) within the better-quality acid grasslands, which often had wavy hair grass as the dominant specie.

Changes in vegetation then increased the sensitivity of the upland landscape to erosion as trees were replaced by heath and then acid grassland, a sequence commonly recorded, for example, in the Howgill Fells (Cundill, 1976). Sensitivity can also be increased when changes in vegetation concentrate grazing on small patches of vulnerable ground, e.g. the expansion of bracken (Taylor, 1986) and mat grass, which are unpalatable to sheep.

The grasslands most vulnerable to erosion are usually on steep slopes with shallow ranker or brown podsol soils. In the Pennines sheep scars are characteristically found at the break of slope between the upper, gentler slopes carrying peat and the steeper slopes with mineral soils. In the Peak District there is an association between the extent of erosion at this convexity and that of the peat upslope (Anderson & Tallis, 1981). It is likely that the peat was exposed by gully incision and scar retreat upslope of the convexity. In parts of the Lake District, the Pennines and the Hebrides peat exposed by overgrazing is expanding rapidly in area at the moment. In exposed situations the very mobile peat surface resists recolonisation by vegetation (Tallis & Yalden, 1983).

On mineral soils scar extension downslope is particularly rapid where the slopes are underlain by scree, and may also act as initiating points for debris flows. In parts of the Lake District such flows seem to have become more prevalent in recent years. Scars formed by animals at convexities in valley floors can also initiate gullying, perhaps this is how the gullies in the New Forest (Tuckfield, 1965) started. Increased grazing pressures could lead to increased runoff (Evans, 1989/90), leading to increased sediment in streams as well as channel erosion, as Dearing, Elner & Happey-Wood (1981) have postulated in north Wales.

Where sheep have thinned the grass and created scars, rabbits have often exacerbated the erosion. Rabbits were introduced after the Norman invasion in 1066, but until the mid-18th century, except for within their extensive warrens, did little damage to the adjacent countryside (Sheail, 1971). The rabbits that escaped from the warrens were kept in check by the lack of food over winter and by predation and did not therefore lower geomorphic thresholds. However, the agricultural revolution ensured that food was available over winter, and the shooting and trapping of predators by gamekeepers meant that breeding was insufficiently checked. There was an explosive expansion of the rabbit population, followed by overgrazing and exposure of bare soil. These soils were liable to blow, especially if they were sandy

(Sheail, 1971). On the southern chalk downlands rabbits were so numerous that farms were abandoned as all crops were eaten down (Sheail, 1971). If the enormous populations of sheep on the downs at that time (e.g. Cobbet, 1830; Defoe, 1724–26; Young, 1813a) were allowed to graze the escarpments and steep valley sides, the combination of rabbits and sheep could have led to severe erosion.

Photographs taken in the 1930s and 1940s show severely eroded steep slopes on chalk. In the early 1950s myxomatosis reduced the rabbit population to about 1% of its former size (Sheail, 1971), and revegetation of bare soil occurred. But in the 1980s the combination of a series of mild winters together with green overwintering crops (autumn-sown cereals and oilseed rape) as a food source has resulted in the massive increase in the rabbit population (Long, 1990). Erosion of chalk escarpments, for example, near Ivinghoe, Buckinghamshire and Ladle Hill, Hampshire has been initiated in the 1980s (Evans, unpublished), especially where sheep were also present.

On lowland heaths, the combination of fire, rabbits and sheep increased the land's sensitivity to erosion, and blowing of the soil often resulted, e.g. in Norfolk (Young, 1804a) and Suffolk (Young, 1813b). Once the heaths were enclosed for agriculture around the turn of the 19th century, or became less intensively used, the land became less sensitive to erosion.

People too can directly alter the sensitivity of the land to erosion by trampling the vegetation cover and exposing soil. Many ancient routeways have eroded deeply, e.g. in Devon and on the chalk, and wherever subsoils are erodible. Tracks recently opened up in the uplands for access for farmers, stalkers, shooters and fishers can become severely eroded (e.g. Duck, 1985; Watson, 1984a). Runoff channelled from roadside drains, cattle or farm tracks onto slopes has caused gullying and extended the stream network (e.g. Fairbairn, 1967; Ovenden & Gregory, 1980).

Footpath erosion, though, is a modern phenomenon of the tourist industry and has been widely reported throughout the uplands of England, Wales and Scotland (e.g. Bayfield, 1973; Shimwell, 1981; Watson, 1984b, 1985). Footpaths were widest where they were on steep unstable slopes or on gentle slopes where wet peat and bog vegetation occurred; deep gullies often formed where erodible, coarse-textured stony head was exposed. Blowouts occurred in sand dunes where the grass became worn away by trampling. In the Cairngorms damage caused by skiing and associated activities (Bayfield, 1974; Watson, 1985) has taken many years to grass over at high altitudes (Bayfield, 1980). Bayfield, Urqhart & Cooper (1981) noted that though walkers damaged lichens and so allowed the weather to act on the soil, trampling damage was much less than that caused by animal tracks and scrapes. All these features could disfigure the landscape, but the area covered by them was tiny compared with the patches of eroding soil initiated by overgrazing or pollution.

The most spectacular erosion in Britain is in the peat uplands, especially in the Peak District. Although formerly considered to be due to natural causes (e.g. Mayfield & Pearson, 1972), the evidence now strongly points to the industrialist and the farmer whose combined actions increased greatly the sensitivity of the peat uplands to erosion.

Some erosion of the moors in the Peak District, and probably of those elsewhere, started between 1000 and 1200 AD when the climate was more favourable to erosion and the margins of the moors were being cleared of forest. However, much of it dates from after 1750 AD with the onset of industrial pollution, which killed the protective mosses (Tallis, 1989/90). Defoe (1724–26), when he ascended The Cheviot in the early 18th century, noted the top was 'a smooth and . . . a most pleasant Plain'; it is not now, the peat is severely eroded. The Peak District, with its lead mills and sulphur mills (Farey, 1811), was also surrounded by industrial towns east and west, as well as within the Pennines, from which more sulphurous fumes belched out. Hence, these were the moors most affected by pollution and where erosion has been most marked (Tallis, 1969).

At the beginning of the 19th century, Farey (1811) noted that the mosses present on the Peak District moors were 'a dead and slowly decaying mass, thinly sprinkled with heaths, [and] aquatic grasses', and the peat was 'crossed in all directions by gullies'. By 1824 the vegetation of the northern hills of the Peak seems to have changed and was described as 'heathy' (Rhodes, 1824), probably mostly heather with cotton grass. This change probably led to the moors becoming less susceptible to erosion.

At present cotton grass and heather moors are resistant to erosion; even during a rare large storm peat headcuts in gullies retreated less than 0.3 m (Evans, unpublished). For gullies to have extended rapidly the vegetation at the onset of erosion probably had to comprise mosses. It has been noted in the field (Evans, unpublished) that Sphagnum was easily stripped from a mossy cushion by flood flow. Incision along depressions must have taken place rapidly, and tapping the pools of the gentle divides further speeded up gullying. The drying out of the pools led to frost, rain, wind and animals disturbing and eroding the surface, so producing the devastated landscape we see now. Present-day erosion rates of the peat in the northern Peak District are sufficient to account for the volumes eroded since the late 1700s (Burt & Labadz, 1990). A similar scenario is envisaged for other upland moors. A combination of circumstances conspired therefore to drastically increase the moors' sensitivity to erosion.

In many parts of the uplands gullies later stabilised and their floors became clothed with vegetation (Figure 12.6) when erosion slowed down on reaching the more resistant mineral

Figure 12.6 Ridge between Littondale and Langstrothdale, Yorkshire Dales, showing stabilised grassed floors of gullies cut through peat to mineral soil; on the crest is a zone of eroding, occasionally hagged peat. Where the streams meet, and along the flanks of the valley, are eroding scars in mineral soil. The scars may have been initiated by slips formed during a large storm, or may be due to overgrazing or, more likely, they are a result of a combination of the two, as sheep are constantly disturbing and extending the bare soil surface

subsoil below. From then on the gully walls slowly retreated, and the rate at which the unstable peat surface became colonised by vegetation depended on the interaction between disturbance by sheep and the weather (Evans, 1989/90).

Industrialisation of Britain led to erosion of slopes where fumes from smelters killed the vegetation cover, e.g. in the Swansea valley, and probably elsewhere. Mining has also disrupted the vegetation cover in many parts of the uplands and caused the formation of mobile screes. Coal mine and other mineral waste spoil heaps have been and are vulnerable to water erosion (Howe, 1955; Miller, 1951). In England and Wales, for example, 0.3% of the surface is covered by restored opencast coal spoil (SSEW, 1983); much of this is at risk of erosion, especially where vegetation takes a long time to grow or where sheep have overgrazed the pastures established on slopes.

AFFORESTATION AND SENSITIVITY TO EROSION

From the time of Elizabeth I, if not earlier, the lack of woodland in the British countryside has been decried, primarily because there was insufficient good-quality wood, especially of oak, for building warships. In some localities there was also a scarcity of fuelwood (e.g. Pitt, 1813; Young, 1813c). Before the 17th century 'plantations were rare and small' (Rackham, 1986), and they had little impact on the landscape until the Forestry Commission was founded in 1919 to produce export substitution softwoods; there was a further boost to afforestation after 1945 when incentives were given to encourage individuals and companies to plant trees. Since the Second World War about 1.4×10^6 ha of conifers have been planted (Best & Coppock, 1962; Forestry Commission, 1989) — about 6% of Britain.

Until after the Second World War most planting was done by hand and the sensitivity of the land to erosion was little altered. Most of the postwar afforestation has been in large blocks in the uplands, often on gently sloping peat moors, which were not sensitive to erosion. With the introduction of large-scale machinery to plough drainage ditches — which was often done in an insensitive manner in a downslope direction, with ditches being led into streams — the sensitivity of the landscape to erosion has been drastically increased (Burt, Donohue & Vann, 1983; Ferguson & Stott, 1987; Murgatroyd & Ternan, 1983; Newson, 1980b; Robinson, 1980; Soutar, 1989; Stretton, 1984; Tuckfield, 1980). Ditches, in or reaching mineral soil, with gradients as low as 2° eroded — and more rapidly than those wholly in peat (Newson, 1980b), as these had smoother channel surfaces and therefore less turbulent flow. Erosion was severe for a number of years until the ditches became vegetated over, and even then sediment loads in streams remained higher than they had been prior to afforestation. Many forestry roads and tracks also rilled and gullied severely and could continue to do so for many years.

SENSITIVITY OF ARABLE LAND TO EROSION

Present-day sensitivity

Arable land is the most sensitive of all landscapes in Britain to erosion because the farmer exposes the soil to the weather most years. Bare soil, especially when it has a smooth surface at drilling time, is vulnerable to wash and windblow. In some years in England and Wales as much as about 25% of the land most sensitive to rilling has eroded (Evans, 1988). Boardman

(Chapter 13 this volume) discusses in detail the sensitivity of arable land to water erosion, with specific reference to the chalk Downlands; here only broad outlines are given.

Landscapes that are most sensitive to erosion are sandlands; fields can erode by wind or water in most years. These light-textured, freely draining soils are easily worked, so a wide range of crops can be drilled at various times of year, thus increasing the risk of erosion (Evans, 1988, 1990b). However, the soils most sensitive to rilling have a high content of coarse silt or fine sand (Evans, 1988, 1990c); fine sands and peats are most vulnerable to blowing (Evans & Cook, 1986). Valley floors, especially on heavier-textured soils, are sensitive to water erosion, whereas on lighter soils slopes below convexities are most at risk (Evans, 1988; Evans & Cook, 1986). Gradients of slopes in water-erodible landscapes are generally greater than 3° and relief within fields is greater than 5 m (Evans, 1990c).

Land drilled in autumn for winter cereals is sensitive to water erosion as soils can be saturated for much of the winter period, thus maximising the chances of runoff. However, rilling occurs disproportionately more and faster in sugar, beet and potatoes (Evans, 1988). This is because these crops take a long time to cover and protect the ground; spring and early summer storms are often more intense than those in winter; and compacted tractor wheelings down which water can flow are more frequent. Crops most sensitive to blowing are those drilled into very smooth tilths, often in spring, and which take a long time to cover the ground, such as sugar beet, carrots and onions.

Recent improvements in agricultural techniques, technology and profitability have increased the sensitivity of the landscape to erosion. For example, winter cereals can now be grown in areas — often wetter — that were not previously considered, because weeds could not be controlled there and profitability was low. Erosion has spread with winter cereals to the west of England, Wales and Scotland (Speirs & Frost, 1985; Watson, 1987). Increasing the areas of fields by amalgamating them has increased the risk of both water (Evans, 1988) and wind erosion. Irrigation can also cause severe erosion.

Sensitivity in the Past

Where erosion is most evident now, so it was in the past (Evans, 1990a). In general, the sensitivity to erosion of the landscape when under arable has changed little. Erosion, mostly by water, must not have been unusual in the 18th and 19th centuries, from evidence given in diaries and reviews of agricultural practices (e.g. Kilvert, 1870—79; Pitt, 1813; Stone, 1794; Vancouver, 1808, 1810; White, 1788; Young, 1804a, 1804b, 1813a, 1813b). In particular, sandlands were known to wash and blow (Stone, 1794; Young, 1804a, 1813a, 1813b), and chalk downland was susceptible to water erosion (Vancouver, 1810). The extent of land down to arable has fluctuated. Arable land has been most extensive and most sensitive to erosion when population pressures have been highest, particularly in Romano-British times, the medieval period just prior to the Black Death, and today (Evans, 1990a).

More land is sensitive to erosion now, probably, than in earlier times, because more is under arable and much of the land brought into arable in the last 250 years is vulnerable to erosion. Although some ridge and furrow indicative of medieval agriculture is still visible in long, unploughed grasslands in the lowlands, especially on heavy soils in the Midlands, and at altitudes higher than those to which arable presently extends (Parry, 1978), much other formerly 'waste' land has been brought into cultivation since 1750.

As the population outstripped Britain's capacity to grow food, sometime between 1750 and 1800, so the 'wastes' that had previously been an integral part of the farming system to maintain fertility of the arable land were enclosed and brought into production, not always successfully. Much of this land was sandy heath, e.g. in the Midlands, Norfolk and Suffolk, and very sensitive to erosion. Much of the arable land in Scotland, some of which is sensitive to erosion (Speirs & Frost, 1985; Watson & Evans, 1991) was probably enclosed around this time. The coarser, less-fertile sands not enclosed for arable during the great enclosure movement, e.g. the Suffolk Sandlings, have been brought into arable since the Second World War.

Much low-lying ground has been drained for agriculture, continuing the trend started in part of the Fens of East Anglia in the mid-17th century. These peat lands often wasted rapidly, a well-known phenomenon in the Fens (Hutchinson, 1980), and in the Lancashire mosses the ground surface fell 1.4 m in a very short time (Holt, 1795). This land was also susceptible to blowing, as were the sandy soils of the Trent lowlands and the more recently drained Vale of York.

Stamp (1955) gave figures for the area of England and Wales covered by arable in the early (18−24%; after Maxey, 1601) and late (24%; after King, 1698) 17th century, as well as the early 19th century (30%; after Comber, 1808). These compare with about 35% now (MAFF, 1989).

After the agricultural revolution of the 18th century, to the high farming period of the 1870s (Coppock, 1971), arable and short-term grass covered about 31% of Britain, compared to about 32% in 1966 (MAFF, 1968) and about 29.5% in the 1980s (MAFF, 1989); it was only about 21% in the depths of the agricultural recession in 1938 (MAFF, 1968). Until after the Second World War much arable land was farmed on a rotational basis, comprising grass, which protected the land from erosion, and crops, which were more sensitive. Since then, some of this grass has been replaced by crops, though the proportion varies throughout the country.

Rates of erosion have probably changed over time, as crops of different sensitivity to erosion have been introduced, or as the proportions of land sown to different crops have varied. The growing of potatoes as a field crop in the 18th and 19th centuries (Salaman, 1949), and sugar beet in the 1920s (MAFF, 1968), would both increase rates of erosion.

Prior to enclosure, arable land was often left fallow over winter and summer one year in two or three, to allow it to recover its fertility. This land was ploughed and harrowed three or four times to control weeds, resulting in a rough surface resistant to erosion. But unstable soils would slake to give a smoother, more erodible surface. Unless weeds protected the soil, much more of the land than now would be vulnerable to summer thunderstorms as presently very little of the land ($<1\%$) is bare fallow during summer. Summer storms could be devastating — washing away the soil of the fallows (White, 1788) or removing turnips from the ground (Kilvert, 1870−71) — and still can be (Howe, 1955; Oakley, 1945).

The crops presently most sensitive to rilling in summer, namely sugar beet, potatoes and vegetables, which are mainly vulnerable to about mid-June, and reseeded grass fields, which are often drilled in late summer, rarely cover more than 10% of the landscape. A storm around Balsham, Cambridgeshire, 19 June 1985, when 91 mm of rain fell in 3 h caused remarkably little erosion. Much of the land was drilled to cereals, which covered the ground well, and erosion was rarely seen in these crops; where it was, rates were low. For example, from the eroded part of a winter wheat field, which had about 80% ground cover, 0.8 m^3 ha^{-1} of soil was washed away. However, in the occasional field of sugar beet, which at that time had a minimum of between 20 and 30% plant cover, up to 100 m^3 ha^{-1} of soil, but often less, was washed off the slopes; soil losses diminish rapidly when cover is $>30\%$ (Elwell & Stocking, 1976).

CONCLUSIONS

The sensitivity of the British landscape to erosion largely depends on its land use, and hence its vegetation cover. As the woods have slowly been cleared over the last 5000 years, to be replaced by pasture and then arable, so the land's sensitivity to erosion has increased.

Climate has changed over time since woodland clearance began (Lamb, 1982). But, except in the uplands in the Little Ice Age when mass movements of vegetated as well as scree slopes could have become more frequent, these changes, in comparison to those in land use, would have had little impact on altering geomorphological thresholds and the sensitivity of the land to erosion. Wetter or more stormy climates could have exacerbated washing of slopes of arable fields, e.g. from about 1000 BC to 0 AD, 1200 to 1300 AD and 1500 to 1750 AD (Lamb, 1982). Climatic change to wetter conditions about 3000 BC and 1000 BC could have triggered peat growth, leading eventually to instability and erosion, but the major phase of moorland erosion was related to pollution caused by the Industrial Revolution.

More of Britain is more sensitive to erosion than ever before. There is a larger area under arable than previously, and under forestry too, which instead of protecting the land as the 'wildwood' (Rackham, 1986) did, has been planted in such a way as to cause erosion. The use of modern technology means that land less suited to agriculture and forestry is so used instead of being under grassland, heath or moor, which are less sensitive to erosion. But even these are under threat from overgrazing, fire and tourism.

Water erosion of undulating arable fields may occur on average about once every 3–10 years, depending on soil type (Evans, 1988), and in some years can affect up to an estimated 6% of lowland England and Wales. Wind erosion of sandy or peaty arable soils may have a similar frequency, but less is known about its extent, frequency and magnitude (Evans & Cook, 1986). Erosion of arable land is generally obliterated by the farmer, so its extent and severity are not often realised. Erosion of exposed slopes and plateaux in the uplands can occur all year round, and may have evolved into spectacular forms. The landscapes most sensitive to erosion in Britain are not extensive: about 24% of England and Wales, for example, is considered to be at moderate or higher risk of erosion (Evans, 1990b).

Much of Britain is not sensitive to erosion, as it is clothed with a strong protective cover of vegetation, or is gently sloping, or has a soil resistant to erosion, or some combination of these. These landscapes cover 38% of England and Wales (Evans, 1990b).

Where slopes are steeper but the land is vegetated, mass movements will occur only when large storms, of prolonged duration and/or great intensity occur. These storms are likely to have a recurrence interval of longer than 50 years. Generally, only very small parts of the landscape will be affected, but the effects may remain obvious for many years. Though the bare soil scars will eventually be recolonised by vegetation, this may take more than 15 years in exposed situations; the scars and footslope slump features, for example, will remain identifiable.

If climate continues to ameleriorate, land use changes could make large parts of Britain more sensitive to water erosion (Boardman et al., 1990). Lowland that is presently under cereals, for instance, could be converted to growing more warmth-loving crops such as maize or vines; erosion under both these crops could be greater than under cereals. Grassland could be converted to arable, especially winter cereals. And if winter rainfall increased, as could the frequency of summer storms, more erosion could take place in winter cereals and spring-sown crops such as sugar beet. Previous work (Evans, 1990c) suggested that as erosion rates increase with increased rainfall (Boardman et al., 1990), the numbers of fields that washed could increase rapidly. A response to warmer summers could be to irrigate a greater proportion of the crops, and this could also increase the extent of erosion. Such intensification of land use could increase the area of land in England and Wales at moderate or higher risk of erosion from 24 to 46% (Evans, 1990b).

Identifying those areas that could be most sensitive to erosion in the future (Evans, 1990b), and the processes and thresholds operating there, could enable us to combat erosion and its impacts.

In lowland arable Britain, areas sensitive to erosion have slowly but surely had their topsoils stripped off. This has resulted in a lowering of the land surface, but with little alteration of its form because farmers plough out rills and infill gullies. In the uplands, channel extension and gullying have altered slope form, but surface lowering will take place rarely and will only affect small areas. Much of the landscape of Britain under its present climate, and that which has prevailed for much of the Holocene, is not very sensitive to erosion and hence its form has changed little, even taking into account the impacts of the land user.

ACKNOWLEDGEMENTS

I thank Dr John Boardman for his helpful advice.

REFERENCES

Acreman, M. C. (1989). *The rainfall and flooding on 19 May 1989 in Calderdale, West Yorkshire*, Report, Institute of Hydrology, Wallingford.

Alam, M. S. and Harris, R. (1987). Moorland soil erosion and spectral reflectance. *International Journal of Remote Sensing*, **8**, 593–608.

Anderson, E. W. and Cox, N. J. (1984). The relationship between soil creep rate and certain controlling variables in a catchment in upper Weardale, northern England. In Burt, T. P. and Walling, D. E. (eds), *Catchment Experiments in Fluvial Geomorphology*. Geobooks, Norwich, pp. 419–430.

Anderson, P. and Tallis, J. (1981). Present moorland vegetation. In Phillips, J., Yalden, D. and Tallis, J. (eds), *Moorland Erosion Study. Phase 1 Report*, Peak Park Joint Planning Board, Bakewell, pp. 52–64.

Anderson, M. G. and Calver, A. (1977). On the persistence of landscape features formed by a large flood. *Transactions of the Institute of British Geographers*, **NS2**, 243–254.

Arkell, W. J. (1955). Geological results of the cloudburst in the Weymouth district, 18th July 1955. *Proceedings of the Dorset Natural History and Archaeological Society*, **77**, 90–95.

Baker, C. F., Morgan, R. P. C., Brown, I. W., Hawkes, D. E. and Ratcliffe, J. B. (1979). *Soil Erosion Survey of the Moel Famau Country Park*, Clwyd County Council Planning and Estates Department, Country Park Research Report, National College of Agricultural Engineering, Occasional Paper no 7.

Baird, P. D. and Lewis, W. V. (1957). The Cairngorm floods, 1956: summer solifluction and distributary formation. *Scottish Geographical Magazine*, **73**, 91–100.

Ballantyne, C. K. (1986a). Late Flandrian solifluction on the Fannich Mountains, Rosshire. *Scottish Journal of Geology*, **22**, 395–406.

Ballantyne, C. K. (1986b). Landslides and slope failures in Scotland: a review. *Scottish Geographical Magazine*, **102**, 134–150.

Ballantyne, C. K. and Eckford, J. D. (1984). Characteristics and evolution of two relict talus slopes in Scotland. *Scottish Geographical Magazine*, **100**, 20–33.

Bayfield, N. G. (1973). Use and deterioration of some Scottish hill paths. *Journal of Applied Ecology*, **10**, 639–648.

Bayfield, N. G. (1974). Burial of vegetation by erosion near ski lifts on Cairngorm, Scotland. *Biological Conservation*, **6**, 246–251.

Bayfield, N. G. (1980). Replacement of vegetation on disturbed ground near ski lifts in the Cairngorm Mountains, Scotland. *Journal of Biogeography*, **7**, 249–260.

Bayfield, N. G., Urqhart, U. H. and Cooper, S. M. (1981). Susceptibility of four species of Cladonia to disturbance by trampling in the Cairngorm Mountains, Scotland. *Journal of Applied Ecology*, **18**, 303–310.

Best, R. H. and Coppock, J. T. (1962). *The Changing Use of Land in Britain*, Faber & Faber, London.

Beven, K., Lawson, A. and McDonald, A. (1978). A landslip/debris flow in Bilsdale, North York Moors, September 1976. *Earth Surface Processes*, **3**, 407–419.

Birks, H. J. B. (1988). Long-term ecological changes in the British uplands. In Usher, M. B. and Thompson, D. B. A. (eds), *Ecological Changes in the Uplands*, Blackwell Scientific, London, pp. 37–56.

Birnie, R. V. and Hulme, P. D. (1990). Overgrazing of peatland in Shetland. *Scottish Geographical Magazine*, **106**, 28–36.

Bleasdale, A. and Douglas, C. K. M. (1952). Storm over Exmoor on August 15, 1952. *Meterological Magazine*, **81**, 353–367.

Boardman, J., Evans, R., Favis-Mortlock, D. T. and Harris, T. M. (1990). Climate change and soil erosion on agricultural land in England and Wales. *Land Degradation and Rehabilitation*, **2**, 95–101.

Bowes, D. R. (1960). A bog burst in the Isle of Lewes. *Scottish Geographical Magazine*, **76**, 21–23.

Burt, T. and Labadz, J. (1990). Blanket peat erosion in the southern Pennines. *Geography Review*, **3**, 31–35.

Burt, T. P., Donohue, M. A. and Vann, A. R. (1983). The effect of forestry drainage operations on upland sediment yields: the results of a storm-based study. *Earth Surface Processes and Landforms*, **8**, 339–346.

Caine, T. N. (1963). Movement of low angle scree slopes in the Lake District, northern England. *Revue De Geomorphologie Dynamique*, **14**, 171–177.

Carling, P. A. (1986). Peat slides in Teesdale and Weardale, northern Pennines, July 1983: Description and failure mechanisms. *Earth Surface Processes and Landforms*, **11**, 193–206.

Cobbett, W. (1830; repr. 1967). *Rural Rides*, Penguin Books, Harmondsworth.

Common, R. (1954). A report on the Lochaber, Appin and Benderloch floods, May 1953. *Scottish Geographical Magazine*, **70**, 6–20.

Conway, V. M. (1954). Stratigraphy and pollen analysis of southern Pennine blanket peats. *Journal of Ecology*, **42**, 117–147.

Coppock, J. T. (1971). *An Agricultural Geography of Great Britain*, Bell, London.

Crisp, D. T., Rawes, M. and Welch, D. (1964). A Pennine peat slide. *Geographical Journal*, **130**, 519–524.

Cundill, P. R. (1976). Late Flandrian vegetation and soils in the Carlingill valley, Howgill Fells. *Transactions of the Institute of British Geographers*, **NS1**, 301–309.

Cundill, P. R. (1977). The distribution, age and formation of blanket peat on the North York Moors. *Proceedings of the North of England Soils Discussion Group*, **9**, 25–29.

Curtis, L. F. (1965). Soil erosion on Levisham Moor, North York Moors. In *Rates of Erosion and Weathering*, Institute of British Geographers, Bristol, pp. 19–21.

Dearing, J. A., Elner, J. K. and Happey-Wood, C. M. (1981). Recent sediment flux and erosional processes in a Welsh upland lake catchment based on magnetic susceptibility measurements. *Quaternary Research*, **16**, 356–372.

Defoe, D. (1724–26; repr. 1927). *A Tour Thro' the Whole Island of Great Britain, Vols 1 and 2*, Davies, London.

Dimbleby, G. W. (1962). The development of British heathlands and their soils, *Oxford Forestry Memoir*, No. 23, Clarendon Press, Oxford.

Doornkamp, J. C. (1990). Landslides in Derbyshire. *East Midland Geographer*, **13**, 33–62.

Duck, R. W. (1985). The effect of road construction on sediment deposition in Loch Earn, Scotland. *Earth Surface Processes and Landforms*, **10**, 401–406.

Duckworth, J. A. (1969). Bowland Forest and Pendle floods. *Association of River Authorities Yearbook*, pp. 81–90.

Elwell, H. A. and Stocking, M. A. (1976). Vegetal cover to estimate soil erosion hazard in Rhodesia. *Geoderma*, **15**, 61–70.

Evans, J. G. (1975). *The Environment of Early Man in the British Isles*, Elek, London.

Evans, R. (1974). A study of selected erosion processes acting on slopes in a small drainage basin. Unpublished Ph.D. Thesis, Sheffield University.

Evans, R. (1977). Overgrazing and soil erosion on hill pastures with particular reference to the Peak District. *Journal of the British Grassland Society,* **32**, 65–76.

Evans, R. (1988). Water Erosion in England and Wales 1982–1984. Report for Soil Survey and Land Research Centre, Silsoe.

Evans, R. (1989/90). Erosion studies in the Dark Peak. *Proceedings of the North of England Soils Discussion Group,* **24**, 39–61.

Evans, R. (1990a). Soil erosion: its impact on the English and Welsh landscape since woodland clearance. In Boardman, J., Foster, I. D. L. and Dearing, J. A. (eds), *Soil Erosion on Agricultural Land,* Wiley, Chichester, pp. 231–254.

Evans, R. (1990b). Soils at risk of accelerated erosion in England and Wales. *Soil Use and Management,* **6**, 125–131.

Evans, R. (1990c). Water erosion in British farmers' fields — some causes, impacts, predictions. *Progress in Physical Geography,* **14**, 199–219.

Evans, R. and Cook, S. (1986). Soil erosion in Britain. *SEESOIL,* **3**, 28–59.

Fairbairn, W. A. (1967). Erosion in the Findhorn valley. *Scottish Geographical Magazine,* **83**, 46–52.

Farey, J. (1811). *A General View of the Agriculture and Minerals of Derbyshire,* MacMillan, London.

Ferguson, R. I. and Stott, T. A. (1987). Forestry effects on suspended sediment and bedload yields in the Balquhidder catchment, central Scotland. *Transactions of the Royal Society of Edinburgh: Earth Sciences,* **78**, 379–384.

Finlayson, B. (1981). Field measurements of soil creep. *Earth Surface Processes and Landforms,* **6**, 35–48.

Forestry Commission (1989). *Forestry Facts and Figures 1988–1989.* Forestry Commission.

Gifford, J. (1953). Landslides on Exmoor caused by the storm of 15th August, 1952. *Geography,* **38**, 9–17.

Harvey, A. M. (1986). Geomorphic effects of a 100 year storm in the Howgill Fells, northwest England. *Zeitschrift für Geomorphologie,* **30**, 71–91.

Harvey, A. M. and Renwick, W. H. (1987). Holocene alluvial fan and terrace formation in the Bowland Fells of northwest England. *Earth Surface Processes and Landforms,* **12**, 249–257.

Harvey, A. M., Oldfield, F. and Baron, A. F. (1981). Dating of post-glacial landforms in the central Howgills. *Earth Surface Processes and Landforms,* **6**, 401–412.

Hemingway, J. E. and Sledge, W. A. (1945). A bog-burst neat Danby-in-Cleveland, *Proceedings of the Leeds Philosophical and Literary Society, Scientific Section,* **4**, 276–284.

Holt, J. (1795; repr. 1969). *A General View of the Agriculture of the County of Lancaster,* David & Charles Reprints, Newton Abbot.

Hoskins, W. G. and Stamp, L. D. (1963). *The Common Lands of England and Wales,* Collins, London.

Howe, G. M. (1955). A South Wales thunderstorm. *Meteorological Magazine,* **84**, 281–286.

Hudleston, F. (1930). The cloudburst on Stainmore, Westmorland, June 18, 1930. *British Rainfall,* 287–292.

Hughes, R. E., Dale, J., Ellis Williams, I. and Rees, D. I. (1973). Studies in sheep population and environment in the mountains of north-west Wales. I. The status of sheep in the mountains of North Wales since Mediaeval times. *Journal of Applied Ecology,* **10**, 113–132.

Hutchinson, J. N. (1980). The record of peat wastage in the East Anglian Fenlands at Holme Post, 1848–1978 AD. *Journal of Ecology,* **68**, 229–249.

Imeson, A. C. (1971). Heather burning and soil erosion on the North Yorkshire Moors. *Journal of Applied Ecology,* **8**, 537–542.

Innes, J. L. (1983). Lichonometric dating of debris-flow deposits of the Scottish Highlands. *Earth Surface Processes and Landforms,* **8**, 579–588.

Innes, J. L. (1985). Magnitude-frequency relations of debris flows in northwest Europe. *Geografiska Annaler,* **67A**, 23–33.

Jacobi, R. M., Tallis, J. H. and Mellars, P. A. (1976). The southern Pennines Mesolithic and the ecological record. *Journal of Archaeological Science,* **3**, 307–320.

Jenkins, A., Ashworth, P. J., Ferguson, R. I., Grieve, I. C., Rowling, P. and Stott, T. A. (1988). Slope failures in the Ochil Hills, Scotland, November 1984. *Earth Surface Processes and Landforms,* **13**, 69–76.

Johnson, R. H. (1980). Hillslope stability and landslide hazard — a case study from Longdendale, north Derbyshire, England. *Proceedings of the Geological Association*, **91**, 315–325.

Kilvert, F. (1870–79; repr. 1977). *Kilvert's Diary. 1870–79*, Plomer, W. (ed.), Cape, London.

Lamb, H. H. (1982). *Climate, History and the Modern World*, Methuen, London.

Learmonth, A. T.A. (1950). The floods of 12th August, 1948, in south-east Scotland. *Scottish Geographical Magazine*, **66**, 147–153.

Long, E. (1990). Rabbits. Revivalists run riot. *Farmers' Weekly*, 6 July, 65–68.

MAFF (1968). *A Century of Agricultural Statistics. Great Britain 1866–1966*, HMSO, London.

MAFF (1989). *Agricultural Statistics UK 1988*, HMSO, London.

Mayfield, B. and Pearson, M. C. (1972). Human interference with the north Derbyshire blanket peat. *East Midland Geographer*, **5**, 245–251.

McVean, D. N. and Lockie, J. D. (1969). *Ecology and Land Use in Upland Scotland*, Edinburgh University Press, Edinburgh.

Miles, J. (1985). The pedogenic effects of different species and vegetation types and the implications of succession. *Journal of Soil Science*, **36**, 571–584.

Miller, A. A. (1951). Cause and effect in a Welsh cloudburst. *Weather*, **6**, 172–179.

Moore, P. D. (1968). Human influence upon vegetational history in north Cardiganshire. *Nature*, **217**, 1006–1009.

Moore, P. D. and Chater, E. H. (1969). The changing vegetation of west-central Wales in the light of human history. *Journal of Ecology*, **57**, 361–379.

Murgatroyd, A. L. and Ternan, J. L. (1983). The impact of afforestation on stream bank erosion and channel form. *Earth Surface Processes and Landforms*, **8**, 357–369.

Newson, M. D. (1980a). The geomorphological effectiveness of floods — a contribution stimulated by two recent events in mid-Wales. *Earth Surface Processes and Landforms*, **5**, 1–16.

Newson, M. (1980b). The erosion of drainage ditches and its effect on bed-load yields in mid-Wales: reconnaissance case studies. *Earth Surface Processes and Landforms*, **5**, 275–290.

Oakley, K. P. (1945). Some geological effects of a 'cloudburst' in the Chilterns. *Records of Buckinghamshire*, **14**, 265–276.

Ovenden, J. C. and Gregory, K. J. (1980). The permanence of stream networks in Britain. *Earth Surface Processes and Landforms*, **5**, 47–60.

Parry, M. L. (1978). *Climate Change, Agriculture and Settlement*, Dawson, Archon Books, Folkestone.

Pennington, W. (1970). Vegetation history in the north-west of England: a regional synthesis. In Walker, D. and West, R. G. (eds), *Studies in the Vegetational History of the British Isles*, Cambridge University Press, Cambridge, pp. 41–79.

Pennington, W., Haworth, E. Y., Bonny, A. P. and Lishman, J. P. (1972). Lake sediments in northern Scotland. *Philosophical Transactions of the Royal Society*, **B264**, 191–294.

Pitt, W. (1813; repr. 1969). *General View of the Agriculture of Worcester*, David & Charles Reprints, Newton Abbot.

Rackham, O. (1986). *The History of the Countryside*, Dent, London.

Richards, K. S. (1981). Evidence of Flandrian valley alluviation in Staindale, North York Moors. *Earth Surface Processes and Landforms*, **6**, 183–186.

Richards, K. S., Peters, N. R., Robertson-Rintoul, M. S. E. and Switsur, V. R. (1986). Recent valley sediments in the North York Moors: evidence and interpretation. In Gardiner, V. (ed.), *International Geomorphology, Part 1*, Wiley, Chichester, pp. 869–883.

Rhodes, E. (1824). *Peak Scenery*, London.

Roberts, R. A. (1959). Ecology of human occupation and land use in Snowdonia. *Journal of Ecology*, **47**, 317–323.

Robinson, M. (1980). The effect of pre-afforestation drainage on the streamflow and water quality of a small upland catchment. Institute of Hydrology Report no. 73.

Rogers, N. W. and Selby, M. J. (1980). Mechanism of shallow translational landsliding during summer rainstorms, New Zealand. *Geografisker Annaler*, **62A**, 11–21.

Salaman, R. N. (1949). *The History and Social Influence of the Potato*, Cambridge University Press, Cambridge.

Selby, M. J. (1976). Slope erosion due to extreme rainfall: a case study from New Zealand. *Geografiska Annaler*, **58A**, 131–138.

Sheail, J. (1971). *Rabbits and their History*, David & Charles, Newton Abbot.

Shimwell, D. (1981). Footpath erosion. In Phillips, J., Yalden, D. and Tallis, J. (eds), *Peak District Moorland Erosion Study. Phase 1 Report*, Peak Park Joint Planning Board, Bakewell, pp. 160–170.

Simmons, I. G. (1963). The blanket bog of Dartmoor. *Report and Transactions of the Devon Association for the Advancement of Science*, **95**, 180–196.

Simmons, I. G. and Cundill, P. R. (1974). Late Quaternary vegetational history of the North York Moors. I. Pollen analyses of blanket peats. *Journal of Biogeography*, **1**, 159–165.

Slaymaker, H. O. (1972). Patterns of present subaerial erosion and landforms in mid-Wales. *Transactions of the Institute of British Geographers*, **55**, 47–68.

Smith, R. T. and Taylor, J. A. (1969). The post-glacial development of vegetation and soils in northern Cardiganshire. *Transactions of the Institute of British Geographers*, **48**, 75–96.

Soutar, R. G. (1989). Afforestation, soil erosion and sediment yields in British fresh waters. *Soil Use and Management*, **5**, 82–86.

Speirs, R. B. and Frost, A. (1985). The increasing incidence of accelerated soil water erosion on arable land in the east of Scotland. *Research and Development in Agriculture*, **2**, 161–167.

SSEW (1983). *Soil Map of England and Wales*, Soil Survey of England and Wales, Harpenden.

Stamp, L. D. (1955). *Man and the Land*, Collins, London.

Statham, I. (1976). Debris flows on vegetated screes in the Black Mountains, Carmathenshire. *Earth Surface Processes*, **1**, 173–180.

Stone, T. (1794). *General View of the Agriculture of the County of Bedford*, Hodson, London.

Steven, H. M. (1951). The forests and forestry of Scotland. *Scottish Geographical Magazine*, **67**, 110–123.

Stewart, A. J. A. and Lance, A. N. (1983). Moor-draining: a review of impacts on land-use. *Journal of Environmental Management*, **17**, 81–99.

Stretton, C. (1984). Water supply and forestry — a conflict of interests: Cray Reservoir, a case study. *Journal of the Institute of Water Engineering and Scientists*, **38**, 323–330.

Tallis, J. H. (1964). Studies on southern Pennine peat. II. The pattern of erosion. *Journal of Ecology*, **52**, 333–344.

Tallis, J. (1969). The blanket bog vegetation of the Berwyn Mountains, north Wales. *Journal of Ecology*, **57**, 765–787.

Tallis, J. (1981). Uncontrolled fires. In Phillips, J., Yalden, D. and Tallis, J. (eds), *Moorland Erosion Study. Phase 1 Report*, Peak Park Joint Planning Board, Bakewell, pp. 176–182.

Tallis, J. H. (1989/90). Aspects of blanket peat erosion. *Proceedings of the North of England Soils Discussion Group*, **24**, 27–38.

Tallis, J. and Johnson, R. H. (1980). The dating of landslides in Longdendale, north Derbyshire, using pollen analytical techniques. In Davidson, D. A. and Lewin, J. (eds), *Timescales in Geomorphology*, Wiley, Chichester, pp. 189–205.

Tallis, J. H. and Switsur, V. R. (1983). Forest and moorland in the south Pennine uplands in the mid-Flandrian Period. I. Macrofossil evidence of the former forest-cover. *Journal of Ecology*, **71**, 585–600.

Tallis, J. H. and Yalden, D. W. (1983). *Peak District Moorland Restoration Project. Phase 2 Report: Re-vegetation Trials*, Peak Park Joint Planning Board, Bakewell.

Taylor, J. A. (1986). The bracken problem: a local hazard and global issue. In Smith, R. T. and Taylor, J. A. (eds), *Bracken: Ecology, Land-use and Control Technology*, Parthenon Press, Carnforth, pp. 21–46.

Thomas, T. M. (1965). Sheet erosion induced by sheep in the Pumlumon (Plynlimon) area, mid-Wales. In *Rates of Erosion and Weathering in the British Isles*, Institute of British Geographers, Bristol, pp. 11–14.

Tinsley, H. M. and Derbyshire, E. (1976). Late-glacial and postglacial sedimentation in the Peris-Padarn rock basin, north Wales. *Nature*, **260**, 234–238.

Tivy, J. (1957). Influences des facteurs biologique sur l'erosion dans les Southern Uplands Ecossais. *Revue De Geomorphologie Dynamique*, **8**, 9–19.

Trow Smith, R. (1957). *A History of British Livestock Husbandry to 1700*, Routledge & Keagan Paul, London.

Tuckfield, C. G. (1965). Rate of erosion by gullying in the New Forest. In *Rates of Erosion and Weathering in the British Isles*, Institute of British Geographers, Bristol, pp. 15–18.

Tuckfield, C. G. (1980). Stream channel stability and forest drainage in the New Forest, Hampshire.

Earth Surface Processes and Landforms, **5**, 317–329.

Turner, J. (1964). The anthropogenic factor in vegetational history. I. Tregaron and Whixall Mosses. *New Phytologist*, **63**, 73–90.

Turner, J., Hewetson, V. P., Hibbert, F. A., Lowry, K. H. and Chambers, C. (1973). The history of the vegetation and flora of Widdybank Fell and the Cow Green Reservoir basin, upper Teesdale. *Philosophical Transations of the Royal Society*, **B265**, 327–408.

Vancouver, C. (1808; repr. 1969). *General View of the Agriculture of the County of Devon*, David & Charles Reprints, Newton Abbot.

Vancouver, C. (1810). *General View of the Agriculture of Hampshire, Including the Isle of Wight*, Richard Phillips, London.

Watson, A. (1984a). A survey of vehicular hill tracks in north-east Scotland for land use planning. *Journal of Environmental Management*, **18**, 345–353.

Watson, A. (1984b). Paths and people in the Cairngorms. *Scottish Geographical Magazine*, **100**, 151–160.

Watson, A. (1985). Soil erosion and vegetation damage near ski lifts at Cairn Gorm, Scotland. *Biological Conservation*, **33**, 363–381.

Watson, A. and Evans, R. (1991). A comparison of estimates of soil erosion made in the field and from photographs. *Soil & Tillage Research*, **19**, 17–27.

Webb, N. (1986). *Heathlands*, Collins, London.

Werritty, A. (1984). Stream response to flash floods in upland Scotland. In Burt, T. P. and Walling, D. E. (eds), *Catchment Experiments in Fluvial Geomorphology*, Geobooks, Norwich, pp. 537–560.

Werritty, A. and Ingram, H. A. P. (1985). Blanket mire erosion in the Scottish Borders in July, 1983. *British Ecological Society Bulletin*, **16**, 202–203.

White, G. (1788; repr. 1951). *The Natural History of Selborne*, Lutterworth Press, London.

Young, A. (1804a; repr. 1969). *General View of the Agriculture of the County of Norfolk*, David & Charles Reprints, Newton Abbot.

Young, A. (1804b; repr. 1971). *General View of the Agriculture of Hertfordshire*, David & Charles Reprints, Newton Abbot.

Young, A. (1813a; repr. 1970). *General View of the Agriculture of the County of Sussex*, David & Charles Reprints, Newton Abbot.

Young, A. (1813b; repr. 1969). *General View of the Agriculture of the County of Suffolk*, David & Charles Reprints, Newton Abbot.

Young, A. (1813c; repr. 1970). *General View of the Agriculture of Lincolnshire*, David & Charles Reprints, Newton Abbot.

13 The Sensitivity of Downland Arable Land to Erosion by Water

JOHN BOARDMAN

School of Geography, University of Oxford, UK

ABSTRACT

The application of the concept of landscape sensitivity to arable landscapes has been neglected. Farmers' decisions change the sensitivity of the land; the selection, timing and siting of the crop, plus the choice of cultivation practices, for example, are important controls on where and when erosion takes place. The distribution of erosion through time is controlled by the interaction between rainfall and the prepared soil surface, the character of which frequently maximises the risk of runoff occurring. On the South Downs, southern England, sites that are at risk of erosion can be identified using soil and morphological criteria. A distinction should be made between the sensitivity of arable fields to the initiation and to the continuation of erosion. Some rainfall-threshold values at which rilling commences have been proposed for Britain. In the future, climatic change may increase the sensitivity of arable landscapes to erosion.

INTRODUCTION

Recent interest in landscape sensitivity (Brunsden & Thornes, 1979) and the associated concern with geomorphological thresholds (Schumm, 1979), has centred almost exclusively on 'natural' landscapes and those subject to grazing pressure or mineral exploitation (Graf, 1979). The theme of complex gully systems developed as a result of landscape destabilisation by overgrazing is particularly important in the debate about thresholds (e.g. Womack & Schumm, 1977). The implications of these concepts for landscape management is discussed by Schumm (1979) with reference to gully erosion and changes in river patterns. Application of the concepts of sensitivity and threshold to the erosion of arable land has been neglected, though many workers implicitly acknowledge the relevance of these ideas.

In this chapter reference is made to the erosion of arable land in Britain particularly to the eastern South Downs where a monitoring project has been running for nine years (Figure 13.1) (Boardman, 1990a). Several recent reviews of erosion in Britain have been published (Speirs & Frost, 1987; Evans, 1988; Boardman, 1990b), and this material will not be repeated except where it is relevant to the issue of landscape sensitivity.

The sensitivity of the land to erosion can change when it is put down to arable. Gross morphological aspects of the landscape do not change in the timespan of modern farming. However, field boundaries are a morphological element that play a crucial role in the control of runoff and storage of sediment (e.g. Boardman, 1986). Widespread concern about the

Landscape Sensitivity. Edited by D. S. G. Thomas and R. J. Allison
© 1993 John Wiley and Sons Ltd

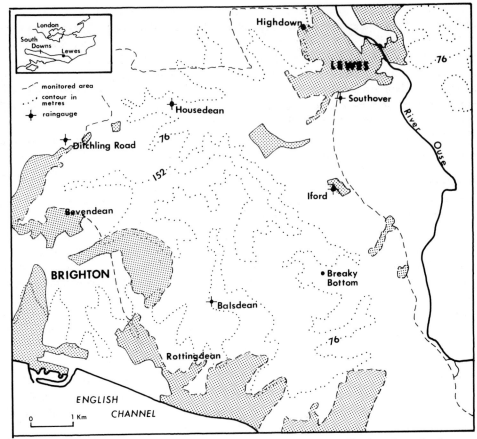

Figure 13.1   The monitored area on the eastern South Downs. The location of rain gauges is shown. A rain gauge at Offham is situated 1 km north of Highdown

loss of field boundaries in the last 50 years in western Europe has focused on the creation of long, uninterrupted slopes and large catchment areas; this change is frequently seen as a factor contributing to the increase in erosion. However, De Ploey (1989) points out that features such as field boundaries may act to concentrate water and thus encourage erosion: their impact is therefore ambivalent and they require further study. Of relevance here is that slope morphology is, at least in part, a product of decisions made by farmers, since the removal of a boundary may link individual slope segments. The changes tend to be abrupt in that removal of hedges, walls or grass banks, which may be very old features of the landscape, can be accomplished in a day.

Over the last 5000 years the sensitivity of soils to erosion has changed as soils have changed. Soil texture changes as a result of water and wind erosion and deposition (e.g. Boardman, 1983). Farming practices may also affect soil texture. In southern England, with a history of up to 5000 years of farming and intermitent erosion, an original thick loess cover has been removed or thinned (Catt, 1978; Evans, 1990a). In some areas such as the South Downs these thin soils (often <20 cm to chalk) are now stony due to the incorporation by ploughing of chalk stones and flints. These soils are less erodible than the former stoneless, silt-rich soils.

A rapid change in the soil's character, which may change its sensitivity to erosion, is the decline in organic matter that follows conversion of grassland to arable. Data in Boardman & Robinson (1985) show typical organic carbon percentages for fields on the South Downs that have been arable since before 1945 of around 3% compared to about 12% for permanent grassland. Greenland, Rimmer & Payne (1975) suggested that aggregates become unstable at organic carbon levels below 2%. There is some evidence (Evans, 1990a) that erosion is more prevalent on soils with low, less than 2%, organic carbon levels, but it is difficult to isolate organic carbon as the controlling variable, as in England and Wales generally (Evans, 1990b), and Somerset specifically (Colborne & Staines, 1985), the lowest values are associated with coarser-textured soils. Some emphasis has been placed on declining levels of organic matter as an explanation for increasing erosion (Hodges & Arden-Clark, 1986), but erodible chalkland soils appear to be well above the threshold value.

Year-to-year variations in the occurrence and rates of erosion at a site are not controlled by slope morphology or soil characteristics. Changes in crop type, timing of cultivations, farming practices and rainfall alter the sensitivity of the land to erosion over the short term. For a given crop, management will be similar year after year. Significant differences in management may occur if bad weather delays the timing of drilling or spraying. A farmer may then take risks, e.g. spraying when the ground is too wet and causing compaction of the soil.

Hence, most elements of the arable landscape change slowly through time; in some localities this even applies to crop type. With the decline in the use of arable rotations, winter cereals are often grown in the same field year after year.

In discussing the sensitivity of arable landscapes to erosion it seems inevitable, as sensitivity appears to vary from year to year, that emphasis must fall on weather patterns as a major explanatory factor. However, a fundamental point should be stressed, namely that on bare ground, a characteristic (though not necessary) element of arable systems, even light rainfall may cause erosion.

CHANGES IN LANDSCAPE SENSITIVITY TO EROSION THROUGH THE CROPPING YEAR

In the previous section reference was made to changes in slope morphology and soils which influence erosion. Most of these changes, with the exception of field boundary removal, occur slowly over long periods of time. Major controls on seasonal erosion of the arable landscape are changes of the field surface that take place over periods of days and weeks, and these interact with rainfall.

In Britain the increase in erosion of arable land that has taken place since the mid-1970s is a function of the change from spring to winter cereals (Boardman, 1984; Boardman & Robinson, 1985; Colbourne & Staines, 1985; Evans & Cook, 1986). Erosion continues to occur on other crops such as sugar beet, potatoes, fruit and vegetables, and to occur at other times especially during summer thunderstorms but, in terms of the area affected, erosion is mostly in winter cereal fields (Evans, 1988). This observation poses the question: why are landscapes where winter cereals are the dominant crop (much of southern and eastern England and eastern Scotland) particularly sensitive to erosion by water?

The principal reason is that the date the crop is drilled falls between the end of September and mid-November, which in many areas coincides with the wettest part of the year. Thus

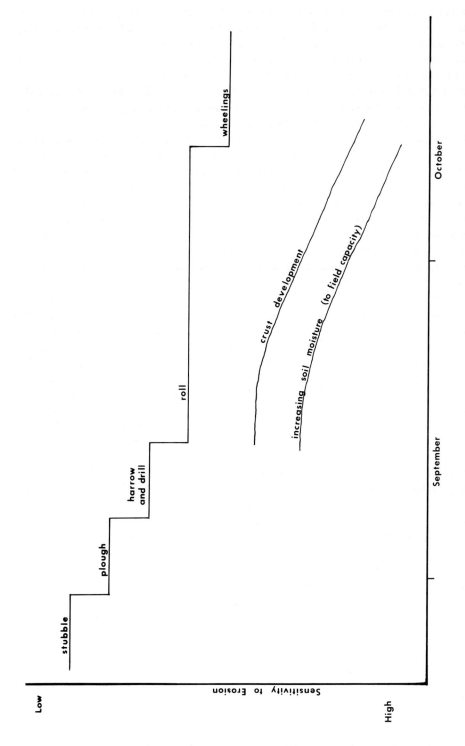

Figure 13.2 Progressive reduction in barriers to erosional change in a winter cereal system: the South Downs. The timing of farming operations and therefore crust development and soil moisture levels will vary somewhat from field to field

the period when soils are bare and smooth is also when there is the highest risk of rainfall. In addition, during this time period, soils approach or attain field capacity, in which condition they remain throughout the winter (Robinson & Boardman, 1988).

The sensitivity of winter cereal fields to erosion is strongly influenced by farming operations. Over a period that varies between two weeks and two months (Figure 13.2), the barriers to change within the arable system are progressively reduced (cf. Brunsden & Thornes, 1979, figure 8). Less energy is required, therefore, to initiate erosion at progressively later points in this time period. The end result is fields that are smooth (low microtopography) rolled and have wheelings on them (compacted zones of low infiltration). If the fields have not been rolled — and this may vary regionally — then producing fine seedbeds also reduces the resistance of the soil surface to erosion. Frost & Speirs (1984) considered the widespread use of powered harrows to produce fine tilths to be part of the explanation for increased erosion in Britain. Observational evidence that erosion frequently begins along vehicle wheel tracks is supported by experimental evidence (e.g. Reed, 1986; Fullen, 1985; Rauws & Auzet, 1989).

The sensitivity of arable land to erosion is strongly related to the susceptibility of the soil to slaking and crust development. The development of a crust reduces significantly the infiltration rate (Table 13.1), and leads to more frequent runoff. On sandy soils in the west Midlands, 10−15 mm of rain falling over 10−20 days caused the formation of 1−4 mm thick crusts. The threshold at which runoff and erosion occurred was reduced to rainfalls of 5−10 mm at intensities of 1.5−2.0 mm h^{-1} (Fullen & Reed, 1986). Heavy rainfall at the beginning of a growing season may lead to crusting and thus ensure that subsequent minor rainfall events produce runoff (Boardman, 1988a). Short periods of high intensity rainfall may also cause crusting and erosion (Fullen & Reed, 1986; Boardman & Spivey, 1987).

The major force inhibiting erosion is the growing crop. Vegetation cover values in excess of 30% inhibit erosion (Elwell & Stocking, 1976). This is an approximate figure dependent on site conditions. Crop growth is largely controlled by temperature and moisture, and winter cereals tend to grow more slowly than spring cereals, in their equivalent early stages of growth, because of the declining temperatures of the autumn and winter months. Rates of growth vary from year to year (Figure 13.3). When autumn is warm the cereals grow quickly and achieve the ~30% threshold value by November. In most winters on the South Downs the majority of fields fall into this category. In cold winters (e.g. 1985−86), growth is slower and the ~30% cover value is not reached until spring (Robinson & Boardman, 1988). The periods of risk are therefore of different lengths. However, warmer winters with shorter periods when the land is at risk of erosion, perhaps six weeks, tend to be wetter and serious erosion

Table 13.1 The development of soil crusts and changing wet soil infiltration: typical values on loamy soils in northern France (adapted from Papy & Boiffin, 1989)

Crusting state	Range of infiltration rates on wet soil (mm h^{-1})
Initial fragmentary structure	30−50
Altered fragmentary state with structural crusts	5−30
Continuous crusts	1−2
Disturbed crusts (climatic and biological factors)	>5?

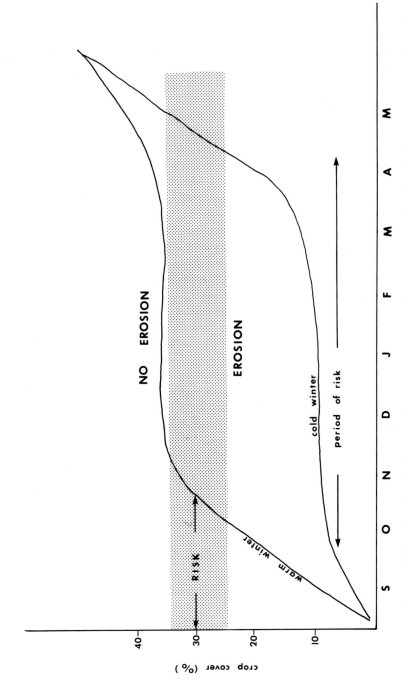

Figure 13.3 Periods of risk in relation to crop cover: winter cereals on the South Downs

can occur, whereas cold winters tend to be dry. Present evidence suggests rainfall thresholds necessary to cause moderate or serious erosion are exceeded about once every five years and are associated with warm, wet, early winter weather.

On the South Downs the temporal pattern of erosion during the last decade has resulted from the dominance of winter cereals and the October−November rainfall peak. The chief variation from this pattern has been relatively dry winters when little erosion occurred. In 1989−90 a different distribution of rainfall with a late winter peak produced a landscape response that was unexpected. Early winter rainfall had led to some erosion of soils under oilseed rape, grass leys and winter cereals (Figure 13.4). At the end of January serious erosion (>10 m^3 ha^{-1} year^{-1}) occurred on ploughed fields that had crusted earlier in the winter. Rates of erosion were an order of magnitude less than the highest rates recorded under cereals (Boardman, 1988a, 1991). Ploughed surfaces, because of their greater roughness, had been regarded as more resistant to erosion than smooth-surfaced cereal fields. This erosion of ploughed fields in January accords with recent findings in northern France where a cumulative rainfall total of about 450 mm was required to reduce a ploughed surface of a loamy soil to a strongly crusted state where runoff readily occurs (Papy & Boiffin, 1989). Rather lower rainfall totals appear to produce the same effect on the South Downs where the initial ploughed surface tends to be free of residue because of stubble burning. Hence, early ploughing of fields in preparation for the drilling of spring cereals puts the fields at risk of erosion if there is heavy late winter rain. The distribution of rainfall is therefore an important control on the sensitivity of different land uses to erosion (Boardman, 1991).

SPATIAL VARIATION IN LANDSCAPE SENSITIVITY TO EROSION

When crop type, management practice and rainfall are uniform, the sensitivity of a landscape to erosion will vary with its landform and soils. The influence on erosion of slope length, angle and soil type are reflected in the variables in the Universal Soil Loss Equation (USLE; Wischmeier & Smith, 1978). Two important morphological parameters are neglected in the USLE. Convex slopes with wide crestal areas acting as collecting grounds for runoff appear to be particularly susceptible to erosion (Evans, 1980a). Many sites of major erosion conform to this pattern (e.g. Boardman & Robinson, 1985). Sites of erosion are also often depressions or valley bottoms where concentrated flow of water occurs (e.g. Evans & Cook, 1986). This is also a pattern widely recognised in western Europe (Auzet *et al.*, 1990).

On the South Downs the spatial distribution of erosion can be predicted on the basis of land use and field morphology; the *timing* of erosion is controlled by the interplay of rainfall, land use decisions and farming practices. High-risk sites may be identified using the following criteria:

- Slope length >200 m; slope angle $>10°$; relief within the field >30 m (Boardman, 1988a, 1988b).
- Valley bottom or depression locations. Figure 13.5 shows that most of the erosion in a sample area occurs in valley-bottom locations and relatively little on the valley sides. The latter act as areas of runoff generation but critical stress levels are only exceeded where flow is concentrated in depressions. In high gradient depressions incision of up to 1 m is typical; in lower gradient sites broad shallow flows incising to 20 cm occur,

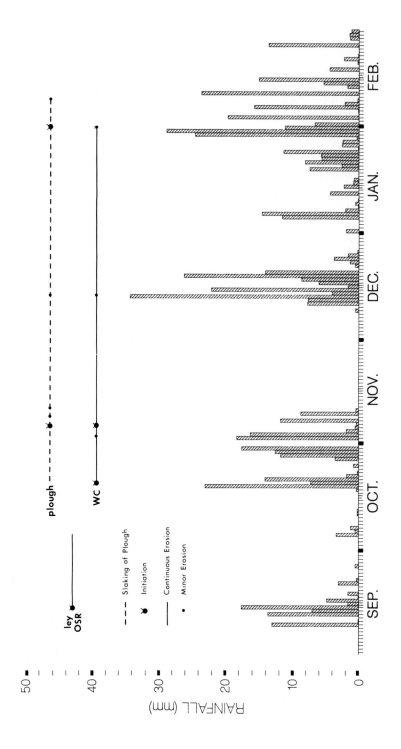

Figure 13.4 Rainfall September 1989—February 1990 at Southover, Lewes, and erosion within the monitored area. Plough: ploughed and cultivated fields; ley OSR; newly drilled grass ley and oilseed rape fields; WC: winter cereal fields (from Boardman, 1991, by permission)

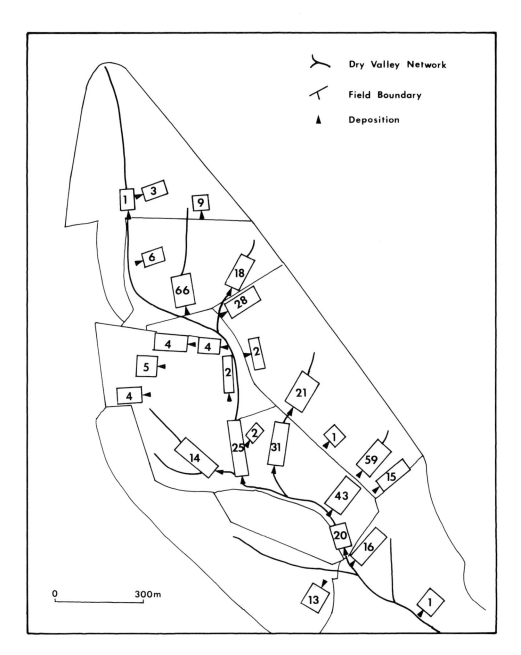

Figure 13.5 An example of the spatial distribution of erosion in a dry valley system on the South Downs. Figures are cumulative erosion 1982–90 in m³. Erosion was predominantly in winter cereals, with small amounts in oilseed rape and in ploughed fields. Triangles indicate soil storage sites, which are frequently controlled by field boundaries

whereas low gradient valley bottoms appear to be stable and ungullied under present conditions (Boardman, 1992).

Elsewhere in Britain, erosional thresholds may be lower on sandy soils and higher on clays in comparison to the silt loams of the South Downs (Evans, 1990b).

LANDSCAPE SENSITIVITY

Initiation of Erosion

An important theme when discussing the changing sensitivity of a landscape to erosional processes has been the initiation of rilling, instability or failure. External destabilising forces such as large rainfalls are important as are internal forces such as reductions in material strength. Changes in surface roughness on arable fields as a result of farming operations (Figure 13.2), reduce the ability of the soil to resist erosion so that progressively smaller forces that occur more frequently initiate erosion.

The magnitude and frequency of the rainfall event likely to cause erosion of arable land needs to be known. Threshold values that have to be surmounted to initiate rilling are listed in Table 13.2. These values are based on observations at a site or a locality and refer to specific soil, management and land use conditions, beyond which they may not be applicable. There is little data on the frequency of occurrence of these events (cf. Evans, 1990b).

Prior to the initiation of rills, aggregate breakdown due to raindrop impact, soil redistribution and sheet wash may occur. These processes appear more important on some soils than others. On non-cohesive sandy soils, sheet flow is more common than on cohesive silty or clayey soils. Sheet wash, a thin flow of water over the surface, is considered important on bare sandy soils in Bedfordshire (Jackson, 1986; Morgan, Martin & Noble, 1987). In some cases large amounts of soil may be moved by splash and sheet flow but distances moved are short and net losses are likely to be small. There are difficulties in making estimates of the effectiveness of this process (Phillips, 1988). Sheet wash is rarely observed on the silt loams of the South Downs, occurring only on patches of sandy soil (Boardman, 1990a).

Table 13.2   Rainfall-erosion thresholds for Britain

>7.5 mm day^{-1}	Evans & Nortcliff (1978) based on Evans & Morgan (1974)
>1 mm h^{-1} during rainfalls of 10 mm on compacted soils	Reed (1979)
KE > 10 index based on >10 mm h^{-1} for >10 min	Morgan (1980)
Successive rainfalls in winter totalling about 20–25 mm	Evans (1980)
>10 mm in summer >20 mm in 3 days in winter	Evans (1981)
15–20 mm rainfall events especially when soil at field capacity	Speirs & Frost (1985)
30 mm in 2 days	Boardman (1990a)

It may be that on all soil types, splash and wash are of approximately equal importance, but on some soils rilling is so clearly dominant that these processes have been ignored. Evans (1990b) reviews the available British data for transport of soil by sheet flow. In all cases, values are < 1 m^3 ha^{-1} year^{-1} and he concludes that, 'generally, wash on most soils under arable crops will transport <0.3 m^3 ha^{-1} year^{-1}, and this probably applies to all sloping arable fields'. Since splash and sheet wash occur prior to, as well as during rilling, the rainfall threshold level at which they commence is lower than that for rill initiation.

On the South Downs the threshold value for the initiation of rilling is 30 mm in two days, and this applies to the risk period September–November inclusive. It has been exceeded at least once in each of the last nine years. It applies to smooth-drilled surfaces which are typical of the arable area at that time of year. Although the threshold value was originally suggested with reference to plots of 200 m length and 7° slope, rill initiation was observed under the same rainfall conditions on a range of field morphologies. The slope length and angle criteria do not seem to be essential elements of the domain to which the threshold value is applicable. Field morphology appears important in controlling the *magnitude* of the erosion.

The rainfall events that have initiated erosion on the South Downs over a nine year period are listed in Table 13.3. The rainfall gauge nearest to the site of erosion has been used. In 1982, extensive rilling occurred at Bevendean on 20 September, midway between two gauges. Even with six gauges within or close to the monitored area the amount falling at a specific site can only be estimated (Figure 13.1). Some rainfall amounts were relatively uniform at several gauges, whereas, on other occasions, there was marked variability. Two sets of dates for 1989 are listed. On 13–14 September rills formed in fields drilled with oilseed rape and grasses (Figure 13.4). At this time there were no fields drilled with winter cereals. The initiation of rilling on winter cereals did not occur until 19–21 October.

There are 17 rainfall events greater than 30 mm in two days in the 1 September–30 November part of a seven year period (Figure 13.6). The rainfall-erosion threshold is likely

Table 13.3 Rainfall thresholds for rilling on the South Downs 1982–1990

1982	19, 20 Sept (D) 17, 12 mm	erosion Bevendean on grass ley
	22, 23 Sept (B) 11.9, 20.5 mm	(Boardman & Robinson, 1985)
1983	13, 14, 15 Oct (S): 18.2, 8.6, 18.6 mm	erosion pre-18 Oct; very little erosion
1984	2, 3 Oct (S): 19, 14.9 mm	erosion pre-24 Nov, probably 2, 3 Oct
1985	8, 9 Nov (O): 10.9, 22.5 mm	rills developed on 9 Nov
1986	14, 15 Oct (S): 27.6, 0.2 mm	erosion pre-2 Dec, probably 14, 15 Oct
1987	6, 7 Oct (B): 12.1, 62.9 mm	extensive and severe erosion (Boardman, 1988a)
1988	8, 9 Oct (H): 39.2, 9.1 mm	virtually no erosion
1989	13, 14 Sept (D): 13.2, 16.5 mm	erosion OSR and grass ley
	19, 20, 21 Oct (S): 22.9, 7.2, 14.0 mm	erosion on winter cereals
1990	26, 27 Oct (H): 30.4, 38.7 mm	erosion 27 Oct

Rain gauges and crops: D, Ditchling Road; B, Balsdean; S, Southover; O, Offham; H, Housedean; OSR, oilseed rape.

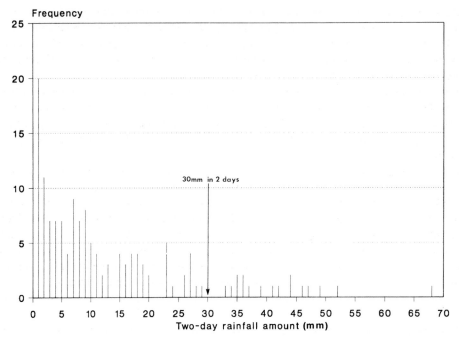

Figure 13.6 The frequency of two day rainfall totals 1 September–30 November 1982–88, Southover rain gauge

to be exceeded every year. The extent and magnitude of erosion will then depend on the size of the initiating event (e.g. 1987) and the magnitude and frequency of succeeding rainfalls.

Continuation of Erosion

Rainfall values that trigger erosion subsequent to initiation have been little discussed. In some analyses it was assumed that erosion events were only triggered by repeated rainfalls of a given magnitude (Evans & Nortcliff, 1978; Boardman, 1983), because the resisting forces at a site, especially surface roughness, remained constant even after the establishment of a rill system: this is untrue. In cases of severe erosion significant hydrological changes occur. For example, on a part of a slope of 2.1 ha at Breaky Bottom, South Downs (Boardman, 1988a), an unrilled area was replaced by one with a drainage density of 650 km km^{-2}.

In an early case study of erosion on arable land in Britain, Foster (1978) recognised the increased efficiency of a rilled or gullied slope compared to a non-eroded one. On 1 October 1976, 64.8 mm of rain led to gully formation. On 2 October there were three minor rain events of 0.5 mm (maximum intensity 2 mm h^{-1}); 0.5 mm (5 mm h^{-1}) and 3.0 mm (>5 mm h^{-1}). Each of these produced further incision. Foster comments:

> the time interval between the start of rainfall and the start of flow in the gully decreased during each successive storm period; this almost certainly reflects the degree of saturation of the topsoil in the unploughed catchment area i.e. the topsoil had to be saturated before overland flow could occur. (Foster, 1978, p. 160)

Once a gully and rill network had been initiated, runoff generated on interrill areas also had a shorter distance to travel to a channel and the network responded quicker to the later rainfall events. Speirs & Frost (1985) commented that on soils that had previously suffered erosion, 'further significant soil losses resulted from rainfalls of less than 10 mm'. Similarly, erosion on the South Downs initiated by a rainfall of about 60 mm on 7 October 1987 continued when daily rainfall amounts were as low as 7 mm (Boardman, 1988a).

It is important to distinguish between erosion and runoff. Where rill systems have been established, then runoff from the rilled slopes is likely to be fast and efficient so that large proportions of the rain falling on the surface will run off rather than infiltrating. This is particularly so when soils have crusted. However, sediment yield is likely to decline as supplies become exhausted.

Landscapes that are sensitive to the initiation of erosion may not always be sensitive to the continuation of erosion; and the reverse may also be the case. Data on erosion rates from localities widely spread throughout England and Wales show that stony silt loams such as those of the South Downs are less erodible than soils in many other areas (Evans, 1990b; Boardman *et al.*, 1990). The threshold for rill initiation is undoubtedly higher on these chalky soils than on sandy and loamy soils of nearby areas. However, the frequency of serious erosion (> 10 m^3 ha^{-1} year^{-1}) suggests that management and field morphological factors are important. Steep, long valley-side slopes with winter cereal crops readily develop rill systems that are often more extensive than those that can form in smaller fields where soils are more erodible.

Little is known of the relationship between energy input (rainfall) and erosion above the threshold of rilling. It could be that the relationship is linear, more soil being removed at higher levels of energy. However, our knowledge of thresholds within systems suggests that this simple relationship is unlikely to hold true. In the monitored area of the South Downs, erosive rainfall in the autumn−winter risk period can be represented by a Rainfall Index (RI; Table 13.4), which over the period 1950−89 has varied from 0 to 16. During the period of monitoring, values of 3−14 have occurred (Boardman, 1990a). For values between 1 and 12 the relationship between RI and total soil loss is approximately linear. Between index values 12 and 14 there is a threshold; above 12, soil losses increase by an order of magnitude. A 3-fold increase is seen also in annual median values of soil loss. The physical explanation for the increase in magnitude of erosion at the higher RI values is not known.

The sensitivity of the landscape to events of different magnitudes and frequencies has been investigated (Wolman & Miller, 1960). In the arable landscape most of the questions remain unanswered through lack of long-term monitoring projects. Without such projects it is difficult to assess the importance of the many so-called 'catastrophic events' that are recorded in case studies. An insight into the importance of such events is the heavy rainfall on the South Downs in October 1987. Much of the erosion was accomplished by a storm with a maximum rainfall

Table 13.4 Values allotted to rainfall events to obtain rainfall index (RI). The values are summed for the winter period (1 September−1 March)

Rain event	Index value
30 mm in 2 days	1
30 mm in 1 day	2
60 mm in 2 days	3
60 mm in 1 day	4

of 63 mm at Balsdean (Figure 13.1), probably falling over 12 h. The return period for such an event is estimated at 25 years and it increases the median erosion rate from under 2 m^3 ha^{-1} $year^{-1}$ to about 5 m^3 ha^{-1} $year^{-1}$. At many sites where serious erosion has occurred the return period has been considerably less (Boardman, 1988a).

Attempts to assess the impact on an arable landscape of high magnitude rainfall events of frequencies > 100 years are more speculative. There does not appear to be an example of such an occurrence in Britain during the present phase of intensive arable farming. Storms such as that of early July 1968 (130 mm fell at Bath in a day), which affected much of the south-west and Midlands, appear to have had little impact on arable fields (Newson, 1975). This may be because crops covered the ground at that time of year or because the effect of such storms on arable land was not reported. Hanwell & Newson (1970, appendix), list 44 impacts of the storm over the Mendips but only at two of these sites is soil erosion on arable fields mentioned. Similarly, the intense August thunderstorm that hit western Derbyshire in 1957 (Barnes & Potter, 1958) appears to have had little effect on arable land, despite giving rise to flooding, channel changes and property damage. This storm, when 152 mm of rain fell on Rodsley in 24 h with estimated intensities of 25 mm h^{-1} sustained for 5 h, is in striking contrast to a rather minor storm that caused serious erosion on arable land in the same area in April 1983 (Boardman & Spivey, 1987). The difference was that in 1983 the rain fell on recently drilled spring cereal and potato fields.

On the South Downs, Potts (1982) describes two large rainfalls that occurred on 20−21 September and 10−11 October 1980 in the Worthing area, where gauges recorded 112 and 133 mm in 24 h. Potts suggests that these represent return periods in excess of 1000 years. The impact of these events on arable land was surprisingly limited and certainly no greater than storms of much lower magnitude in subsequent years. This seems to have been because the area of winter cereals was much lower in 1980 than, for example, in 1990 (Boardman & Evans, unpublished).

In each of these cases, local land use and timing of the storm — and possibly a lack of interest in soil erosion on arable land particularly where no flood impact on property occurred — makes it impossible to draw any conclusions about the sensitivity of the landscape to high-magnitude events. We do not know, for example, what would be the effect of 120 mm in a day falling on the South Downs under current land use conditions in October or November. This would be twice the highest daily fall in the last ten years. Nor is there much information about the relative importance of total amount of rainfall as opposed to high intensities. There have been suggestions that amount is more relevant than intensity (Evans & Nortcliff, 1978), but the effect of very short periods of high-intensity rain, which normally go unrecorded, may be important (Boardman & Spivey, 1987).

LANDSCAPE SENSITIVITY AND CLIMATE CHANGE

The sensitivity of landscape to future changes of climate (the 'greenhouse effect') has been the subject of recent meetings (e.g. Boer & De Groot, 1990), though even at those meetings devoted specifically to agriculture there has been little discussion of soil erosion (e.g. Bennett, 1989). There are good reasons for this lack of progress, the major one being the failure of General Circulation Models (GCMs) to provide detailed and reliable data regarding future climate at a regional scale. There is also the problem that GCMs provide more reliable temperature than rainfall data. Prediction of soil erosion rates requires considerable climatic

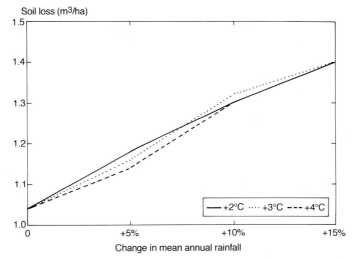

Soil loss (m3/ha)

Figure 13.7 Simulated mean annual soil loss for the Woodingdean site D under conditions of increased temperature and winter rainfall (from Boardman *et al.*, 1990, by permission)

detail, especially of precipitation. Seasonal changes in rainfall are also poorly understood.

Accepting these constraints, and simulating a series of possible climatic scenarios, we can model future soil erosion using the EPIC model (Williams & Renard, 1985). For a winter cereal field on the South Downs, where erosion rates have been monitored, Boardman *et al.* (1990) show increases in erosion from the present median rate of 1.04 to 1.40 m^3 ha^{-1} year^{-1} with a 15% increase in rainfall in the winter months. An increase in temperature has little effect on erosion (Figure 13.7). Future changes in erosion rates in other parts of the country may be predicted based on the known relationship between present rates in selected localities (Evans, 1988; Boardman *et al.*, 1990). Changes in thunderstorm frequency, rainfall intensity and the use of irrigation under future climates are largely speculative but will have an important influence on the sensitivity of landscapes to erosion.

Predictions of erosion rates on arable land in Britain in the future have so far been based on several simplistic assumptions regarding climate and land use change. It has also been assumed that climatic variability will be the same as it is now, even though mean values will change: this is unlikely to be the case (Wigley, 1989). Changes in the frequency of high-magnitude events will have to be taken into account in future modelling exercises.

CONCLUSION

Arable farming systems have certain characteristics that distinguish them from more 'natural' landscapes in their sensitivity to erosion. Forms such as rills are removed in the preparation of the land for the next crop. On a given field, erosion is generally confined to a period of less than three months since the growing crop inhibits soil loss. Farming operations such as the production of fine tilths and the incidence of wheelings increase the sensitivity of fields to erosion. Unfortunately, many decisions taken by British farmers in order to maximise crop yields serve also to increase the likelihood of erosion. While it is all too easy to blame erosion on the vagaries of rainfall, the conditions of the soil surface in generating runoff also needs to be taken into account when explaining the sensitivity of the arable landscape to water erosion. Some questions regarding the sensitivity of British arable landscapes to erosion remain unanswered. Fortunately, the absence of high-magnitude, low-frequency rains falling on arable fields when they are at their most vulnerable to erosion has ensured that the role of such events remains a matter for speculation.

ACKNOWLEDGEMENTS

I would like to thank Drs R. Evans and R. Parkinson for useful comments on the paper. David Favis-Mortlock provided Figures 13.6 and 13.7. Southern Water Authority and, latterly, the National Rivers Authority, kindly made available rainfall data.

REFERENCES

Auzet, A. V., Boiffin, J., Papy, F. and Maucorps, J. (1990). An approach to the assessment of erosion forms and erosion risk on agricultural land in the northern Paris Basin, France. In Boardman, J., Foster, I. D. L. and Dearing, J. A. (eds), *Soil Erosion on Agricultural Land*, Wiley, Chichester, pp. 383–400.

Barnes, F. A. and Potter, H. R. (1958). A flash flood in western Derbyshire. *East Midland Geographer*, **10**, 3–15.

Bennett, R. M. (ed.) (1989). The 'greenhouse effect' and UK agriculture. CAS Paper 19, Centre for Agricultural Strategy, University of Reading.

Boardman, J. (1983). Soil erosion at Albourne, West Sussex, England. *Applied Geography*, **3**, 317–329.

Boardman, J. (1984). Erosion on the South Downs. *Soil and Water*, **12**(1), 19–21.

Boardman, J. (1986). The context of soil erosion. *SEESOIL*, **3**, 2–13.

Boardman, J. (1988a). Severe erosion on agricultural land in East Sussex, UK, October 1987. *Soil Technology*, **1**, 333–348.

Boardman, J. (1988b). Public policy and soil erosion in Britain. In J. M. Hooke (ed.), *Geomorphology in Environmental Management*, Wiley, Chichester, pp. 30–50.

Boardman, J. (1990a). Soil Erosion on the South Downs: a review. In Boardman, J., Foster, I. D. L. and Dearing, J. A. (eds), *Soil Erosion on Agricultural Land*, Wiley, Chichester, pp. 87–105.

Boardman, J. (1990b). *Soil Erosion in Britain: Costs, Attitudes and Policies*, Social Audit Paper no 1, Education Network for Environment and Development, University of Sussex.

Boardman, J. (1991). Land use, rainfall and erosion risk on the South Downs. *Soil Use and Management*, **7**(1), 34–38.

Boardman, J. (1992). Current erosion on the South Downs: implications for the past. In Boardman, J. and Bell, M. (eds), *Past and Present Soil Erosion*, Oxbow Books, Oxford, pp. 9–19.

Boardman, J. and Robinson, D. A. (1985). Soil erosion, climatic vagary and agricultural change on the Downs near Lewes and Brighton, autumn 1982. *Applied Geography*, **5**, 243–258.

Boardman, J. and Spivey, D. (1987). Flooding and erosion in west Derbyshire, April 1983. *East Midland Geographer*, **10**(2), 36–44.

Boardman, J., Evans, R., Favis-Mortlock, D. T. and Harris, T. M. (1990). Climate change and soil erosion on agricultural land in England and Wales. *Land Degradation and Rehabilitation*, **2**(2), 95–106.

Boer, M. M. and De Groot, R. S. (1990). *Landscape Ecological Impact of Climatic Change, Proceedings of a European Conference*, Lunteren, The Netherlands, IOS Press, Amsterdam.

Brunsden, D. and Thornes, J. B. (1979). Landscape sensitivity and change. *Transactions of the Institute British Geographers*, **NS4**, 463–484.

Catt, J. A. (1978). The contribution of loess to soils in lowland England. In Limbrey, S. and Evans,

J. G. (eds), *The Effect of Man on the Landscape: The Lowland Zone*, Research Report no. 21, Council for British Archaeology, London, pp. 12–20.

Colbourne, G. J. N. and Staines, S. J. (1985). Soil erosion in south Somerset. *Journal of Agricultural Science*, **104**, 107–112.

De Ploey, J. (1989). Erosional systems and perspectives for erosion control in European loess areas. *Soil Technology*, **1**, 93–102.

Elwell, H. A. and Stocking, M .A. (1976). Vegetal cover to estimate soil erosion hazard in Rhodesia. *Geoderma*, **15**, 61–70.

Evans, R. (1980a). Characteristics of water-eroded fields in lowland England. In De Boodt, M. and Gabriels, D. (eds), *Assessment of Erosion*, Wiley, Chichester, pp. 77–87.

Evans, R. (1980b). Mechanics of water erosion and their spatial and temporal controls: an empirical viewpoint. In Kirkby, M. J. and Morgan, R .P. C. (eds), *Soil Erosion*, Wiley, Chichester, pp. 109–128.

Evans, R. (1981). Assessments of soil erosion and peak wastage for parts of East Anglia, England. A field visit. In Morgan, R. P. C. (ed.), *Soil Conservation: Problems and Prospects*, Wiley, Chichester, pp. 521–530.

Evans, R. (1988). *Water Erosion in England and Wales 1982–1984*, Soil Survey and Land Research Centre, Silsoe.

Evans, R. (1990a). Soil erosion: its impact on the English and Welsh landscape since woodland clearance. In Boardman, J., Foster, I. D. L. and Dearing, J. A. (eds), *Soil Erosion on Agricultural Land*, Wiley, Chichester, pp. 231–254.

Evans, R. (1990b). Water erosion in British farmer's fields — some causes, impacts, predictions. *Progress in Physical Geography*, **14**(2), 197–219.

Evans, R. and Cook, S. (1986). Soil erosion in Britain. *SEESOIL*, **3**, 28–58.

Evans, R. and Morgan, R. P. C. (1974). Water erosion of arable land. *Area*, **6**(3), 221–225.

Evans, R. and Nortcliff, S. (1978). Soil erosion in north Norfolk. *Journal of Agricultural Science*, **90**, 185–192.

Foster, S. (1978). An example of gullying on arable land on the Yorkshire Wolds. *Naturalist*, **103**, 157–161.

Frost, C. A. and Speirs, R. B. (1984). Water erosion of soils in south-east Scotland — a case study. *Research and Development in Agriculture*, **1**(3), 145–152.

Fullen, M. A. (1985). Compaction, hydrological processes and soil erosion on loamy sands in east Shropshire, England. *Soil and Tillage Research*, **6**, 17–29.

Fullen, M. A. and Reed, A. H. (1986). Rainfall, runoff and erosion on bare arable soils in east Shropshire, England. *Earth Surface Processes and Landforms*, **11**(4), 413–425.

Graf, W. L. (1979). Mining and channel response. *Annals of the Association of American Geographers*, **69**(2), 262–275.

Greenland, D. J., Rimmer, D. and Payne, D. (1975). Determination of the structural stability class of English and Welsh soils, using a water coherence test. *Journal of Soil Science*, **26**, 294–303.

Hanwell, J. D. and Newson, M. D. (1970). The great storms and floods of July 1968 on Mendip. *Wessex Cave Club Occasional Publication*, Series 1, no. 2.

Hodges, R. D. and Arden-Clark, C. (1986). *Soil Erosion in Britain: Levels of Soil Damage and their Relationship to Farming Practice*, Soil Association, Bristol.

Jackson, S. J. (1986). Soil erosion survey and soil erosion risk in mid-Bedfordshire. *SEESOIL*, **3**, 95–105.

Morgan, R. P. C., Martin, L. and Noble, C. A. (1987). Soil erosion in the United Kingdom: a case study from mid-Bedfordshire. *Occasional Paper* no. 14, Silsoe College, Cranfield Institute of Technology.

Newson, M. D. (1975). *Flooding and Flood Hazard in the United Kingdom*, Oxford University Press, Oxford.

Papy, F. and Boiffin, J. (1989). The use of farming systems for the control of runoff and erosion. *Soil Technology*, **1**, 29–38.

Phillips, G. J. (1988). An assessment of soil erosion on selected soils of the Sussex Weald. Unpublished B.Sc. Dissertation, Brighton Polytechnic.

Potts, A. S. (1982). A preliminary study of some recent heavy rainfalls in the Worthing area of Sussex. *Weather* **37**(8), 220–227.

Rauws, G. and Auzet, A. V. (1989). Laboratory experiments on the effects of simulated tractor wheelings on linear soil erosion. *Soil and Tillage Research,* **13**, 75–81.

Reed, A. H. (1979). Accelerated erosion of arable soils in the United Kingdom by rainfall and run-off. *Outlook on Agriculture,* **10**, 41–48.

Reed, A. H. (1986). Soil loss from tractor wheelings. *Soil and Water,* **14**(4), 12–14.

Robinson, D. A. and Boardman, J. (1988). Cultivation practice, sowing season and soil erosion on the South Downs, England: a preliminary study. *Journal of Agricultural Science,* **110**, 169–177.

Schumm, S. A. (1979). Geomorphic thresholds: the concept and its application. *Transactions of the Institute of British Geographers,* **NS4**, 485–515.

Speirs, R. B. and Frost, C. A. (1985). The increasing incidence of accelerated soil water erosion on arable land in the east of Scotland. *Research and Development in Agriculture,* **2**(3), 161–167.

Speirs, R. B. and Frost, C. A. (1987). Soil water erosion on arable land in the United Kingdom. *Research and Development in Agriculture,* **4**(1), 1–11.

Wigley, T. M. L. (1989). The effect of changing climate on the frequency of absolute extreme events. *Climate Monitor,* **17**, 44–55.

Williams, J. R. and Renard, K. G. (1985). The physical components of the EPIC model. *Proceedings of the International Conference on Soil Erosion and Conservation,* Honolulu, Hawaii.

Wischmeier, W. H. and Smith, D. D. (1978). Predicting rainfall erosion losses, *USDA Agricultural Research Service Handbook* 537.

Wolman, M. G. and Miller, J. P. (1960). Magnitude and frequency of forces in geomorphic processes. *Journal of Geology,* **68**, 54–74.

Womack, W. R. and Schumm, S. A. (1977). Terraces of Douglas Creek northwestern Colorado: an example of episodic erosion. *Geology,* **5**, 72–76.

14 Catchment Sensitivity to Land Use Controls

TIM P. BURT
School of Geography, University of Oxford, UK

A. LOUISE HEATHWAITE
and
STEPHEN T. TRUDGILL
Department of Geography, University of Sheffield, UK

ABSTRACT

The pathways taken by hillslope water draining to a stream determine many of the characteristics of the landscape, the ways in which land can be used and the practices required to sustain this use. At the same time, land use may significantly influence catchment hydrology. It has been obvious for many years that catchments are susceptible to land use change, in terms of both the quantity and quality of runoff produced. Following a brief review of hydrological processes in headwater catchments, this chapter examines the sensitivity of runoff production to disturbance, using examples mainly from the authors' own work. Although the general direction of change is now well known, it remains difficult to determine which catchments (and indeed which sites within a catchment) are most sensitive to these changes. Certain locations may be particularly susceptible to change: it is necessary to identify these areas, to understand their functional role and to recognise their importance within strategies for the wise management of drainage basins.

HYDROLOGICAL PROCESSES IN HEADWATER CATCHMENTS

When rain and meltwater reach the surface of the ground, they encounter a filter that is of great importance in determining the path by which hillslope runoff will reach a stream channel (Dunne, 1978). Hillslope hydrology is concerned primarily with flow processes within the soil and over the soil surface, but in certain locations groundwater flow may also be of interest. Precipitation (or meltwater) is partitioned between various flow routes, which attenuate and delay the flow to different extents, so that a knowledge of the relevant mechanisms is important (Kirkby, 1988). Dunne (1978) notes that the paths taken by the water determine many of the characteristics of the landscape, the uses to which land can be put, and the strategies required for wise land management. In addition, the quality of input water may be greatly transformed depending on its hydrological pathway. However, the point must also be made that, though the hydrological characteristics of a catchment may determine the way in which land is used, land use and techniques of land management may, in turn, significantly influence the hydrological response of the catchment.

Landscape Sensitivity. Edited by D. S. G. Thomas and R. J. Allison
© 1993 John Wiley and Sons Ltd

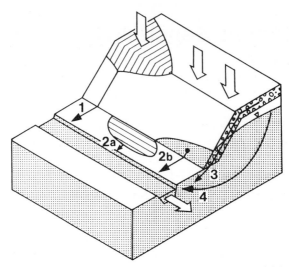

Figure 14.1 Schematic diagram showing hydrological pathways. (1) Infiltration-excess overland flow. (2) Saturation-excess overland flow: (a) direct runoff; (b) return flow. (3) Subsurface stormflow. (4) Groundwater discharge

Figure 14.1 illustrates the mechanisms by which precipitation is delivered to a stream channel. There may be overland flow across the surface, subsurface flow in the near-surface soil and regolith, and deeper groundwater flow through bedrock. If the precipitation falling directly on channels and lakes is ignored (this is usually less than 5% of the precipitation), then it is fair to say that almost all the streamflow reaches the channel by one of these routes. Nearly all the storm runoff is generated by surface and near-surface processes. Baseflow may be produced either by drainage of the soil or by groundwater discharge, or both. Groundwater has often been ignored by hillslope hydrologists (who have been most concerned with storm runoff), but the distinction between shallow and deep subsurface flow is not always a clear one and, in any case, groundwater quality may be an important facet of catchment hydrology.

In general terms, two models of storm runoff generation may be identified. The partial area model (Betson, 1964) recognises that the occurrence of infiltration-excess or 'Hortonian' overland flow is more localised than Horton argued. Even so, for those areas where this type of runoff does occur, Horton's (1933) classic theory of surface runoff may still be applied: the sole source of storm runoff is the excess water which is unable to infiltrate into the soil. The variable source area model (Hewlett, 1961) was proposed to account for the production of storm runoff in areas where infiltration-excess overland flow was rare or absent. In this scheme, subsurface runoff is viewed as the major mechanism providing runoff, both in its own right and because of its influence in generating localised areas of (saturation-excess) overland flow. Thus, the difference between the source area models lie in the mechanisms and pathways of runoff production that cause the expansion and contraction of the source area. These matters are fully reviewed in Anderson & Burt (1990).

Climate, soil and bedrock control which type of runoff mechanism will dominate in a particular catchment. Topography provides a secondary influence at the subcatchment or hillslope scale, especially with regard to the accumulation of soil water and the maintenance of zones of permanent soil saturation which produce overland flow. Vegetation cover may also be an important factor, affecting both the type of storm runoff generated as well as the balance between evaporation and runoff. Anderson & Burt (1990) attempted to summarise the origin of stormflow in relation to catchment characteristics (soil thickness and hydraulic conductivity, vegetation cover and valley floor slope). The production of baseflow is largely controlled by bedrock characteristics, though baseflow may interact with stormflow production to the extent that effluent discharge from an aquifer may help maintain soil saturation in appropriate topographic positions (Figure 14.1).

In relation to catchment sensitivity, the important point to consider is the extent to which natural climatic variations and human impact can modify the pattern of runoff experienced in a given drainage basin. This assessment must be made, not only in respect of the quantity of runoff produced, but also with regard to its quality. Insensitive catchments might be regarded as those where the runoff regime is well buffered from disturbing influences, whereas a sensitive catchment is one where the type and quality of runoff generated is easily changed. Sensitivity may also be considered spatially in relation to the location of source areas of runoff within the basin. The approach taken here will be to focus on individual runoff processes, using case studies to illustrate the type of disturbance that may occur. We focus deliberately on rural headwater catchments in humid temperate areas since this is the type of basin in which the majority of our own research has been conducted. Burt (1992) has considered the hydrology of such basins in some detail, including their definition: he assumes that the pattern of outflow discharge from headwater catchments is strongly related to runoff production at the hillslope scale and that water quality in such catchments is largely determined by non-point rather than by point source inputs.

THE SENSITIVITY OF RUNOFF PRODUCTION TO DISTURBANCE

Infiltration-excess Overland Flow

Surface runoff production from partial areas is largely controlled by infiltration capacity. Low infiltration and high rates of surface runoff and erosion are often associated with bare soils, which may occur naturally in areas where lack of precipitation, throughout the year or seasonally, leads to an incomplete vegetation cover, or under agriculture where natural vegetation cover has been removed. Church & Woo (1990) note that spatial variation in rainfall intensity is much less than the range of soil permeabilities, so that soil type may be a more important control of runoff generation than climate. Where rainfall intensities commonly exceed infiltration capacity, infiltration-excess overland flow will be the dominant storm runoff mechanism. Where the infiltration capacity of the soil is higher than the rainfall rate, subsurface stormflow and saturation-excess overland flow will prevail. An important threshold exists therefore: a soil whose surface condition is easily disturbed will be most likely to shift from one flow domain to another. This can be illustrated using results from Heathwaite, Burt & Trudgill (1990) who used a rainfall simulator to measure infiltration and surface runoff on silty clay loam soils at Slapton, Devon, England. Although normally of high infiltration capacity, these soils are easily compacted by overgrazing or by heavy farm machinery. The results presented in Table 14.1 show that as the bulk density of the soil surface increases, so the infiltration capacity drops dramatically. Compared to old pasture and woodland, the infiltration capacity of these intensively farmed soils is easily lowered and brought within the range of rainfall intensities commonly experienced in this area. In the very wet winter of 1989−90, all fields planted with winter barley contained rills: rolling the seedbed compacted

Table 14.1 Rainfall simulation of the effect of land use on surface runoff from hillslope plots (after Heathwaite, Burt & Trudgill, 1990, with additional data from Burt *et al.*, 1983)

Land use	Rainfall intensity (mm h^{-1})	Infiltration capacity (mm h^{-1})	Bulk density (g cm^{-3})
Temporary grass	12.50	12.33	0.96
Barley	12.50	11.04	1.08
Rolled, bare ground	12.50	4.00	0.93
Lightly grazed permanent pasture	12.50	5.85	1.12
Heavily grazed permanent pasture	12.50	0.10	1.18
Permanent pasture[a]	—	9	—
		(range 3.6–36)	
Freshly ploughed soil[a]	—	50	—
Woodland soil[a]	—	180	—

[a] Determined using a ringe infiltrometer.

the soil surface, reducing the infiltration capacity to around 4 mm h^{-1}. On a number of occasions, hourly rainfall totals exceeded this level, inducing widespread runoff and rill erosion. This analysis was confined to hourly intensities; no doubt further significant erosion was achieved during short bursts of very intense rain. The Slapton catchments represent an area where subsurface stormflow dominates the runoff regime (Burt & Butcher, 1985). However, because the soils are sensitive to compaction, infiltration-excess overland flow may easily be produced if soils are mismanaged. Not surprisingly, as agriculture intensifies, observations of surface wash and rill erosion have become much more common in this region.

Of particular concern is the effect on infiltration of a thin surface zone of reduced permeability. As noted in the example given above, this can arise due to compaction by animals or machinery. More importantly perhaps, a surface crust can develop during a rainstorm as a consequence of a breakdown in soil structure by impacting raindrops, by disintegration of surface clods and soil aggregates when wetted with free water, or by clogging of pores by suspended soil particles transported in infiltrating water (Romkens, Prasad & Whisler, 1990). Such effects become most pronounced where the vegetation cover has been removed. Carson & Kirkby (1972, pp. 216–219) considered rates of soil wash erosion in relation to mean rainfall per rain day and the daily rainfall able to infiltrate under different vegetation cover. They show that the effect produced by the removal of vegetation is least where the rain per rain day is large (e.g. monsoonal climate in India) since infiltration-excess overland flow would already be common, and rates of wash erosion high, even under forest cover. At the other extreme, where rain per rain day is low (e.g. south-west England), even removal of grassland would increase rates of wash erosion by several orders of magnitude. Nevertheless, rates of wash erosion on bare ground in south-west England would remain well below those in India, because of the much higher rainfall intensities experienced in the latter region. Church & Woo (1990) argued that the soil may be a more sensitive control of surface runoff than climate. Carson and Kirkby's approach suggests paradoxically that, on a global scale at least, the effect of vegetation removal will appear most dramatic in cool temperate regions since these are areas of relatively low mean rainfall per rain day.

The impact of forestry practices on runoff are discussed in the next section. However, in upland Britain these effects are complicated by drainage prior to planting. The normal method used in the poor peaty soils is deep ribbon ploughing at right angles to the contour. This greatly extends the drainage network. The effect is to increase peak discharge and reduce time-to-peak of flood hydrographs in small catchments (Robinson, 1986; see also the discussion of Acreman's, 1985, work below). Significant erosion of the furrows may also result, especially where sensitive soils are ploughed (Burt, Donohoe & Vann, 1983a; Soutar, 1989). Improved silvicultural practices recommended by the Forestry Commission (1988), such as leaving unploughed a buffer zone between land to be planted and the stream channel, will reduce, but not necessarily prevent, this problem from occurring in the future.

If the spatial pattern of erosion is to be fully described, then both the location of partial source areas and on-site erosion rates must be predicted. Even where this is achieved, the problem of identifying possible linkages between the source of erosion and the stream remains. Sources distant from the stream may connect via roads and tracks. At the same time, significant amounts of sediment may be deposited downslope, so that the delivery of sediment is much less than the on-site erosion rate. Walling (1990) concludes that it remains difficult to predict how much of any soil eroded will be delivered to the channel of the neighbouring stream. Thus, while the sensitivity of certain soils and land use practices to surface wash erosion is now well established, the *spatial* sensitivity of the process remains uncertain. One is left with the simplistic conclusion that wash erosion in fields adjacent to the stream channel is probably more of a problem in terms of stream sediment load than erosion in distant fields. We concur with Walling (1990) who argued that more field investigations of sediment delivery systems are needed, in order to provide a basis for an improved understanding of the processes and spatial linkages involved.

SUBSURFACE RUNOFF AND VARIABLE SOURCE AREAS

Land use may control the production of subsurface runoff through its influence on the water balance of a slope and the leaching of solutes from the soil. The volume and timing of runoff differ significantly between forested and unforested slopes. This matter has been the subject of much research (see the review by Bosch & Hewlett, 1982), most notably in the United States at the Coweeta Hydrological Laboratory in North Carolina (Swank & Crossley, 1988) and in the United Kingdom at the Plynlimon experimental catchment in mid-Wales (Calder, 1990). It is most likely that such differences relate to changes in evaporative loss, notably via interception, rather than to modification of soil properties. Douglass & Swank (1972) showed that runoff volume increased significantly in summer and autumn when a forested catchment at Coweeta was clearcut. Only at the end of winter, when soils in the forest had been recharged also, were runoff volumes similar to those in the cleared area. Burt & Swank (1992) used flow frequency analysis to show, for forest to grass conversion, that flow rates at all flow frequencies were higher than predicted in winter, except when there was a dense grass growth. The marked increase in winter baseflow suggests that any soil moisture deficit is quickly removed under grass, compared with forest, so increasing flow in the early part of the winter. This effect would be particularly encouraged since, under grass, low flows are also much higher than expected, implying that soils under grass remain wetter at that time also. Table 14.2 shows the effect of afforestation with conifers at Plynlimon: mean runoff on the Severn (forested) was decreased by one-quarter compared with the Wye (open moorland), while evaporation from the Severn more than doubled, mainly as a result of additional interception loss. Clearly then, catchments are sensitive to afforestation or

Table 14.2 Mean annual values of precipitation (mm) and runoff (mm) for two Welsh basins, 1970–1980 averages (after Calder, 1990)

	Severn (forested)	Wye (moorland)
Precipitation (P)	2300	2417
Runoff (Q)	1501	2047
$P - Q$	799	370
$(P_s - Q_s) - (P_w - Q_w)$	+429	

Subscripts: S = Severn; w = Wye.

deforestation. Such changes in the water balance probably represent the most important hydrological effect of land use change in rural catchments. Only urbanisation can produce comparable changes in water yield.

There have, however, been few studies of the spatial impact of forestry. Bosch & Hewlett (1982) established that the percentage area reforested or deforested correlated significantly with the annual change in runoff. It is clear from research conducted at various sites in the United Kingdom by the Institute of Hydrology that, at the basin scale, sensitivity to this sort of change depends on the duration of canopy wetness. Therefore not all basins will respond to the same degree to afforestation. Use of simulation models such as PROSPER (Huff & Swank, 1985) or MANTA (Sellers & Lockwood, 1981) should allow us to extend results gained from experimental basins more generally, in order to evaluate the sensitivity of particular basins to such change. What seems not to have been studied is whether the location of the land use changes can affect the water yield or flood hydrograph of the main basin (Burt, 1989). Douglass & Swank (1975) improved a forest clearance — water yield relationship for 22 forest cutting experiments in the Appalachians by adding an energy variable, which itself depends on the slope, aspect and latitude of each catchment. Dunford & Fletcher (1947) showed that removal of streamside vegetation in a Coweeta catchment caused diurnal variations in streamflow to disappear temporarily, presumably because the transpiration demand was lost. However, within a few weeks after cutting, the rapid recovery and regrowth of vegetation caused these fluctuations to reappear in modified form. Acreman (1985) showed how the unit hydrograph of a basin in southern Scotland changed in form as fragmentary afforestation took place within the area. Acreman concluded that, though the small-scale effects of afforestation and preplanting ditching were obvious, the cumulative regional effects of sub-basin land use changes produce a complex modification of the runoff regime by varying the differential timing of flood peaks from different parts of the basin. One solution to such problems might be to use a distributed runoff model in order to establish how sensitive streamflow is to the location of land use changes involving forestry operations.

In low-lying areas, underdrainage is often installed to lower the water table. Open ditches and buried tile drains are both commonly used. The effect on runoff may be quite considerable. There has been a long debate about the impact of drainage on stormflow. Drainage may induce storage and thus reduce flood peaks, or it may conduct water more rapidly out of the basin. Robinson & Beven (1983) showed a seasonal control of flood discharge in tile-drained land. In summer and autumn, when shrinkage cracks in the clay soil have opened wide, runoff may be more rapid from drained land, with macropore flow rapidly reaching the drain. In winter, the undrained clay produces surface saturation more rapidly and peak discharges are higher from undrained than drained areas.

Whether catchments produce more stormflow as a result of drainage remains an open question. What is clear is that tile drains encourage more subsurface flow. This in turn may result in more nitrate leaching, especially since drained land may be farmed more intensively with larger numbers of livestock and/or heavier applications of fertiliser on the drained fields. There is now little doubt that the problem of high concentrations of nitrate in rivers and groundwater has resulted from the progressive intensification of agricultural practices and an increasing reliance on the use of nitrogen fertiliser (Royal Society, 1983). Numerous plot experiments have shown that increased fertiliser applications are likely to leak to larger leaching losses. This problem may be compounded on underdrained clay soils since large quantities of subsurface runoff are generated rapidly following drainage. It is often argued that, for a basin developed in mixed lithology like the Upper Thames, limestone areas are most vulnerable to nitrate losses because of the intensive arable agriculture found on such land. Once drained, clay vales may become equally susceptible.

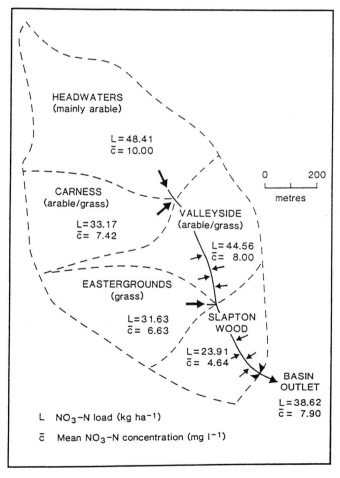

Figure 14.2 The spatial pattern of nitrate leaching in the Slapton Wood catchment. Large arrows indicate point sources of nitrate (springs); small arrows imply non-point inputs for that section of the stream

It has also be possible to relate land use and nitrate leaching at a subcatchment scale. Burt & Arkell (1987) showed that the largest losses of nitrate in the Slapton Wood catchment were from the intensively farmed arable headwaters. There were lower losses from areas of lightly grazed permanent pasture and smallest losses from the woodland section (Figure 14.2). One interpretation of the results shown in Figure 14.2 is gained by invoking the ergodic hypothesis. This suggests that, under certain circumstances, sampling in space can be equivalent to sampling in time and that space–time transformations are possible. The wood/grass/arable gradation observed within the Slapton Wood catchment may therefore resemble the temporal change that has occurred in many agricultural basins as farming has become more intensive over the last half century. In the absence of observations of nitrate concentrations before 1970, we have used an export coefficient model (Johnes & O'Sullivan, 1989) to link land use changes since 1905 to stream nitrate loads (Heathwaite & Burt, 1991) and so to compare with the above ergodic interpretation. Predicted nitrogen losses for the Slapton catchments for the period 1905–85 are shown in Table 14.3. A clear trend of increasing nitrate loss is shown with organic nitrogen as the main source of nitrogen, primarily as a result of larger numbers of livestock in the catchment. These predictions relate only to *potential* losses, since the model is lumped. *Actual* losses will be modified by the location of different land uses in relation to the stream system. For example, land use in riparian zones is likely to be more significant than at sites more distant from the stream. This is why the drainage of land bordering stream channels is potentially such a problem for pollution of fresh water. We conclude that, in general terms, catchments with most intensive land use will be most likely to generate large nutrient losses. In the United Kingdom this will encompass not only the arable land of the south-east but also increasingly heavily fertilised grassland in the west. Within a given basin, the fields most sensitive to such losses are likely to be found in riparian zones. This land was once characterised by high water tables. While drainage allows intensive agriculture and prevents widespread surface runoff, it may well be that important functions associated with the floodplain ecotone (Naiman & Decamps, 1990), providing a barrier to the movement of nutrients and sediment from farmland to the stream channel, have been lost. The role of such ecotones in relation to nitrate leaching is further explored in Haycock & Burt (this volume).

CATCHMENT SENSITIVITY: LOCATION AND SCALE

It has been obvious for many years that drainage basins are susceptible to land use change, both in terms of the quantity and quality of runoff produced. In that sense our review has established nothing new. What is more difficult is to determine which catchments (and indeed which sites within a catchment) are most sensitive to these changes. At the catchment scale, sensitivity might best be investigated using computer simulation models, particularly at the planning stage. For example, USLE could provide initial estimates of soil loss for basins of known climate, soil type, topography and land use. PROSPER or MANTA would give water balances for afforested catchments using information about rainfall duration and intensity. Finally, an export coefficient model could indicate the effect of changing land use on water quality. Such lumped models, though lacking the theoretical structures found in physically based distributed models, are at least amenable to application to large areas without the need for excessive periods of field calibration. Thus they are relatively cheap and efficient (of time and resources) to use.

Table 14.3 Land use, livestock numbers and nitrogen losses in the Slapton catchments, 1905–1985

	1905	1935	1945	1950	1955	1960	1965	1970	1975	1980	1985
(a) Major land uses (hectares)											
Total grass	2997	2902	2329	2656	2910	3135	2795	3259	3231	3095	3207
Total cereals	866	851	1082	990	752	809	955	755	708	752	845
Other arable	631	389	592	456	504	524	412	526	292	290	210
(b) Livestock numbers											
Cattle	1793	1695	2331	3300	3658	4489	4792	6178	6541	4760	6725
Pigs	694	1695	255	574	2263	2074	3507	3371	1971	1105	1096
Sheep	6044	5394	3042	4894	7826	10448	13419	14730	12739	12155	14867
Poultry	?	19226	8740	19924	19944	17183	9116	10989	9329	1367	1697
(c) Fertiliser application (tonnes)											
Inorganic N	55.0	50.5	215.6	132.4	182.1	270.8	305.5	447.1	483.3	566.8	690.8
Organic N	213.8	267.9	198.8	288.0	361.6	430.0	497.3	599.8	586.5	443.6	598.7
(d) Predicted losses of nitrogen (tonnes)											
Loss from farmland											
Inorganic	3.09	2.93	13.30	8.81	12.16	18.91	19.97	28.27	35.66	39.00	48.89
Organic	34.81	43.17	32.25	46.70	58.35	69.64	80.37	97.04	95.13	72.26	100.26
Subtotal	37.90	46.10	45.55	55.51	70.51	88.55	100.34	125.31	130.79	111.26	149.15
Human contribution	6.35	6.35	6.18	6.12	6.22	6.22	6.57	6.87	7.28	7.91	8.53
Total	48.10	56.40	55.68	65.58	80.68	98.89	110.86	136.13	142.02	123.12	161.63
(e) Predicted NO_3-N concentrations (mg l^{-1}) for the Slapton catchments											
	1.98	2.33	2.29	2.71	3.32	4.08	4.57	5.62	5.86	5.07	6.67
(f) Observed NO_3-N concentrations (mg l^{-1}) for the Slapton Wood catchment											
	—	—	—	—	—	—	—	5.14	6.48	6.85	8.32

(a)–(d) from Acott (1989); (f) from Burt et al. (1988).

At the subcatchment scale, there is growing interest in the sensitivity of particular locations to disturbance, especially where land adjoins the river channel. That rivers are no longer bordered by undrained meadows or woodland may create particular difficulties for stream water quality. The recent volume edited by Naiman & Decamps (1990) focuses on the functional properties of such ecotones. It shows how the flow of water, nutrients and other materials from catchment slopes into streams is mitigated by the presence of riparian ecotones, which can retain these pollutants or retard their movement. Where non-point discharges are the major source of river pollution, it may be much easier and, of more immediate benefit, to manipulate riparian land use, rather than to attempt in-stream enhancement activities (Risser, 1990). This presupposes that riparian ecotones do indeed 'function' in the manner described. We conclude that further field research is required to establish the level of alleviation possible by positive management of floodplain ecotones, how such functions vary seasonally and how their position within the catchment might alter their importance as sites of sediment and nutrient retention.

Since changes in land use can affect the volume and quality of runoff produced from small basins, it follows that these changes may prove significant at the larger scale. However, there is surprisingly little evidence that can be used to relate small catchment response, where hillslope runoff generation is the dominant hydrological mechanism, to the flood response of large basins, where channel storage and routing must also be considered. In large basins the effects of travel time and channel storage prevent the establishment of a clear relation between runoff from small basins and the response of the entire basin. It is important to establish the size of basin within which headwater hydrographs can be used to interpret the total basin response. By implication, if the response of a given basin is effectively the sum of inputs from individual source areas, then the specific location of the source areas may also prove important in that case (Burt, 1989). Even less seems to be known about the links between the nature of runoff from small tributaries and water quality in the main river. Walling (1983) noted that detailed understanding of sediment delivery to the basin outlet depends in part upon a knowledge of sediment source areas within the basin. Similar comments could be made with respect to solute delivery.

Certain locations may be particularly sensitive in that changes in land use here may cause significant changes to runoff and water quality downstream. Monitoring in small catchments may therefore have relevance to and be representative of the behaviour of larger basins (Heathwaite, Burt & Trudgill, 1989). Sensitive areas include riparian zones and those areas contributing overland flow. It is crucial to identify these areas, to understand their functional role and to recognise their importance within strategies for drainage basin management. Burt & Haycock (1991) have identified the role of floodplains as buffer zones between farmland and rivers. Their work suggests that the use of undrained floodplains as barriers to nitrate movement could allow modern farming and water supply to coexist in the same basin. Such buffer zones would also trap eroded soil (and associated nutrients) and additionally could provide habitats of high conservation value. Burt & Haycock (1991) suggest that floodplain land should be deliberately taken out of production in order to avoid some of the pollution problems associated with modern agriculture. This strategy of protecting sensitive locations may well have more general relevance for the wise management of catchment areas.

REFERENCES

Acott, T. (1989). An investigation into nitrogen and phosphorus loading on Slapton Ley, 1905–1985. Unpublished B.Sc. Dissertation, Department of Environmental Sciences, Plymouth Polytechnic.

Acreman, M. C. (1985). Effects of afforestation on the flood hydrology of the Upper Ettrick valley. *Scottish Forestry*, **39**(2), 89–99.

Anderson, M. G. and Burt, T. P. (1990). *Process Studies in Hillslope Hydrology*, Wiley, Chichester.

Betson, R. P. (1964). What is watershed runoff? *Journal of Geophysical Research*, **69**, 1541–1552.

Bosch, J. M. and Hewlett, J. D. (1982). A review of catchment experiments to determine the effect of vegetation changes on water yield and evapotranspiration. *Journal of Hydrology*, **55**, 3–23.

Burt, T. P. (1989). Storm runoff generation in small catchments in relation to the flood response of large basins. In Beven, K. and Carling, P. (eds), *Floods*, Wiley, Chichester, pp. 11–36.

Burt, T. P. (1992) The hydrology of headwater catchments. In Calow, P. and Petts, G. E. (eds), *Rivers Handbook*, Blackwell, Oxford, pp. 3–28.

Burt, T. P. and Arkell, B. P. (1987). Temporal and spatial patterns of nitrate losses from an agricultural catchment. *Soil Use and Management*, **3**, 138–142.

Burt, T. P. and Butcher, D. P. (1985). Topographic controls of soil moisture distribution. *Journal of Soil Science*, **36**, 469–476.

Burt, T. P. and Haycock, N. E. (1991). Farming and nitrate pollution. *Geography*, **76**, 60–63.

Burt, T. P. and Swank, W. T. (1992). Flow frequency responses to hardwood-to-grass conversion and subsequent succession. *Hydrological Processes*, **6**, 179–188.

Burt, T. P., Donohoe, M. A. and Vann, A. R. (1983). The effects of forestry drainage operations on upland sediment yields: the results of a storm-based study. *Earth Surface Processes and Landforms*, **8**, 339–346.

Burt, T. P., Butcher, D. P., Coles, N. and Thomas, A. D. (1983). The natural history of the Slapton Ley nature reserve XV: hydrological processes in the Slapton Wood catchment. *Field Studies*, **5**, 731–752.

Burt, T. P., Arkell, B. P., Trudgill, S. T. and Walling, D. E. (1988). Stream nitrate levels in a small catchment in south west England over a period of 15 years (1970–1985). *Hydrological Processes*, **2**, 267–285.

Calder, I. R. (1990). *Evaporation in the Uplands*, Wiley, Chichester.

Carson, M. A. and Kirkby, M. J. (1972). *Hillslope Form and Process*, Cambridge University Press.

Church, M. and Woo, M.-K. (1990). Geography of surface runoff: some lessons for research. In Anderson, M. G. and Burt, T. P. (eds), *Process Studies in Hillslope Hydrology*, Wiley, Chichester, pp. 299–326.

Douglass, J. E. and Swank, W. T. (1972). Streamflow modification through management of eastern forests. USDA Forest Service Research Paper SE-94.

Douglass, J. E. and Swank, W. T. (1975). Effects of management practices on water quality and quantity: Coweeta Hydrologic Laboratory, North Carolina. In *Muncipal Watershed Management Proceedings*, United States Department of Agriculture, Forest Service General Technical Report NE-13, pp. 1–13.

Dunford, E. G. and Fletcher, P. W. (1947). Effect of removal of streambank vegetation upon water yields. *Transactions of the American Geophysical Union*, **28**, 105–110.

Dunne, T. (1978). Field studies of hillslope flow processes. In Kirkby, M. J. (ed.), *Hillslope Hydrology*, Wiley, Chichester, pp. 227–293.

Forestry Commission (1988). *Forests and Water: Guidelines*. Forestry Commission, Edinburgh.

Haycock, N. E. and Burt, T. P. (1993). Catchment sensitivity to nitrate leaching: the effectiveness of riparian zones in protecting stream ecosystems. This volume, Chapter 16.

Heathwaite, A. L. and Burt, T. P. (1991). Predicting the effect of land use on stream water quality. IAHS Publication **203**, 209–218.

Heathwaite, A. L., Burt, T. P. and Trudgill, S. T. (1989). Runoff, sediment and solute delivery in agricultural drainage basins: a scale-dependent approach. IAHS Publication no. 182, pp. 175–191.

Heathwaite, A. L., Burt, T. P. and Trudgill, S. T. (1990). Land-use controls on sediment production in a lowland catchment, south-west England. In Boardman, J., Foster, I. D. L. and Dearing, J. A. (eds), *Soil Erosion on Agricultural Land*, Wiley, Chichester, pp. 69–86.

Hewlett, J. D. (1961). Watershed management, Report for 1961, South Eastern Forest Experimental Station, US Forest Service, Asheville, North Carolina.

Horton, R. E. (1933). The role of infiltration in the hydrologic cycle. *Transactions of the American Geophysical Union,* **14**, 446–460.

Huff, D. D. and Swank, W. T. (1985). Modelling changes in forest evaporation. In Anderson, M. G. and Burt, T. P. (eds), *Hydrological Forecasting,* Wiley, Chichester, pp. 125–152.

Johnes, P. J. and O'Sullivan, P. E. (1989). Nitrogen and phosphorus losses from the catchment of Slapton Ley, Devon — an export coefficient approach. *Field Studies,* **7**, 285–309.

Kirkby, M. J. (ed.) (1978). *Hillslope Hydrology,* Wiley, Chichester.

Kirkby, M. J. (1988). Hillslope runoff processes and models. *Journal of Hydrology,* **100**, 315–339.

Naiman, R. J. and Decamps, H. (1990). *The Ecology and Management of Aquatic-Terrestrial Ecotones,* Parthenon Press, UNESCO, Paris.

Risser, P. G. (1990). The ecological importance of land-water ecotones. In Naiman, R. J. and Decamps, H. (eds), *The Ecology and Management of Aquatic-Terrestrial Ecotones,* Parthenon Press, UNESCO, Paris, pp. 7–21.

Robinson, M. (1986). Changes in catchment runoff following drainage and afforestation. *Journal of Hydrology,* **86**, 71–84.

Robinson, M. and Beven, K. J. (1983). The effect of mole drainage on the hydrological response of a swelling clay soil. *Journal of Hydrology,* **64**, 205–223.

Romkens, M. J. M., Prasad, S. N. and Whisler, F. D. (1990). Surface sealing and infiltration. In Anderson, M. G. and Burt, T. P. (eds), *Process Studies in Hillslope Hydrology,* Wiley, Chichester, pp. 127–172.

Royal Society (1983). *The Nitrogen Cycle of the United Kingdom.* Royal Society, London.

Sellers, P. J. and Lockwood, J. G. (1981). A computer simulation of the effects of differing crop types on the water balance of small catchments over long periods. *Quarterly Journal of the Royal Meteorological Society,* **107**, 395–414.

Soutar, R. G. (1989). Afforestation and sediment yields in British fresh waters. *Soil Use and Management,* **5**, 82–86.

Swank, W. T. and Crossley, D. A. (1988). *Forest Hydrology and Ecology at Coweeta.* Ecological Studies 66, Springer-Verlag, New York.

Walling, D. E. (1983). The sediment delivery problem. *Journal of Hydrology,* **65**, 209–237.

Walling, D. E. (1990). Linking the field to the river: sediment delivery from agricultural land. In Boardman, J., Foster, I. D. L. and Dearing, J. A. (eds), *Soil Erosion on Agricultural Land,* Wiley, Chichester, pp. 129–152.

15 Catchment Controls on the Recent Sediment History of Slapton Ley, South-west England

A. LOUISE HEATHWAITE
Department of Geography, University of Sheffield, UK

ABSTRACT

Palaeolimnological evidence and long-term land use, water quality and sediment records are used to examine the sensitivity of a lake basin to catchment change over the past 90 years. The analysis is based on chemical fractionation and ^{210}Pb dating of a sediment core from Slapton Ley. The technique isolates the authigenic, biogenic and allogenic inputs to the lake sediments. It is combined here with a 90 year record for land use change in the catchment and current knowledge on the sediment and nutrient production from different land uses in order to assess the extent to which physical and chemical changes in the lake sediments reflect catchment changes. The results suggest that since 1945, erosion of detrital material from the catchment has increased, which may reflect a postwar increase in the area of arable and temporary grassland. The associated increase in the input of nitrogen and phosphorus to the lake, estimated at 38 tonnes N and 0.4 tonnes P in 1905 and rising to 288 tonnes N and 2 tonnes P in 1990, has accelerated the rate of eutrophication in Slapton Ley.

INTRODUCTION

Palaeolimnological studies can be used to reveal the history of sedimentation and eutrophication in lake basins. They are a means of examining catchment controls, particularly land use changes, on rates of sediment and solute delivery in stream inputs. From the catchment perspective, a number of studies have used export-coefficients to predict the effects of land use on stream water quality and sediment load (Johnes & Burt, 1990; Owens, Edwards & Van Keuren, 1989; Todd *et al.*, 1989). Where historical land use records are available, it is possible to use export coefficients as a basis for detecting the effects of past changes. If combined with other sources of catchment information, such as lake sediment history or long-term stream water quality and sediment records, a reasonable assessment of the impact of land use changes on receiving waters can be made. There are two limitations with such an approach. First, export coefficients are not physically based (Heathwaite & Burt, 1991). In particular, they fail to incorporate the physical and biochemical modifications of sediment and solute export, which form part of the hydrological pathway from the hillslope to the stream. Secondly, palaeolimnological evidence may be isolated from catchment controls because sediment and solute loads may be modified by instream processes before they reach the lake.

Landscape Sensitivity. Edited by D. S. G. Thomas and R. J. Allison
© 1993 John Wiley and Sons Ltd

It is possible that only a small fraction of the catchment load may reach the lake (Walling, 1990) or it may arrive in a substantially altered form. This would make qualification of the link between lake and catchment difficult.

Bearing in mind these limitations, this chapter describes the results of the chemical fractionation of a sediment core from Slapton Ley, south-west Devon, and examines how the information provided may be used to evaluate catchment processes when linked to long-term land use records. Chemical analysis of lake sediment cores involves several difficulties, primarily as a result of the varied and not easily identified sources for the particulate material that constitutes most of the sediment matrix. The chemical fractionation procedure used here isolates the authigenic, biogenic and allogenic fractions of the sediment. It therefore allows a more precise identification of the sediment source than bulk (total) inorganic chemical determinands (Mackereth, 1965). The chemical analysis was coupled with ^{210}Pb dating of the near-surface sediments, which as well as enabling comparison with other long-term information, such as land use records, also allows expression of the chemical results in terms of sediment influx (mg cm^{-2} year^{-1}). This removes the problem of interdependence between different chemical elements, which can occur when results are expressed in terms of concentration (mg g^{-1}).

The changes in the lake sediment history will be examined using the evidence for long-term land use change and more recent data for stream water quality (Heathwaite & Burt, 1991; Heathwaite, Burt & Trudgill, 1989). The current inorganic solute and sediment load from the catchment is of the order: 288 tonnes N, 2 tonnes P and 1440 tonnes suspended sediment (Heathwaite & Burt, 1991). This is a considerable increase from the N and P catchment load predicted for 1905 load of 38 tonnes N and 0.4 tonnes P (Acott, 1989). It is not surprising, then, that Slapton Ley is now considered to be a hypertrophic lake (Heathwaite & O'Sullivan, 1991).

THE SITE

Slapton Ley is a coastal lake, 10 km south-west of Dartmouth, Devon, UK (National Grid Reference: SX 825479). It is divided into the Higher Ley, a 39 ha reedbed system, and the 77 ha Lower Ley, which is a shallow freshwater lake with a maximum depth of 2.8 m. The lake is the sink for solute and sediment inputs from the surrounding 46 km^2 of arable and grassland catchment, which can be divided into four subcatchments, whose current annual sediment and solute loads are summarised in Table 15.1. The topography consists of wide

Table 15.1 Slapton catchments inorganic load (kg ha^{-1}) (October 1987−September 1988)

			Land use		Inorganic load (kg ha^{-1})			
Catchment	Area	Runoff	Arable	Grass	Ammonium	Nitrate	Phosphate	Suspended sediment
	(ha)	(mm)	(%)	(%)		Kg ha^{-1}		
Gara	2362	920	11.9	81.2	1.79	68.03	0.38	503.34
Slapton Wood	93	581	36.1	32.1	0.16	63.60	0.21	66.15
Start	1079	950	34.2	52.6	1.39	103.93	0.58	224.89
Stokeley Barton	153	148	66.0	28.8	0.23	19.51	0.21	23.37

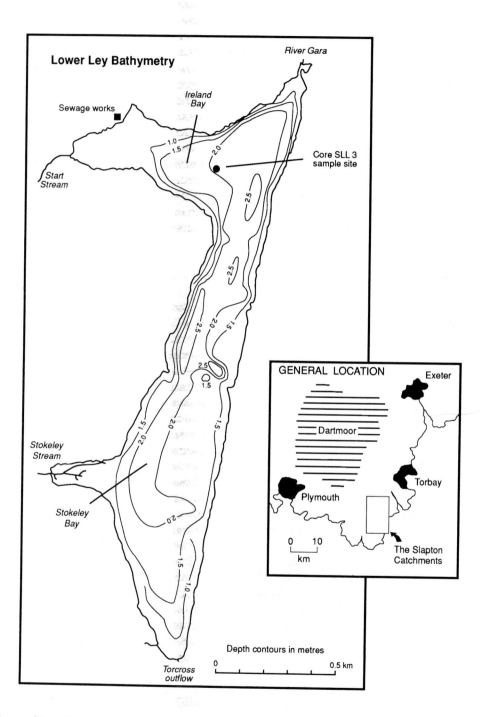

Figure 15.1 The Lower Ley showing lake bathymetry, main inflows and outflow and sample site for core SLL3

plateaux of low gradient dissected by narrow, deep valleys. The area is underlain by impermeable Lower Devonian slates and shales, with freely draining acidic and nutrient-poor silty clay loam soils. Land use in the largest (Gara) subcatchment is dominated by permanent pasture and temporary grass (Table 15.1). In the Start catchment further south, lower altitudes result in a higher proportion of arable land. Mean annual rainfall for the period 1961–1988 is 1039 mm; mean annual runoff is 650 mm. The area has a mean annual temperature of 10.5°C.

The sediments of Slapton Ley have been previously studied by Crabtree & Round (1967) and Morey (1976). Their results suggest that since its formation by onshore movement of shingle, roughly 1000 BP, the Ley has been a shallow, eutrophic lake dominated by *Fragilaria* diatoms and epiphytic and epipelic taxa, (O'Sullivan and Heathwaite, in press). There is little evidence for prolonged marine incursions.

METHODS

Five cores were collected in July 1987 from beneath a 2 m water column in the area of the Lower Ley known as Ireland Bay (Figure 15.1) using a 1 m 'mini' Mackereth corer (Mackereth, 1969). This site was chosen in order to facilitate comparisons between this investigation and that of Crabtree & Round (1967). Cores were stored intact at 5°C until subsampling took place.

The pattern of sedimentation in the lake at the coring site was evaluated by whole core magnetic susceptibility (K) using a Bartington MS1B susceptibility bridge and a loop sensor designed specifically for use with Mackereth cores. Of the five cores taken from Slapton Ley, one (SLL3) was selected for chemical analysis together with ^{210}Pb by gamma spectrometry using a well-type coaxial, low-background, intrinsic germanium detector fitted with a NaI(Tl) escape suppression shield (Appleby *et al.*, 1986). Prior to chemical analysis, subsamples of the core were extracted in twenty 1 cm slices from the top 20 cm of the core

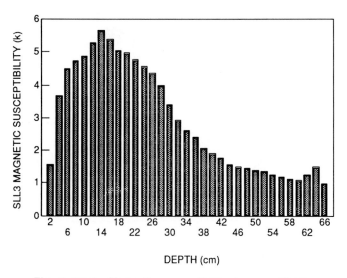

Figure 15.2 Magnetic susceptibility for core SLL3

and 2 cm slices from 20 cm downwards. The sediment below 40 cm was discarded. There were a number of reasons for this. First, this research is concerned only with the very recent history (the last 150 years) of Slapton Ley, since the long-term history has been covered to some extent by Crabtree & Round (1967) and by Morey (1976). Secondly, the ^{210}Pb profile for an adjacent core in the Lower Ley indicated that the top 20 cm of sediment had been deposited since *ca.* 1830. Sediment below this depth was therefore likely to be beyond the range of the ^{210}Pb dating technique. Finally, in order to interpret chemical and sedimentary trends in the Ley, information on catchment changes is needed. Reliable land use records only extend back to the 1850s and the focus is, therefore, on the period with the most accurate catchment data and dating timescale.

The core slices were subsampled for the determination of wet volume, dry mass content, bulk density and ash content. The carbonate content of the samples was calculated from weight loss between ignition at 550°C, and 925°C. The sequential chemical fractionation procedure used here followed that of Engstrom & Wright (1984) and is described fully in Heathwaite & O'Sullivan (1991).

RESULTS AND DISCUSSION

Sediment Chronology and Magnetic Susceptibility

The spatial pattern of lake sedimentation in the Lower Ley has been studied using multiple coring and whole core susceptibility, which enabled the construction of isopachs of sediment thickness (O'Sullivan and Heathwaite, in press). From the ^{210}Pb results, peak magnetic susceptibility (K) in all cores is dated as approximately 10 years BP. Whole core initial apparent reversible magnetic susceptibility for core SLL3 is shown in Figure 15.2. The K peak around

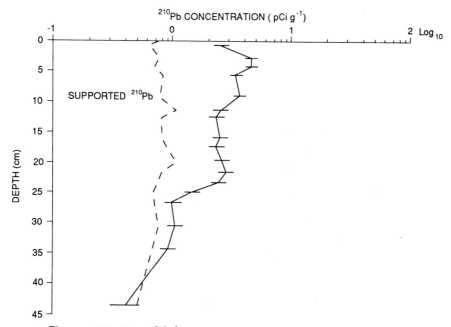

Figure 15.3 Total ^{210}Pb concentration with depth for core SLL3

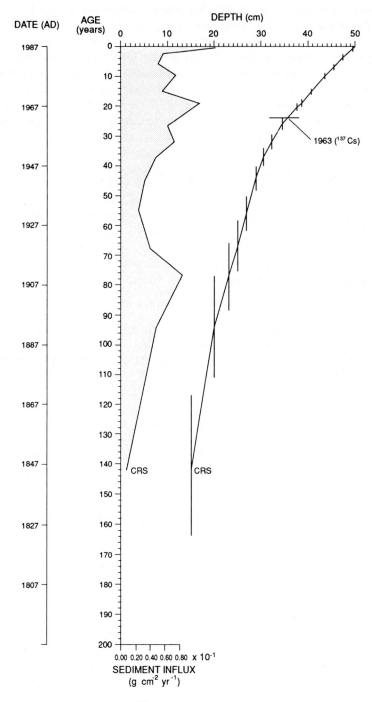

Figure 15.4 ^{210}Pb chronology for core SLL3

12–14 cm and minor peak at 62–64 cm suggests that it is a typical 'Ireland Bay' core which correlates with the multiple cores taken from this area.

Figure 15.3 illustrates the variation in ^{210}Pb concentration with depth for core SLL3. The unsupported ^{210}Pb inventory for core SLL3 was 7.73 pCi cm^{-2}, giving a constant influx of 0.24 pCi cm^{-2} year^{-1}. Owing to the non-monotonic nature of the unsupported ^{210}Pb profile shown in Figure 15.3, chronology could only be derived using the CRS model (Appleby et al., 1986). The time–depth relationship is shown in Figure 15.4. The mean sedimentation rate is 0.248 cm year^{-1} but there is a significant increase in sediment accumulation since ca. 1950, with a maximum in 1968. This peak was confirmed by ^{137}Cs determination.

Physical and Chemical Properties of the Lake Sediments

Figure 15.5 illustrates the total sediment influx and physical properties of core SLL3. There is a clear general upward trend in sediment influx, with four peaks ca. 1822 (36–38 cm), 1908 (27–28 cm), 1968 (12 cm) and 1987 (sediment–water interface). These are discussed below in relation to land use changes in the catchment. A major change in sediment quality at 24–28 cm depth is also shown. Here the total lake sediment influx declines but the proportion of organic matter is increasing. The ^{210}Pb chronology suggests that this change occurred around 1906–1926. Prior to this period, the lake sediments were primarily dominated by mineral matter.

The chemical properties of the core expressed as concentration and influx are shown in Figures 15.6 and 15.7, respectively. The complete curve represents total concentration (Figure 15.6) or total influx (Figure 15.7). The authigenic fraction of this total is represented by light shading and the allogenic fraction by dark shading. The only exception is silica, where light shading reflects the biogenic Si fraction.

Determinands can be grouped into five main categories: (1) major elements: K, Mg, Si and Al; (2) carbonate elements: Ca and CO_3; (3) mobile elements: Mn and Fe; (4) nutrient elements: N and P; and (5) trace elements: Pb and Zn. Of the major elements, K and Al are mainly located in the allogenic fraction (57–97% respectively). This fraction only passes into solution at the last stage of the chemical procedure. It therefore reflects mineral matter embedded in the crystal lattice and a largely 'unaltered' catchment source. Both elements increase in concentration above 24–28 cm depth (Figure 15.6), where the physical composition of the lake sediments also changes. The influx of K and Al increases towards the sediment surface (Figure 15.7), though both show a distinct 'trough' around 16 cm depth, which is also recorded for allogenic Si.

Magnesium, a major element, together with the mobile elements (Mn and Fe), nutrient elements (N and P) and carbonate elements (Ca and CO_3), are all primarily found in the authigenic fraction. Essentially this means that whatever their original source, they reach the lake sediment as a result of chemical fixation within the lake itself. Ca and Mn profiles are similar, and are almost exclusively authigenic. Their influx is high below 20 cm depth and negligible towards the current sediment surface. Below 20 cm, the Mg profile is similar to that of Mn and Ca; towards the sediment surface it resembles that of Fe. The trace elements, together with P and Fe, all have fairly erratic influx, but in general increase towards the current sediment surface. For Si, the rate of influx of both biogenic and allogenic fractions increases above 21 cm (Figure 15.7). In the upper 6 cm, the biogenic Si fraction dominates.

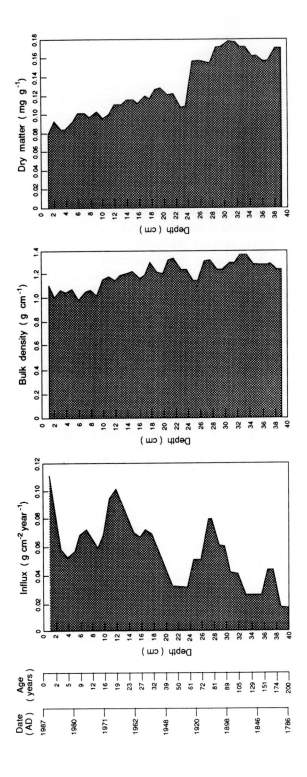

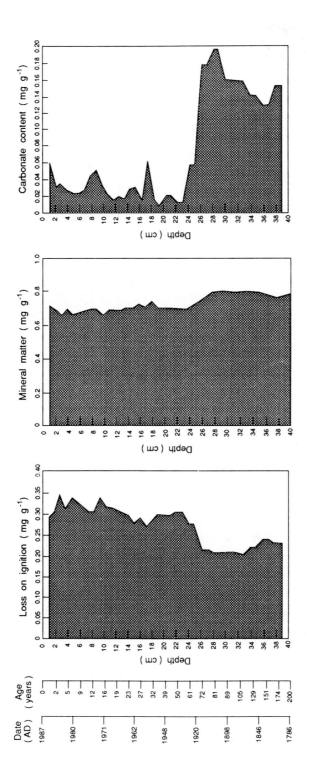

Figure 15.5 Sediment influx and physical properties of core SLL3

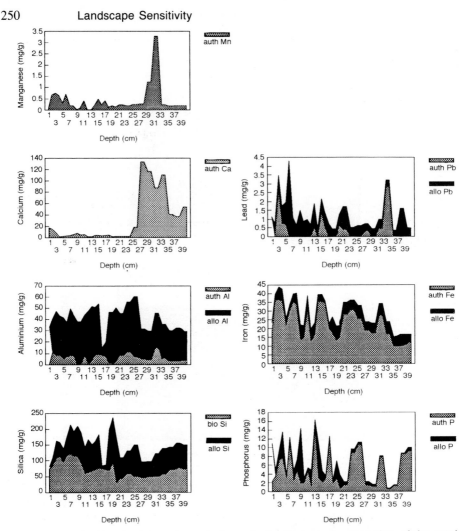

Figure 15.6 Authigenic, biogenic and allogenic fractions as a proportion of the total element concentration (mg g^{-1}) for the Slapton Ley sediments, core SLL3

Lake Sedimentation and Catchment Erosion

The dry mass content of multiple cores from the Lower Ley (Mulder, 1984) suggests that approximately 8 tonnes year^{-1} dry matter is deposited on the lake bed. This is equivalent to an accumulation rate of 9 mm year^{-1}. Relating this to catchment influx, suggests a contribution of around 610 tonnes year^{-1} and an erosion rate of 132 kg ha^{-1}. More recently, the sediment accumulation rate estimated using ^{210}Pb dating of core SLL3 gives a mean sedimentation rate of 2.48 mm year^{-1} over the period 1850–1987. It is clear from Figure 15.4 that the rate of sedimentation has increased since *ca.* 1950 from 2.4 mm year^{-1} in 1948 to a peak of 10.2 mm year^{-1} in 1968. The current annual sedimentation rate is around 8–11 mm year^{-1} at the (1987) sediment–water interface. Using the average dry mass content of the upper lake sediments gives a total sediment input around 534–734 tonnes over the 77 ha lake area.

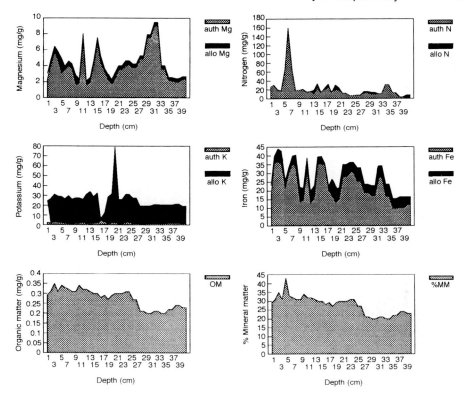

From Table 15.2(a), it is clear that the total catchment suspended sediment load, at 1441 tonnes, is considerably higher than that recorded for the lake sediments. However, this sediment load does not take into account the effect of the Higher Ley reedbed in reducing the sediment load reaching the Lower Ley. Current evidence suggests that the Higher Ley is important in modifying both the sediment and solute input to the Lower Ley (Heathwaite & Burt, 1991). The Gara and Slapton Wood subcatchments pass through the Higher Ley, merging to form a single input to the Lower Ley. Table 15.2(b) illustrates the effect of this hydrological pathway in substantially lowering the sediment input. It is interesting that the modified figure of 365 tonnes year^{-1} sediment input to the Lower Ley is now lower than that predicted from the core. The lake estimate does not allow for any redistribution of sediment within the lake. Slapton Ley is a shallow, turbulent lake. Langmuir cells incorporating the whole water column can often form and may redistribute the lake sediments, possibly giving rise to a higher sedimentation rate in the Ireland Bay area from where this core was taken. Evidence for lake turbulence is reflected in caesium results for core SLL3 (not shown), where ^{137}Cs was found below the 1954 sediment depth dated using ^{210}Pb.

On a longer timescale, there is clear evidence in the chemical fractions of core SLL3 for increased catchment erosion, particularly since the 1950s. This is found in the association between increased influx of K, Al and Si and mineral matter which, according to Engstrom & Wright (1984), can be used to indicate that the amount of eroded material reaching the Lower Ley has increased over the past 150 years. It is interesting to note that on a concentration basis, K, Al and Si are not correlated with mineral matter (Figure 15.6). According to Mackereth's (1966) hypothesis, this would suggest that they are not indicators of erosion.

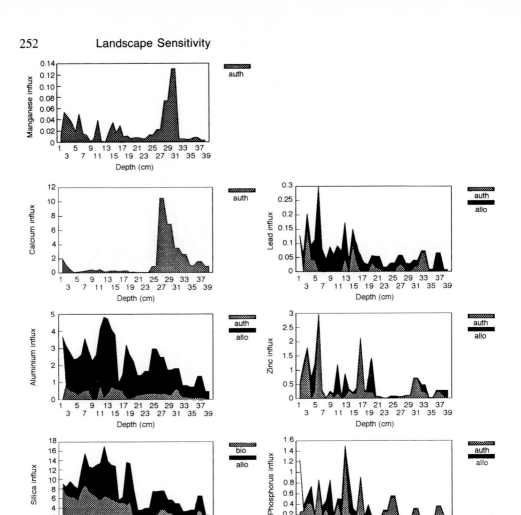

Figure 15.7 Authigenic, biogenic and allogenic fractions as a proportion of the total element influx (mg cm^{-2} year^{-1}) for the Slapton Ley sediments, core SLL3

However, one of the problems in using mineral matter as the basis for this interpretation is that loss on ignition at 550°C may leave variable proportions of authigenic oxides and biogenic silica in the ash. Without chemical fractionation of core sediments, this component cannot be accounted for.

It may be possible to relate the increase in lake sedimentation (Figure 15.5) to catchment changes, using long-term land use records combined with information on sediment and solute production from different land use types. Rainfall simulation studies (Heathwaite, Burt & Trudgill, 1990a, 1990b) indicated that overgrazed permanent grassland is an important source of sediment, total nitrogen and total phosphorus. This is mainly due to increased surface runoff as a result of compaction in heavily grazed areas. Runoff from overgrazed grassland is at least double that from lightly grazed areas, and over 12 times that of ungrazed areas. Where the vegetation cover had been removed by severe poaching, sediment and nutrient production

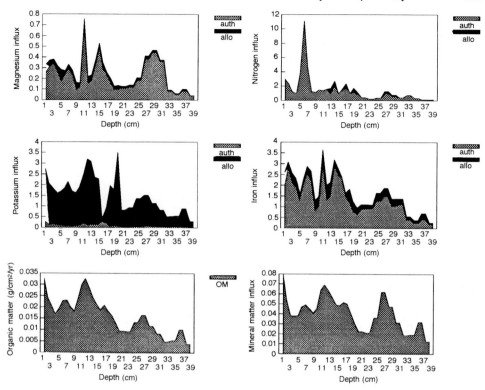

Table 15.2 Suspended sediment delivery from the Slapton catchment (October 1987–September 1988)

Catchment	Area[a] (ha)	Total (tonnes year^{-1})	Load (kg ha^{-1})
(a) Total catchment contribution			
Gara	2362	1189	503
Slapton Wood	93	6	66
Start	1079	243	225
Stokeley Barton	153	4	23
Total	36871	1441	391
(b) Contribution to the Lower Ley			
Higher Ley[b]	24552	118	48
Start	1079	243	225
Stokeley Barton	153	4	23
Total	3687	365	99

[a] Gauged catchment area only.

[b] Catchment area of Higher Ley (Gara + Slapton Wood subcatchments).

(a) Land use (ha)

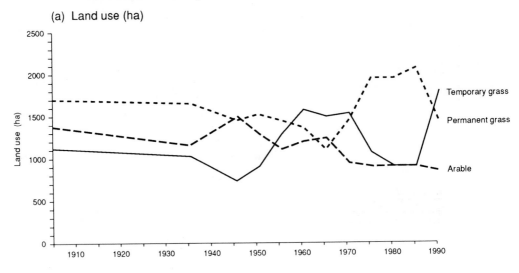

(b) Stocking density

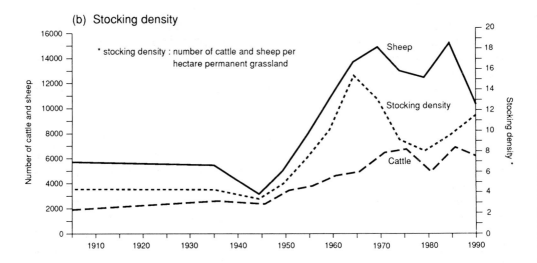

Figure 15.8 Land use, livestock and inorganic nitrogen application in the Slapton catchments (1905–1990)

increased further. For sediment, production in runoff from poached areas was 30 times that of areas with intact vegetation cover. These results suggest that changes in the stocking density of grazed permanent grassland could be an important source of sediment production and hence stream sediment loads. Permanent grassland forms at least 60% of riparian land use (Heathwaite, Burt & Trudgill, 1990a). The remaining area is primarily temporary grassland (19%) or woodland (16%). The proximity of permanent grassland to the stream will increase its relative contribution to the stream sediment and nutrient load. This is because the hydrological pathways and potential modifications of these sources through, for example,

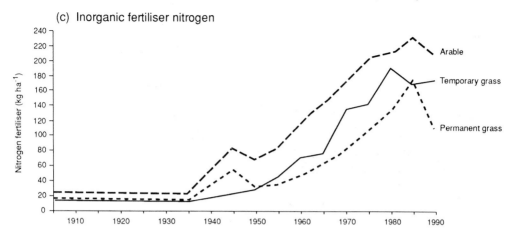

(c) Inorganic fertiliser nitrogen

trapping of sediment on vegetation or adsorption of dissolved phosphorus on sediment particles, will be shorter. Other sources of stream sediment and nutrients are likely to include conversion of land from grassland to arable use. However, because arable land tends to be located on plateaux that are some distance from the stream, modification of the sediment and nutrient produced might be anticipated.

Figure 15.8(a) illustrates the catchment land use over the period 1905–1990. There are three main periods of change that may affect stream sediment and nutrient loads. First, the increase in the proportion of arable land between 1935 and 1945. This occurred at the expense of grassland and would result in an increase in the area of ploughed land in the catchment, so a increase in the stream sediment load might be anticipated. Secondly, the substantial increase in the area of temporary grassland over the period 1945 to 1960. This is matched by a decline in both arable and permanent grassland. Owing to the areal extent of the transfer of land into temporary grassland at this time it is likely that some permanent grassland in riparian zones was ploughed. In addition to the potential increase in sediment production from these areas, the high nitrogen release from ploughed long-term grassland is well known (Young, 1986). Thirdly, the significant increase in the area of temporary grassland in the last five years. As there is little change in the area of arable land, this must reflect the ploughing and reseeding of areas of permanent grassland. Again, the dominance of this land use in riparian zones is likely to increase the stream sediment and solute load, and may in part explain the high loads shown for suspended sediment and nitrate in Table 15.1 for the water year 1987–1988.

Any change in the area of permanent grassland affects the stocking density (Figure 15.8b), which has repercussions for the rate of sediment and solute production from overgrazed grassland (Heathwaite, Burt & Trudgill, 1990a). The stocking density was relatively stable until 1945, after which it increased significantly, peaking in 1960. This is a result of the decline in the available area of permanent grassland and an increase in the number of cattle and sheep in the catchment. The 1960s peak in stocking density, according to the results of the rainfall simulation experiment (Heathwaite, Burt & Trudgill, 1990b), may have significantly increased the sediment and solute load of streams flowing into the Lower Ley.

Comparing these land use changes to the lake sediment record shows four peaks in sediment accumulation, dated as *ca*. 1822, 1908, 1968, 1987 (Figure 15.5). The 1822 and 1908 peaks are unreliable for reasons discussed earlier, and are also beyond the range of the land use records. The 1968 peak marks the end of a period of increased sediment accumulation which began around 1945. It may be a result of land use change from grassland to arable around 1935, followed by the switch to temporary grass around 1945 (Figure 15.8a). Both changes would increase the proportion of ploughed land in the catchment. Land use change is coupled with an increase in the stocking density (Figure 15.8b) between 1945 and 1965, which would again increase sediment and solute export. The major peak in sediment influx at the 1987 sediment−water interface may reflect the recent conversion of permanent to temporary grassland, particularly where riparian areas are ploughed.

Further corroborative evidence for the increase in soil erosion in the catchment since the 1950s is provided by Harris (1988). Cores collected from the Slapton Wood stream were evaluated using their magnetic properties and dated using ^{137}Cs. They suggested that a peak in sediment delivery from channel sources occurred in the 1950s. A further peak was found in the late 1970s, which was linked to hillslope sources. This occurred when stocking densities in the catchment were high (Figure 15.8b), which could increase the load from hillslope sediment sources as a result of trampling and poaching by cattle.

Lake Eutrophication

The increase in catchment erosion and association nutrient influx as a result of land use changes has had a demonstratable effect on the nutrient status of the Lower Ley (Heathwaite & O'Sullivan, 1991; Heathwaite & Burt, 1991). Biogenic silica is often used as an index of primary productivity (Schelske *et al.*, 1987). The proportion of biogenic Si in the lake sediments steadily increases towards the sediment surface from *ca*. 1948, until, in the upper 5 cm of the core (1982−1987), it forms 93% of total Si. Influx of total phosphorus also increases towards the sediment surface; authigenic P is more abundant in the lower core below 21 cm (*ca*. 1942), but since 1950, the proportion of allogenic P has increased. This suggests that prior to 1950, the Lower Ley was eutrophic, but less influenced by its catchment than at present. This is supported by diatom analysis of the lake sediments (O'Sullivan and Heathwaite, in press) where a change in the flora over this period, from abundant *Fragilaria* to centric planktonic diatoms, especially *Cyclostephanos dubius* occurred. This appears to be a response to higher nutrient influx of, for example, catchment minerotrophic P, since 1950. This also ties in well with land use changes in the catchment and a general increase in the sediment loading.

Table 15.3 indicates the predicted inorganic and organic nitrogen and phosphorus losses from the Slapton catchment over the period 1905−1985. The figures were based on export coefficients for the different land uses. There is a clear trend of increasing nutrient export, particularly nitrogen, over the 80 year period. After 1950, the export of these nutrients accelerated and, as a consequence, the Lower Ley became more eutrophic. This is indicated by the increased influx of biogenic Si to the lake sediments.

Organic nitrogen and phosphorus comprise the main nutrient export fraction, and probably reflect the increase in livestock in the catchment over this period (Figure 15.8b). However, a large increase in the rate of fertiliser nitrogen application also occurred (Figure 15.8c), which would also contribute to the stream solute load (Table 15.1). Note that the figures given in Table 15.3 are *potential* nitrogen and phosphorus losses; *actual* losses will be modified

Table 15.3   Predicted losses[a] of nitrogen and phosphorus from the Slapton catchments

Year	N loss from farmland			P loss from farmland		
	Inorganic	Organic	Total[b]	Inorganic	Organic	Total[b]
1905	3.1	34.8	37.9	0.45	0.85	1.30
1935	2.9	43.2	46.1	0.42	1.16	1.58
1945	13.3	32.2	45.6	0.42	0.73	1.15
1950	8.8	46.7	55.5	0.42	1.08	1.50
1955	12.2	58.4	70.5	0.42	1.48	1.90
1960	18.9	69.6	88.6	0.45	1.74	2.19
1965	20.0	80.4	100.4	0.44	2.07	2.51
1970	28.3	97.0	125.3	0.43	2.41	2.84
1975	35.7	95.1	130.8	0.42	2.23	2.65
1980	39.0	72.6	111.3	0.41	1.63	2.04
1985	48.9	100.3	149.2	0.41	2.20	2.61

All units in tonnes year^{-1}.

[a] After Acott (1989).

[b] Totals assume fixed losses from woodland (2.27). Settlement (0.12) and precipitation (1.56 tonnes year^{-1}).

by land use location in relation to the stream system. For example, land use in riparian zones is likely to be more important than land use in plateau areas. Furthermore, nitrogen and phosphorus losses will be modified by physical and biochemical processes, such as deposition with sediment on the hillslope, assimilation by microorganisms and adsorption of dissolved forms onto suspended sediment. Table 15.4 gives a comparison of the predicted and observed stream nitrate loads for the Slapton Wood catchment, which has a 20 year record (Burt *et al.*, 1988). A significant increase in the stream nitrate load is shown. However, the predicted concentration of inorganic N is an underestimate, although total N is close to observed values. This suggests that either total N is readily transformed to inorganic (nitrate) N in the stream,

Table 15.4   Predicted and observed long-term nitrate concentrations in the Slapton catchments

Year	Predicted total N concentration[a] (all catchments) (mg l^{-1})	Predicted inorganic NO$_3$-N concentration[a] (all catchments) (mg l^{-1})	Observed inorganic NO$_3$-N concentration (Slapton Wood) (mg l^{-1})
1905	1.98	—	—
1935	2.33	—	—
1945	2.29	0.55	—
1950	2.71	0.36	—
1955	3.32	0.50	—
1960	4.08	0.78	—
1965	4.57	0.83	—
1970	5.62	1.17	5.14
1975	5.86	1.47	6.40
1980	5.07	1.60	6.85
1985	6.67	2.01	8.32

[a] After Acott (1989).

or hillslope physical and biochemical processes modify the potential nitrogen export before it reaches the stream (Heathwaite & Burt, 1991).

It is possible to examine the significance of catchment inputs to the Lower Ley using the 1976 drought, which resulted in high authigenic influx to the lake sediments (Figure 15.7). Authigenic N, for example, has a major peak at 6 cm depth, dated as 1977−1979. Here the concentration increased from less than 40 mg g^{-1} N to over 160 mg g^{-1} N (Figure 15.6), and influx increased from 1−2 mg N cm^{-2} year^{-1} to 11 mg N cm^{-2} year^{-1}. Water quality records for this period suggest that the stream nitrate load following the 1976 drought increased from approximately 1.1 tonnes year^{-1} to over 5.3 tonnes year^{-1}, and the mean nitrate concentration increased from 6.93 mg l^{-1} to 8.29 mg l^{-1}, though a sustained peak of 14 mg l^{-1} was recorded in the autumn following the drought. This event is clearly recorded in the lake sediment of the Lower Ley. It may also have triggered a further shift in the trophic status of the lake (Heathwaite & O'Sullivan, 1991). Today Slapton Ley is regarded as a hypertrophic lake.

CONCLUSIONS

Sequential inorganic chemical analysis of core SLL3 has revealed a number of points concerning the history of Slapton Ley and its catchment over the last 150 years. A major change in the composition of the lake sediments began around 1935−1945, when the proportion of allochthonous material increased. Prior to this date, it appears that the lake was a shallow but clear water environment, dominated by macrophytes and epiphytic diatoms. After 1945, increased catchment erosion, probably as a result of the transfer of land from permanent pasture to arable or temporary grassland and increased stocking densities, accelerated the rate of lake sedimentation and increased the influx of sedimentary N and P. As a result the lake became more eutrophic. A peak in sedimentary authigenic N is correlated with the 1976 drought. The lake appears to have been triggered by this event to its current hypertrophic state.

REFERENCES

Acott, T. (1989). Nitrogen and phosphorus loading on Slapton Ley, 1905−1985. Unpublished B.Sc. dissertation, Plymouth Polytechnic.

Appleby, P. G., Nolan, P., Gifford, D. W., Godfrey, M. J., Oldfield, F., Andersen, N. J. and Batterbee, R. W. (1986). ^{210}Pb dating by low background gamma counting. *Hydrobiologia*, **141**, 21−27.

Burt, T. P., Arkell, B. P., Trudgill, S. T. and Walling, D. E. (1988). Stream nitrate levels in a small catchment in southwest England over a period of 15 years (1970−1985). *Hydrological Processes*, **2**, 267−284.

Crabtree, K. and Round, F. E. (1967). Analysis of a core from Slapton Ley. *New Phytologist*, **66**, 255−270.

Engstrom, D. R. and Wright, H. E. (1984). Chemical stratigraphy of lake sediments as a record of environmental change. In Haworth, E. Y. and Lund, W. G. (eds), *Lake Sediments and Environmental History*, Leicester University Press, pp. 11−69.

Harris, T. R. J. (1988). Tracing sediment source linkages in a small catchment using mineral magnetism: the effects of land use change. Unpublished BA dissertation, University of Oxford.

Heathwaite, A. L. and Burt, T. P. (1991). Predicting the effect of land use on stream water quality. In Peters, N. E., Schobel, F., Sturm, M. and Walling, D. E. (eds), *Sediment and Water Quality in a Changing Environment*, IAHS, Publication 203, pp. 209−218.

Heathwaite, A. L. and O'Sullivan, P. E. (1991). Sequential inorganic chemical analysis of a core from Slapton Ley, Devon, UK. *Hydrobiologia*, **214**, 125−135.

Heathwaite, A. L., Burt, T. P. and Trudgill, S. T. (1989). Runoff, sediment and solute delivery in agricultural drainage basins — a scale dependent approach. In Ragone, S. (ed.), *Regional Characterization of Water Quality*, IAHS Publication no. 182. IAHS Press, Wallingford, pp. 175−191.

Heathwaite, A. L., Burt, T. P. and Trudgill, S. T. (1990a). The effect of agricultural land use on nitrogen, phosphorus and suspended sediment delivery to streams in a small catchment in south west England. In Thornes, J. B. (ed.), *Vegetation and Erosion*, Wiley, Chichester, pp. 161−179.

Heathwaite, A. L., Burt, T. P. and Trudgill, S. T. (1990b). Land use controls on sediment delivery in lowland agricultural catchments. In J. Boardman, I. D. L. Foster and J. Dearing (eds), *Soil Erosion on Agricultural Land*, Wiley, Chichester, pp. 69−87.

Johnes, P. E. and Burt, T. P. (1990). Modelling the impact of agriculture on water quality in the Windrush catchment — an export coefficient approach in a representative basin. In Hooghart, J. C., Posthumus, C. W. S. and Warmerdam, P. M. M. (eds), *Hydrological Research Basins and the Environment*, T.N.O., Vol. 44, pp. 245−252.

Johnes, P. E. and O'Sullivan, P. (1989). The natural history of Slapton Ley Nature Reserve XVIII: Nitrogen and phosphorus losses from the catchment — an export coefficient approach, *Field Studies*, **7**, 285−309.

Mackereth, F. J. J. (1965). Chemical investigations of lake sediments and their interpretation. *Proceedings of the Royal Society of London*, **B161**, 295−309.

Mackereth, F. J. J. (1966). Some chemical observations on post-glacial lake sediments. *Philosophical Transactions of the Royal Society of London*, **250**, 167−213.

Mackereth, F. J. J. (1969). A short-core sampler for sub-aqueous deposits. *Limnology Oceanography*, **14**, 145−151.

Morey, C. R. (1976). The natural history of Slapton Ley Nature Reserve IX: The morphology and history of the lake basins. *Field Studies*, **4**, 353−368.

Mulder, T. B. (1984). A preliminary investigation of the sediment dynamics of Slapton Ley, South Devon, using magnetic susceptibility. Unpublished B.Sc. dissertation, Plymouth Polytechnic.

O'Sullivan, P. E. and Heathwaite, A. L. (in press). The natural history of Slapton Ley Nature Reserve XXI: the palaeolimnology of the uppermost sediments of the lower ley, with interpretations based on ^{210}Pb dating and the historical record. *Field Studies*.

Owens, L. B., Edwards, W. M. and Van Keuren, R. W. (1989). Sediment and nutrient losses from an unimproved, all-year grazed watershed. *Journal of Environmental Quality*, **18**, 232−238.

Schelske, C., Conley, D. J., Stoermer, E. F., Newberry, T. L. and Campbell, C. D. (1987). Biogenic silica and phosphorus accumulation in sediments as an index of eutrophication in the Laurentian Great Lakes. In Loffler, H. (ed.), *Palaeolimnology IV*, Junk, The Hague.

Todd, D. A., Bedient, P. B., Haasbeek, J. F. and Noell, J. (1989). Impact of land use and NPS loads on lake quality. *Journal of Environmental Engineering*, **115**, 633−649.

Van Vlymen, C. D. (1979). The water balance of Slapton Ley. *Field Studies*, **5**, 59−84.

Walling, D. E. (1990). Linking the field to the river: sediment delivery from agricultural land. In Boardman, J., Foster, I. D. L. and Dearing, J. (eds), *Soil Erosion on Agricultural Land*, Wiley, Chichester, pp. 129−152.

Warren, K. (1990). Slapton Ley catchment land use. Unpublished undergraduate dissertation, University of Sheffield.

Young, C. P. (1986). Nitrate in groundwater and the effects of ploughing on release of nitrate. In Solbe, J. F. (ed.), *Effect of Land use on Freshwaters*, Water Research Centre, Medmenham, pp. 221−237.

16   The Sensitivity of Rivers to Nitrate Leaching: The Effectiveness of Near-stream Land as a Nutrient Retention Zone

NICK E. HAYCOCK
Silsoe College, Bedford, UK

and

TIM P. BURT
School of Geography, University of Oxford, UK

ABSTRACT

This chapter considers whether aquatic−terrestrial ecotones may be used as nutrient retention zones for non-point source pollution. More specifically, it addresses the question whether near-stream land can be suitably managed so as to mitigate the worst excesses of nitrate leaching from agricultural land before nitrate-rich water drains into a river or lake. Results are presented for a small undrained floodplain in the Cotswold Hills, England. Even for a narrow floodplain, the nutrient retention function of near-stream land is demonstrated. It is suggested that near-stream land should be taken out of production in order to make rivers less sensitive to the pollution problems associated with modern intensive agriculture.

THE ECOLOGICAL IMPORTANCE OF RIPARIAN ECOTONES

Ecotones, or boundary zones, are important regulators of the movement of energy and material across landscapes. The recent volume, *The Ecology and Management of Aquatic−Terrestrial Ecotones* (Naiman & Decamps, 1990), focuses on the dynamic nature and management potential of ecotones occurring at land−freshwater boundaries. This chapter is concerned with one aspect of this topic: can aquatic−terrestrial ecotones be used as nutrient retention zones for non-point source pollution? More specifically, can near-stream land be suitably managed so as to mitigate the worst excesses of nitrate leaching from agricultural land before nitrate-rich water drains into a river or lake? A satisfactory response to these questions depends on establishing that terrestrial−aquatic ecotones can function effectively as nutrient retention zones. The results presented here provide such evidence for a small floodplain in an upland rural catchment in the Cotswold Hills in the Midlands of England. Our basic argument is that agricultural systems are too strongly coupled to aquatic ecosystems. It is argued here

Landscape Sensitivity. Edited by D. S. G. Thomas and R. J. Allison
© 1993 John Wiley and Sons Ltd

that terrestrial—aquatic ecotones may be positively managed to desensitise rivers by effectively separating agricultural land from the stream environment.

Like many concepts in ecology, the ecotone was originally conceived by Clements (1897) as a boundary between two community types. Leopold (1933) saw the ecotone as a zone of transition particularly rich in species numbers; his idea of an 'edge effect' remains well known. Decamps & Naiman (1990) note that, until recently, ecological research focused on the functioning of individual ecosystems, avoiding questions of spatial variability and therefore the edges or boundaries constituting the ecotones. They identify two parallel developments that have resulted in a re-examination of the ecotone concept over the last 15 years. First, freshwater ecologists began to realise that the well-being of fresh waters depends largely on the terrestrial environment (Hynes, 1975). Secondly, growing interest in disturbance and patch dynamics has led to numerous studies oriented towards interactions and exchanges across heterogenous landscapes, influences of heterogeneity on biotic and abiotic processes, and management of that heterogeneity.

Risser (1990) notes that the traditional definition of an ecotone — as a zone of transition between vegetation types — fails to incorporate two important ideas explicitly. First, an ecotone is now regarded as a dynamic rather than a static zone, possessing properties of its own. Secondly, ecotones are now considered integral parts of the landscape, deriving their properties from their position in the landscape. Given these broader ideas, Risser suggests that the following definition of an ecotone is more appropriate (Holland, 1988):

> An ecotone is a zone of transition between adjacent ecological systems, having a set of characteristics uniquely defined by space and time scales and by the strength of the interactions between adjacent ecological systems.

Thus, an ecotone is now considered to be a functional entity whose attributes and interconnections depend on its transitional position between adjacent ecological systems. The ecotone so defined is not only a boundary but also an area of interaction with two or more adjacent systems (Pinay et al., 1990).

The terrestrial—aquatic ecotone provides a connection zone between a terrestrial ecosystem and the river channel. In temperate catchments fluxes will usually be from land to water, though seepage into the channel bank or overbank flooding may reverse this. Such connectivity may provide the opportunity for near-stream ecotones to function as natural sinks for sediment and nutrients emanating from farmland. This in turn raises the possibility of establishing and maintaining buffer zones at the bottom of agricultural fields adjacent to streams to trap both soil particles and nutrients and to protect bodies of water from possible eutrophication (Odum, 1990). It is clear, for most hydrological situations, that the terrestrial—aquatic ecotone coincides with the floodplain, in other words, the zone between the hillslope and river channel where water tables remain high for much of the time, mainly because of slope drainage but also partly because of overbank flooding and/or effluent seepage from the channel. We tend to avoid the term 'riparian' to describe this ecotone since traditionally this has included only a very narrow strip bordering the channel. Our concept of a riparian ecotone would encompass a much broader zone, in some cases the entire floodplain (if undrained), though very narrow where fields come right to the edge of the channel.

Several recent studies have suggested that near-stream ecotones may indeed function as nutrient retention zones, with the implication of sustaining this capacity over a period of many years (Lowrance, Todd & Asmussen, 1984; Peterjohn & Correll, 1984; Jacobs & Gilliam,

1985; Pinay & Decamps, 1988). However, none of these studies has involved sufficiently intensive field monitoring, in space or time, to establish details of this ecotone function. The aim of the investigation reported here was to monitor the passage of nitrate-rich groundwater as it crosses a floodplain and thus to evaluate the nitrate reduction capacity of the floodplain. A four-dimensional approach (three-dimensional space plus time) is shown to be necessary for a meaningful understanding of the changing nitrate concentration of subsurface water as it moves across the floodplain to the channel. Most of the results presented relate to the 1989−90 winter. Winter is the time of highest nitrate concentrations in stream water and thus the period when the nutrient adsorption capacity of riparian ecotones may be crucially important. The results for the floodplain section are set in context using information from a spatial water sampling programme. The protection afforded by a riparian ecotone is made clear by comparison with point source inputs where no such barrier exists between source and stream.

RESEARCH DESIGN

Two narrow floodplain sections of the River Leach, a headwater tributary of the Thames, were selected as study sites (National Grid Reference: SP 134132). This stretch of floodplain fulfilled two requirements. First, the floodplain is undrained: consequently, effluent groundwater cannot short-circuit the floodplain via tile drains or open ditches. Moreover, the level of the water table will be maintained at a higher level in undrained soils, which may enhance the rate of operation of processes such as denitrification. Secondly, selection of two sites allowed comparison between a grazed pasture and a plantation of poplar trees (Populus *italica*). Slopes above the floodplain are underlain by Great Oolite limestone. Soils are brown rendzinas of the Sherborne series, being freely draining, slightly stony, calcareous clay loams (USDA classification: Typic Udorthent; FAO: Calcaric Regosol — Avery, 1990). The floodplain lies on impermeable Fullers' Earth clay on which are soils characteristic of the Evesham series, greyish calcareous soils with blocky gleyed subsurface horizons (USDA: Vertic Haplaquept; FAO: Calcaric Gleysol — Avery, 1990). The floodplain soils appear to be a mixture of colluvium and alluvium. They appear reasonably free draining, and contain gravelly horizons at a depth approximately equal to the current stream-bed elevation. Saturated hydraulic conductivity was measured using a ring permeameter. Minimum values obtained were 0.00089 mm s^{-1}, equivalent to 32 mm h^{-1}. These results indicate a permeable soil well capable of conducting large volumes of subsurface flow.

At each site a grid of unlined boreholes (depths 0.7−3.1 m) were augered on the floodplain and on the adjoining slope using a 6 in. bucket auger. These holes allowed water table elevation to be measured by hand and water samples to be taken. The grassland site had a grid of 24 boreholes, measures 24 m × 16 m, with six holes upslope and four holes upvalley. At the poplar site there was a grid of 14 boreholes, with seven upslope and two upvalley, the instrumented area measuring 32 m × 18 m respectively. At one borehole in the grassland, water table elevation was measured continuously using at Ott stage recorder and water samples are taken automatically using a Rock and Taylor pump water sampler. Other measurements and sampling of boreholes was done by hand on a weekly basis. River discharge was measured using a rated section within the poplar plantation, stage again being recorded using at Ott recorder. (At the time of writing, a satisfactory rating curve for this section was not yet available.) Here, only results for the grassland site are presented: those for the popular plantation mirror these very closely. Topographical surveys were conducted using a Kern EDM.

Water samples were collected routinely once a week at several sites within (or close to) the Leach valley: several nearby springs which issue directly into the river (SS, V, TH), on the main stream above (SWU) and below (SWL) the Northleach sewage works and at the rated section (RS — there a Rock and Taylor sampler collected samples once a day), the sewage effluent itself (SWP), from a borehole (BH) in the Inferior Oolite (limestone beds below the Fullers' Earth aquitard), from a spring below an ornamental woodland, at least 120 years old at Chedworth Roman Villa (CRV), and from the River Coln immediately downstream of the Chedworth spring. Together, these samples provide a broader context for the floodplain study and allow some assessment of in-stream nitrate reduction.

Water samples were preserved with mercuric chloride (1 ml of 40mg l^{-1} $HgCl_2$ to 100 ml of sample) and stored in cool boxes for a period of 7−14 days until analysis. Analysis of the inorganic *nitrogen* content of water samples is divided into nitrate (NO_3-N), nitrite (NO_2-N), which together sum to give oxidised nitrogen (TON-N), and ammonium (NH_4-N) species. These determinations are conducted on a continuous flow Chemlab System 4 autoanalyser, using standard colorimetric methods (Chemlab Laboratory Instruments, 1985).

ECOTONE FUNCTIONS OF THE LEACH FLOODPLAIN

Hydrology

Stage and nitrate concentration for the River Leach are shown in Figure 16.1 for the period May 1989−April 1990; also shown on Figure 16.1 are nitrate concentrations for a nearby spring which flows directly into the stream. The period up to mid-November was marked by a significant rainfall deficit. Streamflow remained at a very low level and nitrate concentrations at the spring fell gradually throughout the summer, remaining above the EC maximum acceptable concentration of 11.3 mg l^{-1} NO_3-N until late in the autumn. Stream nitrate concentrations were significantly below those of the spring throughout the summer, presumably because of in-stream uptake by macrophytes (this matter is discussed further below). Modest rainfall in October and November served to recharge the soil moisture deficit and stream discharge began to rise slowly. Heavy rainfall in December, January and February caused a sharp rise in streamflow and a concomitant increase in nitrate concentrations of both spring and stream. Individual flood hydrographs in the winter period have a distinctive double peak. Such hydrographs have been discussed in detail elsewhere (Burt & Butcher, 1985). The first peak is largely the result of surface runoff though some rapid subsurface runoff may be generated at this time too. The delayed peak is produced entirely by subsurface runoff. These results for the Leach show that a shallow aquifer may produce a subsurface stormflow response that is similar both in timing and magnitude to 'throughflow' hydrographs generated in deep permeable soil which overlie impermeable bedrock.

Figure 16.2 shows a topographical survey of the grassland site. This shows that the present stream channel does not occupy the lowest part of the floodplain. Map evidence in the Northleach museum suggests that the channel may have been relocated in the 18th century. This means that when the water table is high, within a few centimetres of the soil surface, the flow direction of water draining from the hillslope is not straight across the floodplain to stream. Instead flow lines curve down-valley, implying that water eventually drains into the stream some distance (< 100 m) lower down the valley. This situation is perhaps not

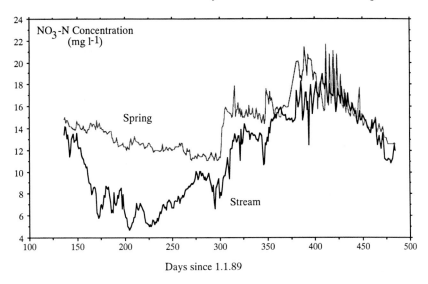

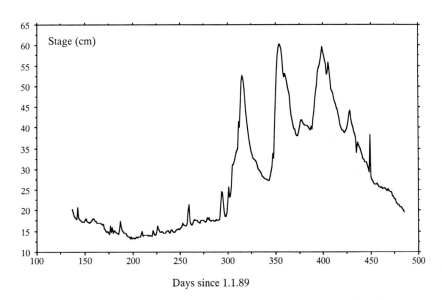

Figure 16.1 The nitrate concentration of spring and stream water for sites in the River Leach catchment, May 1989–April 1990 inclusive. Also shown is river stage for the same period

so unusual, however, and might occur wherever the down-valley gradient exceeds the cross-floodplain gradient (cf. Anderson & Kneale, 1982). During the dry summer of 1989, the floodplain was very dry. Few boreholes contained water, except those closest to the stream, showing that flow direction was influent from channel to floodplain, rather than vice versa. However, from mid-November onwards, sufficient recharge had taken place that drainage from the aquifer through the floodplain soil to the stream was re-established.

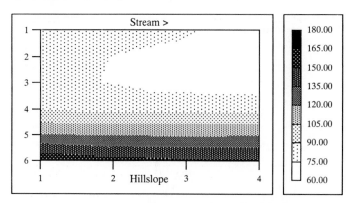

Figure 16.2   The topography of the study site. Contours give heights (cm) above an arbitrary datum, a point on the bed of the present stream channel

Figure 16.3(a) uses box–whisker plots to show the distribution of water table elevation through the grassed floodplain section for the period mid-November 1989 to the end of February 1990. The results are averaged by row, so that down-valley components of flow are not included. Each 'box' shows the median level and interquartile range, while the 'whiskers' show the 10 and 90% deciles. Row 4 is shown to be consistently the site of lowest water level on the floodplain. During the period of high flow, water flows continuously towards this point from the hillslope, driven by a high hydraulic gradient. Given the relatively high hydraulic conductivity of the soil, calculations show that the amounts of subsurface flow through the floodplain soil, when the water table is high, are significant and quite sufficient to account for increases in stream discharge along the stream section. Figure 16.4(a) maps mean water table elevations for January 1990, a period of maximum flow when the water table was almost at the soil surface. Since the elevation of the water table is equivalent to the hydraulic potential (or 'hydraulic head'), lines drawn orthogonal to contours of water table elevation indicate flow direction through an isotropic medium. The flow lines drawn on Figure 16.4(a) indicate that flow direction is from hillslope (aquifer) to floodplain and then curve sharply downvalley to trend obliquely towards the channel. Figure 16.3(a) shows that, for the instrumented plot, only from rows 1 and 2 does water flow directly to the channel; otherwise the zone around row 4 represents the focus for drainage, from where water moves down-valley. We do not believe that this situation is unusual: many floodplains have complex topography within which the direction of flow for water draining to the channel may well be tortuous rather than direct. The important point is that floodplains are characterised (at least seasonally) by high water tables and low hydraulic gradients.

Nitrate Retention

Results plotted on Figure 16.3(b) show nitrate concentrations across the grassed section, again averaged by row. These results cover the winter period, when the nitrate concentration of water draining from the aquifer is known to be high (cf. Figure 16.1). The pattern is one of consistently low nitrate concentrations on the floodplain with high concentrations only found in those boreholes close to the aquifer. We interpret these results as showing a sharp loss of nitrate as groundwater drains into the floodplain soil. We believe this loss is caused

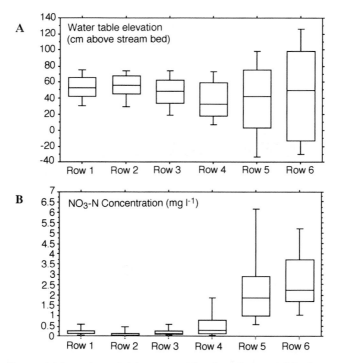

Figure 16.3 Box−whisker plots of (a) water table elevation and (b) nitrate concentration for boreholes across the study site (row 1 is closest ot the stream). See text for details

mainly by denitrification, though there may be some assimilation of nitrogen by soil bacteria also. Denitrification can be rapid where anaerobic conditions prevail and where a carbon source is available to fuel the process. We ourselves have not yet made any measurements of denitrification rates at the site, but results presented recently by Ambus & Christensen (1990) showed that denitrification in a riparian meadow can rapidly remove nitrate from agricultural drainage. It might be argued that the results presented on Figure 16.3(a) simply show a dilution of aquifer drainage by floodplain water. However, two points may be set against this argument. First, high flow rates of water leaving the aquifer mean that large quantities of floodplain water would be needed for dilution. This might be possible on a large alluvial floodplain but seems improbable on the narrow Leach floodplain. Volumes of runoff are also too large for precipitation to provide a means for dilution. Secondly, for dilution to occur, there must be a supply of water of low nitrate concentration. All sources of water in the catchment show *high* concentrations, with the exception of floodplain water and precipitation. Any water draining into the floodplain from the river will have a high concentration (Figure 16.1) and could only dilute if it first looses its own nitrate content. This may well occur since we have evidence from summer periods that drainage from channel to floodplain is accompanied by rapid loss of nitrate load. While some dilution may take place, our calculations suggests that flow volumes from aquifer to floodplain are too large for a simple mixing process to explain the nitrate profile shown on Figure 16.3(b). We interpret the reduction in nitrate concentration as an active denitrification loss as groundwater enters the floodplain soil.

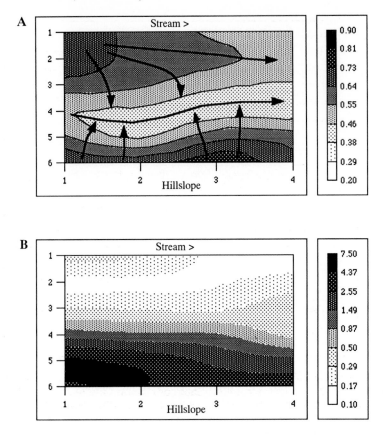

Figure 16.4 Maps of (a) mean water table elevation and (b) the mean nitrate concentration of borehole water for January 1990. (a) includes flow lines drawn at right-angles to the equipotential isolines

On Figure 16.4(b) the mean nitrate concentration of borehole water is mapped for January 1990, a time of maximum subsurface runoff. As noted above, water table levels indicate a strong hydraulic gradient from aquifer to floodplain at this time. Despite the high rate of flow, nitrate concentrations fall greatly across the floodplain, from a range of $4-8$ mg l^{-1} NO_3-N at the upslope row of boreholes, to below 2 mg l^{-1} NO_3-N at row 4 where flow begins to move down valley. At this time, most of the boreholes on the floodplain contain water with concentrations below 1 mg l^{-1} NO_3-N. These results imply that the nitrate retention capacity of the floodplain remains large, even at times of maximum subsurface flow. It should be noted that these results were obtained during a mild winter. At the nearby Radcliffe Meteorological Station, Oxford, soil temperatures at 300 mm remained in the range $5.3-8.0°C$ during this time. If the floodplain had remained saturated during the summer, these results imply that the nitrate retention function of the floodplain soils might be even more effective at such times. Even so, since winter is the time of maximum nitrate concentration in river water, the fact that a floodplain can fulfil this function even during winter is of some potential importance to water supply companies.

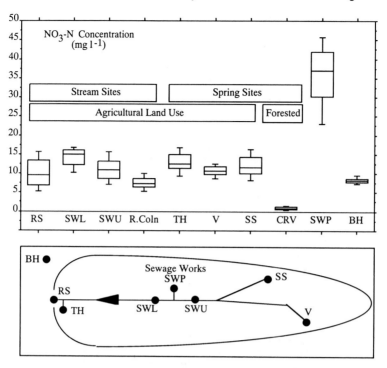

Figure 16.5 Box–whisker plots of nitrate concentrations at a number of sampling sites within, or close to, the Leach catchment. Each 'box' shows the median level and interquartile range while the 'whiskers' show the 10 and 90% deciles. The schematic map indicates the relative position of the sampling sites: those for Chedworth Roman Villa and the River Coln are not included

Spatial Variations in Nitrate Concentrations

Figure 16.5 uses box–whisker plots to show NO_3-N concentrations at a number of sampling sites within, or close to, the Leach catchment. With the exception of the site immediately downstream of the sewage works (SWL), all stream sites have lower concentrations than the springs. This may happen for at least two reasons. First, in-stream losses occur in summer (Figure 16.1). Thus, the stream itself has some capacity for nutrient retention, though this function is seasonal and problems may arise in winter when macrophytes, having been fertilised by the high nitrate concentration of the stream water, decay. Secondly, the mitigating effects of nitrate retention by the floodplain, as discussed above, may help reduce the nitrate concentration of the stream water, especially in winter when the aquifers are discharging.

The spring below long-established woodland at Chedworth has consistently low nitrate levels, and as such provides a benchmark against which all other sites may be compared. Many plot experiments have shown that, once land comes under the plough, nitrates losses increase significantly, even if fertiliser applications remain low. The Inferior Oolite borehole (BH), being relatively isolated from the land surface, shows lower concentrations than springs from the Great Oolite, which is unconfined. The point source input from Northleach sewage treatment works (SWP) is high in concentration but low in volume; it only affects the concentration of the stream locally, immediately downstream of the outfall.

DISCUSSION AND CONCLUSIONS

Further field measurements are required to establish the rates of denitrification operating within these floodplain soils, especially when the water table is high, and to confirm the hydraulic continuity between aquifer, floodplain and stream; both these aspects are the subject of ongoing research. Nevertheless, results to date already demonstrate what other researchers have suggested from more cursory field studies: that aquatic—terrestrial ecotones are capable of functioning effectively as nutrient retention zones.

The Leach is a headwater tributary of the Thames: the floodplain is narrow (<25 m), if present at all. Generally therefore, for this stretch of the Leach at least, aquifers discharge directly into the stream and there is little opportunity for the mitigating function of the floodplain to operate. For this reason, spring and stream concentrations are very similar during the winter period. In other words, the Leach is too well coupled to its aquifer. Further downstream, where floodplains are wider, underdrainage has often been installed so that aquifer drainage will by-pass the floodplain, better drainage allowing arable crops to be cultivated rather than the traditional meadow or woodland usage of such land. The results of Ambus & Christensen (1990) show that, if drains are broken at the hillslope edge of the floodplain so that water must flow over or through the floodplain soils to reach the stream, denitrification is rapid and the quality of agricultural drainage water is greatly and rapidly improved. Our results also show that the nitrate loss may take place within a very small distance, just a few metres. It may be that denitrifying bacteria operate best at the junction of anaerobic/aerobic zones where both carbon and nitrogen are abundant: such conditions are found around row 4 (Figure 16.4) where we believe the former channel of the Leach to have been located. However, Ambus & Christensen's results show that our site is not a special case: floodplain soils are typically carbon-rich and waterlogged. Since the grassland and poplar sites function in a similar way, we cannot determine whether, woodland provides a more effective nutrient buffer than grassland.

It follows that the sensitivity of a river to nitrate leaching might be reduced if it can be decoupled from the agricultural land. Where floodplains have been underdrained, one option might be to block drains and so force water to move slowly through floodplain soils to the stream. Such strategic removal of near-stream land from cultivation might require compensation to be paid to farmers for loss of production, but might well prove a more cost-effective remedy nationally for controlling nitrate levels in river water than alternative measures. (No attempt has been made to calculate the cost of such a scheme.) In addition, re-establishment of water meadows, whether lightly grazed or ungrazed, would bring the benefits of enhanced conservation and amenity value. Where point sources enter the stream directly, our results imply that a narrow zone of seepage between spring and stream could be of great benefit. So-called tertiary treatment of sewage effluent is already common in some countries: effluent is passed through wetland ecosystems before entering the river. It would be valuable to know whether wetlands constructed around springs might serve a similar purpose in dealing with the worst effects of agricultural drainage. Certainly tertiary treatment of (point source) sewage effluent would seem worthwhile even for small rural works; the removal of other nutrients, particularly phosphate, would also be highly beneficial here.

Even on the upper Leach, where the floodplain is narrow, the nutrient retention function of near-stream land has been demonstrated. Thus, a policy of setting aside agricultural land in sensitive locations is not restricted to the floodplains of large alluvial rivers, but has relevance too for the uplands. This is where relatively unpolluted water — to dilute pollution entering the river further downstream — is commonly thought to originate. Our results show that headwater streams in agricultural areas may be highly polluted too, often above legal limits. The concept of setting aside land from agricultural use in order to protect groundwater sources has already been implemented in the United Kingdom (MAFF, 1989). We advocate an extension of this scheme to include protection of surface waters. The use of undrained floodplains as barriers to nitrate movement could allow modern farming and water supply to coexist in the same basin. Such nutrient retention zones might also trap eroded soil and associated pollutants, and would provide habitats of high conservation value. Burt, Heathwaite & Trudgill (Chapter 14 this volume) stress that particular locations within a basin may be especially sensitive to land use change, and that the effect of changes in small basins may well prove significant at the larger scale. Our new suggestion here to 'set aside' near-stream land might solve some of the pollution problems associated with modern agriculture and make rivers less sensitive to land use changes within their catchment area.

REFERENCES

Ambus, P. and Christensen, S. (1990). Denitrification in a riparian meadow removes nitrate from agricultural drainage. Poster paper, 6th Workshop on Nitrogen in Soils, The Queen's University, Belfast.

Anderson, M. G. and Kneale, P. (1982). The influence of low-angled topography on hillslope soil-water convergence and stream discharge. *Journal of Hydrology*, **57**, 65–80.

Avery, B. W. (1990). *Soils of the British Isles*, CAB International, Wallingford.

Burt, T. P. and Butcher, D. P. (1985). On the generation of delayed peaks in stream discharge. **78**, 361–378.

Chemlab Laboratory Instruments (1985). *System 4 Sampler, User Manual*, Chemlab Instruments, Hornchurch, Essex.

Clements, F. E. (1897). Peculiar zonal formations in the Great Plains. *American Naturalist*, **31**, 968.

Decamps, H. and Naiman, R. J. (1990). Towards an ecotone perspective. In Naiman, R. J. and Decamps, H. (eds), *The Ecology and Management of Aquatic–Terrestrial Ecotones*, Parthenon Press, UNESCO, Paris, pp. 1–5.

Holland, M. M. (compiler) (1988). SCOPE/MAB technical consultations on landscape boundaries: report of SCOPE/MAB workshop on ecotones. *Biology International, Special Issue*, **17**, 47–106.

Hynes, H. B. N. (1975). The stream and its valley. *Verhandlungen Internationale Vereinigung Limnologie*, **19**, 1–15.

Jacobs, T. C. and Gilliam, J. W. (1985). Riparian losses of nitrate from agricultural drainage waters. *Journal of Environmental Quality*, **2**, 480–482.

Leopold, A. (1933). *Game Management*, Charles Scribner's Sons, New York.

Lowrance, R. R., Todd, R. L. and Asmussen, L. E. (1984). Nutrient cycling in an agricultural watershed. *Journal of Environmental Quality*, **13**, 22–32.

MAFF (1989). Nitrate sensitive area scheme. Consultation document, Ministry of Agriculture, Fisheries and Food (UK), 9 May.

Naiman, R. J. and Decamps, H. (eds) (1990). *The Ecology and Management of Aquatic–Terrestrial Ecotones*, Parthenon Press, UNESCO, Paris.

Odum, W. E. (1990). Internal processes influencing the maintenance of ecotones: do they exist? In Naiman, R. J. and Decamps, H. (eds), *The Ecology and Management of Aquatic–Terrestrial Ecotones*, Parthenon Press, UNESCO, Paris, pp. 91–102.

Peterjohn, W. T. and Correll, D. L. (1984). Nutrient dynamics in an agricultural watershed: observations on the role of riparian forest. *Ecology*, **65**, 1466–1475.

Pinay, G. and Decamps, H. (1988). The role of riparian woods in regulating nitrogen fluxes between the alluvial aquifer and surface waters: a conceptual model. *Regulated Rivers*, **2**, 507–516.

Pinay, G., Decamps, H., Chauvet, E. and Fustec, E. (1990). Functions of ecotones in fluvial systems.

In Naiman, R. J. and Decamps, H. (eds), *The Ecology and Management of Aquatic—Terrestrial Ecotones*, Pathenon Press, UNESCO, Paris, pp. 141—169.

Risser, P. G. (1990). The ecological importance of land-water ecotones. In Naiman, R. J. and Decamps, H. (eds), *The Ecology of Management of Aquatic—Terrestrial Ecotones*, Parthenon Press, UNESCO, Paris, pp. 7—21.

17 Environmental Responses and Sensitivity to Permanent Cattle Ranching, Semi-arid Western Central Botswana

JEREMY S. PERKINS
Department of Environmental Science, University of Botswana

and

DAVID S. G. THOMAS
Department of Geography, University of Sheffield, UK

ABSTRACT

Expansion of all-year round livestock production into arid and semi-arid African environments devoid of surface water has been made possible by sinking boreholes to tap groundwater supplies. In Botswana, this has led to the expansion of the cattle industry into the Kalahari. Concern about consequential environmental degradation and desertification has ensued, leading to soil erosion, vegetation change and range degradation studies. Much of the concern has centred on the Kalahari — a fragile and sensitive environment where undesirable changes are readily effected and less readily resolved due to long environmental recovery rates. This chapter reports the findings of a study of the impacts associated with a series of borehole-dependent cattle ranches established at intervals since the 1970s. Results suggest that sensitivity to degradation is both complex and affected by factors other than simply time and stocking levels, including topography and time in relation to a quasi-18 year drought cycle. Consequently while vegetation appears to be sensitive to disturbance, it also recovers relatively quickly after surplus rainfall years; soil erosion shows little response to environmental disturbance due to topography and sediment character while soil chemistry falls somewhere between the two. It is concluded that issues of sensitivity and degradation are more complex than widely recognised, requiring caution when longer-term trends are extrapolated.

INTRODUCTION

Increased agricultural utilisation of lands marginal in relation to both areas of traditional sedentary human occupation and to traditionally productive areas has become a characteristic of many post-colonial African nations. Of note have been the rate and intensity of expansion into arid and semi-arid environments — areas generally neglected except by low usage density hunter-gatherer and pastoral-nomad groups. Lack of surface water, poorly developed and mineral-deficient soils and high seasonal and year-to-year rainfall variability are all factors that restrict the utilisation options available and are widely regarded as rendering the physical environment susceptible to disturbance and degradation. Terms such as fragility, desertification

Landscape Sensitivity. Edited by D. S. G. Thomas and R. J. Allison
© 1993 John Wiley and Sons Ltd

and degradation all widely appear in the literature dealing with the impact of land use, particularly pastoralism, on semi-arid environments. In other words, such environments are thought to display a *high sensitivity* to disturbance and a low or even virtually non-existent *recovery rate* once disturbance has occurred (e.g. United Nations, 1977).

Such views are now beginning to be critically assessed (e.g. Warren & Agnew, 1988; Abel & Blaikie, 1989), with two major lines of dispute levelled at the generalisations. First, different arid and semi-arid regions display major differences in climatic regime, environmental history, vegetation communities, soils and geology. Consequently, care needs to be exercised in adapting findings from one area as a generalisation applicable to others. Secondly, the physical environment is frequently regarded as an undifferentiated whole in studies of degradation and desertification, but individual components of the physical environment (soil erosion rates, soil quality, vegetation densities, community structures, etc.) may well respond to and recover from disturbance and degradation at different rates by displaying differing sensitivities. This chapter explores these issues through the results of an investigation into the environmental impact of borehole-dependent cattle ranching in the Kalahari of Botswana.

THE KALAHARI

The Kalahari Desert or thirstland can be defined in a variety of ways using a range of criteria (Thomas & Shaw, 1991). For the purpose of this study it can be designated as an area centred upon western and central Botswana but extending into neighbouring territories (Figure 17.1). It is characterised by semi-arid conditions (criteria of Meigs, 1953) and very limited relative relief. Geologically, it is regarded as an extensive basin infilled with Jurassic to recent sediments dominated by the surficial, nutrient deficient, Kalahari Sand (Thomas & Shaw, 1990, 1991). The vegetation is species-poor and can be typified as a shrub and low tree savanna with a ground layer of tufted perennial grasses (Weare & Yalala, 1971).

The Kalahari, and its sand cover, occupies about 75% of Botswana and can be contrasted with the eastern quarter of 'hardveld'. Mean annual rainfall decreases in a south-westerly direction from about 600 to 150–200 mm, but in all places is highly seasonal, with a summer wet season, and variable, with inter-annual variability as high as 45% at Tshabong in the south-western Kalahari (Bhalotra, 1985a). The 18 year cycle of rainfall fluctuations identified for the whole of the southern African summer rainfall zone (Tyson, Dyer & Mametse, 1975; Tyson, 1986) has been confirmed in a Botswanan context by Bhalotra (1985b) and Tyson (1979). This gives rise to extended periods of above and below average rainfall with, for example, a period of 'good' rainfall in the 1970s separating severe droughts in the 1960s and in the first six years of the 1980s (Perkins, 1990). Additionally, the Kalahari has no permanent surface water and is only touched by perennial rivers on its northern and southern margins. These factors, together with the high infiltration rates afforded by sandy sediments even under high rainfall intensities, make the Kalahari a location where natural water availability is very limited and at best seasonal, confined to accumulations during the rainy season in shallow, natural depressions or pans.

CATTLE RANCHING IN THE KALAHARI

The environmental characteristics described above have traditionally rendered the Kalahari unsuitable for pastoralism, though archaeological evidence now demonstrates that limited use did occur by Bantu cattle herder groups over the last 2000 years and that interactions between pastoralists and indigenous hunter-gatherers took place (Denbow, 1984; Denbow & Wilmsen, 1986).

From the 1950s onwards the Kalahari was increasingly viewed as a large, virtually untapped, grazing resource (Debenham, 1952). Exploitation of this resource has been made possible by the sinking of boreholes to tap groundwater reserves permitting, for the first time, year-round or permanent grazing. This process continued through into the period following independence, intensifying in the 1970s when it was carried out in conjunction with a number of successive programmes to exploit the Kalahari for 'ranch-style' pastoralism. The growth of the Kalahari cattle industry has generated both social and environmental issues, the latter including conflicts between wildlife and cattle utilisation (e.g. Thomas, 1988) and the now widespread belief that livestock grazing is causing severe range degradation and desertification (e.g. Cooke, 1985). A critical issue concerns the sensitivity of the Kalahari to this change in resource use, with the remainder of this paper used to summarise the findings of a detailed study of the environmental impacts of borehole-dependent cattle ranching in one area of the Kalahari.

STUDY SITES AND METHODS

The Tribal Grazing Lands Policy (TGLP) of the Botswana government was instigated in 1975, subsuming earlier centrally implemented ranching schemes. Among its multifarious aims TGLP established commercial leasehold ranches in various parts of the Kalahari, one block of which, in the eastern sandveld of the Kalahari in Central District, was selected for the study (Figure 17.1). The area in which the ranches occur has a mean annual rainfall of 400–500 mm and a natural vegetation equivalent to the 'northern Kalahari tree and bush savanna' of Weare & Yalala (1971). This consists of low shrubs, commonly of *Acacia*, *Grewia* and *Terminalia* species dotted over grassy plains dominated by perennial *Pogonarthria*, *Eragrostis* and *Stipagrostis* species.

The block of eight ranches is bordered to the north and west respectively by the Makoba and Makorro veterinary cordon fences. Each ranch has an area of approximately 6400 ha and is served by a diesel-pumped borehole supplying water for 12 months a year. The ranches studied were typical of those found in the Kalahari and indistinguishable in their operation from the traditional hardveld 'cattleposts', with the absence of paddocks and usually perimeter fencing precluding any rotational system of grazing, such that the borehole is effectively 'the herder' (Jerve, 1982). Furthermore, the ranches within the block were not all established at the same time, ranging in starting date from 1970 (pre-TGLP) to 1988, permitting a temporal component to the investigation.

The study aimed to determine the nature, degree and extent of environmental changes associated with cattle ranching. Specifically, as a working hypothesis, it was assumed that the extent of change would increase with age since establishment and with increasing stocking levels. Such changes might include reduced soil nutrient status, increased soil compaction and reduced infiltration, increased erosion (by wind or water), a decline in grass cover and an associated increase in the woody layer (bush encroachment). Effects were also expected to decrease with distance from the watering point. A borehole that had been drilled but not equipped was used as one control, and the area between the double Makoba fence, which was effectively unavailable for grazing, as another.

Fieldwork, conducted in 1988–89, was based on a 'piosphere' approach (Andrew, 1988),

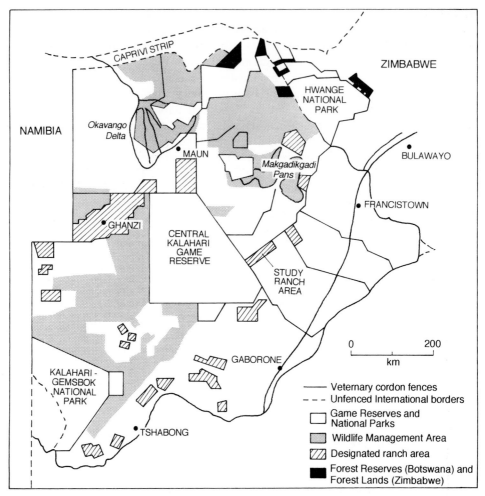

Figure 17.1   The Kalahari in Botswana showing the ranch block investigated in this study

radiating out from the water point on each ranch. A line transect was run out from each borehole and set distances of 0, 25, 50, 200, 400, 800, 1500 m, 3 and 5 km were utilised for sampling. The increasing distance between points reflected the deceasing Herbivore Use Intensity (HUI) (Georgiadis, 1987) with distance away from the borehole — as expected where a single point forms the focus of all animal activity on the ranch.

At each point soil sampling, measurements of infiltration capacity using 7.5 cm diameter rings, and vegetation surveys were carried out. Surface and 30 cm depth soil samples were analysed in the laboratory for organic and nutrient content. Vegetation was investigated using a 25 × 25 m quadrat, subdivided into a grid with 2.5 m intervals. A point-centred quadrat technique (Mentis, 1984) was used to determine species composition and shoot frequency and density. Dung counts (Georgiadis, 1987) and cattle track densities were established as a field based index of HUI, while stocking rate data were derived from centrally held veterinary records.

RESULTS

Detailed results for each ranch are presented in Perkins (1991) with the salient features presented below.

Soil and Geomorphic Data

Tables 17.1 and 17.2 contrast the variation in soil nutrient data with increasing distance away

Table 17.1   Soil chemical data for a borehole utilised since 1972

Dist. (m)	Depth (cm)	pH	Organic matter (%)	CEC (me/100 g)	P (mg/100 g)	N	K	Na	Ca	Mg
							Exchangeable cations (p.p.m.)			
0	0–5	9.65	6.40	10.04	14.60	2.14	243.82	48.60	167.10	29.53
	5–30	8.45	4.40	7.64	11.80	3.91	169.38	41.52	134.05	27.01
25	0–5	7.00	2.00	5.34	27.88	5.68	80.56	43.56	97.00	22.65
	5–30	6.40	1.20	2.85	11.78	3.04	82.24	32.40	38.53	10.55
50	0–5	6.90	2.00	2.43	8.78	2.55	14.71	1.08	44.90	11.08
	5–30	5.80	0.60	2.07	3.73	4.15	11.48	1.33	15.25	4.33
200	0–5	5.85	0.40	1.87	1.74	1.23	3.98	0.78	8.15	2.08
	5–30	5.60	0.80	2.31	0.91	0.46	5.92	0.63	13.58	3.78
400	0–5	5.90	0.40	1.45	1.65	0.28	3.61	0.84	7.80	2.08
	5–30	5.35	0.40	2.48	0.82	0.51	4.59	0.01	6.63	2.18
800	0–5	5.60	0.60	3.02	1.32	0.31	3.67	0.16	8.23	2.83
	5–30	5.10	0.60	2.74	0.74	0.57	3.87	0.16	11.25	5.50
1500	0–5	5.25	0.80	2.89	1.74	0.75	6.74	0.17	17.95	4.65
	5–30	5.70	0.80	1.50	0.74	0.73	5.56	0.35	19.28	8.13
3000	0–5	6.10	0.60	1.74	1.24	0.28	3.13	0.16	13.23	3.68
	5–30	6.00	0.80	1.96	0.24	0.31	5.12	0.45	27.05	8.75

Table 17.2   Soil chemical data for an ungrazed control borehole

Dist. (m)	Depth (cm)	pH	Organic matter (%)	CEC (me/100 g)	P (mg/100 g)	N	K	Na	Ca	Mg
							Exchangeable cations (p.p.m.)			
25	0–5	6.30	0.20	3.42	1.49	0.09	2.49	0.26	12.43	2.15
	5–30	6.05	1.00	3.48	0.91	0.11	4.07	0.65	22.03	6.00
50	0–5	6.00	0.60	3.77	1.32	0.28	3.26	0.19	18.00	2.83
	5–30	6.00	0.60	4.65	1.15	0.39	4.72	0.22	21.43	5.68
200	0–5	5.95	0.40	3.42	0.66	1.03	3.73	0.25	11.10	2.38
	5–30	5.90	0.60	6.20	1.07	0.75	3.82	0.31	16.03	3.33
400	0–5	5.55	0.80	7.62	0.57	0.20	4.03	0.75	11.48	2.75
	5–30	5.40	0.80	2.83	1.15	0.31	3.73	0.40	11.10	3.43
800	0–5	5.70	0.60	3.07	0.91	0.10	4.75	0.21	14.23	3.43
	5–30	5.65	0.60	3.70	0.91	0.15	4.75	0.11	16.70	4.55
1500	0–5	6.00	0.60	7.96	1.24	0.10	3.13	0.15	13.33	2.55
	5–30	6.10	0.60	5.90	0.57	0.16	3.73	0.39	20.23	4.23

from the water trough of the oldest borehole, operational from 1972, with that for the utilised control site. Theoretically the input of dung and urine coincident with the daily concentration of livestock around the water trough would be expected to result in the centripetal movement of nutrients from the area over which livestock graze (Tolsma, Ernst & Verwey, 1987), observed to be up to 8 km in this study. Localised nutrient concentration is quite striking (Figures 17.2 and 17.3), with enrichment occurring within the immediate vicinity of the water trough (0−50 m), before returning to what may be considered to be the 'background' nutrient levels of the Kalahari sands. These are given by the control site in Table 17.2, and as noted in the southern Kalahari (Buckley, Gubb & Wasson, 1987; Buckley, Wasson & Gubb, 1987) are extremely low, even in comparison with other areas noted for their infertility, such as central Australia. This may explain why notable changes in soil chemistry do not appear to be wide ranging (Tables 17.1 and 17.2; Figures 17.2 and 17.3): with such a low starting point, even if degradation was occurring, changes would not necessarily be readily apparent.

It should be emphasised that the results for the soil physical properties measured in the field (infiltration capacity, soil surface resistance and subsoil compaction), while significantly

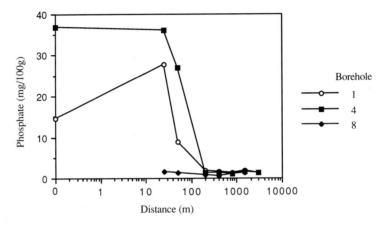

Figure 17.2   Variation in phosphate with distance from the water trough

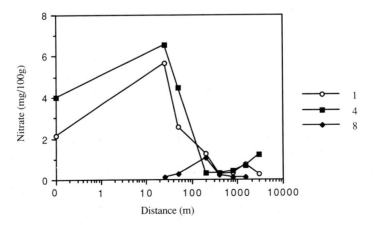

Figure 17.3   Variation in nitrate with distance from the water trough

different at the water trough itself (0 m), did not vary in a manner consistent with HUI beyond this, and so are not reported here. To some extent this finding provides further evidence for highly localised impact on soil properties and may be attributable to the fact that Kalahari sand typically has a negligible silt-clay fraction.

It is often assumed that 'overgrazing' gives rise to an increased potential for soil erosion, both by wind and water. Wind-blown sand was a striking feature in the late dry season of 1988, with increased wind activity clearly evident in the form of surface rippling and in some cases the formation of low dunes (< 1 m high). However, this evidence was confined to the 0−400 m zone around the water trough, forming what might be regarded as a discrete circle of devastation around the watering point, or a 'sacrifice zone' (Stoddart, Smith & Box, 1975). One reason why erosion is spatially restricted is that the types of changes occurring in vegetation communities may actively mitigate against soil loss (see below), though this warrants further study.

Vegetation Communities

The main vegetation changes along HUI gradients in the ranch survey area can be summarised into five main zones, characterised by the following:

1. high nutrient concentration (0−50 m);
2. surface rippling/sand dunes and negligible herbaceous cover (0−400 m);
3. palatable and nutritious species, the 'Digitaria' community (200−800 m);
4. bush encroachment (200−2000 m);
5. the 'unpalatable grazing reserve'.

The maximum extent of each zone is given above and it is important to emphasise that the suggestion is not one of the existence of discrete vegetation communities, but rather a continuum of change along HUI gradients, in which zones 2−4 especially can vary in their significance. Sites with negligible grass cover were omitted from the vegetation analysis, which typically meant those within 0−400 m of the water trough, but also cases where grass cover was enveloped in the canopies of thorn bushes and inaccessible to both measurement and grazers. However, it should be noted that a diverse array of often palatable grass species exploited such protection, with their presence perhaps significant in contributing to the seed bank of the high HUI zone.

A more precise ecological interpretation of the ranch survey area data is provided by Canonical Correspondence Analysis (CCA), available as part of the CANOCO programme (ter Braak, 1988). Through a combination of ordination and multiple regression, CCA directly relates variations in floristic composition to supplied environmental variables. Figure 17.4 illustrates this for grass density data, with the environmental variables 'stocking rate' (SR) and 'cattle tracks' the most significant in accounting for variance in the species data set. The latter related to cattle track density in the field and so provided a simple index of HUI, while SR was estimated from 1988 veterinary data, with absolute variations in the number of cattle ranging from 490 to 725 head or 300 to 480 LSU (livestock units — criteria of Carl Bro., 1982). For a 6400 ha ranch this corresponds to 13−21 LSU ha^{-1}, though the values used in all analyses were calculated on a per annulus basis for the area between sampling points, and so more accurately reflected the pronounced variation in available grazing area with distance away from the water trough (Figure 17.5), which is such a basic feature of piosphere structure (Andrew, 1988).

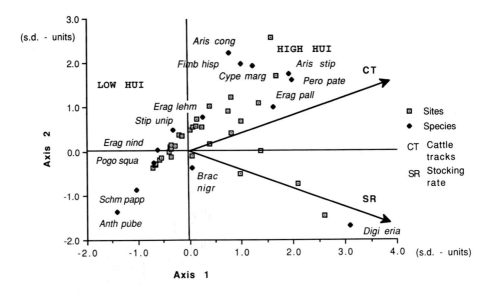

Figure 17.4 Canonical Correspondence Analysis biplot for grass density data

Interpretation of Figure 17.4, known as a biplot, is the same as in any ordination diagram, with sites that lie close to a species point in the diagram likely to have a high abundance of that species (Jongman, ter Braak & van Tongeren, 1987). The environmental variables are represented by arrows, with the arrowhead representing maximal intensities of that environmental factor and the relative length of an arrow its importance. Orthogonal projection of species points (and sites) onto each arrow provides an inferred ranking of species (or sites) in relation to that particular environmental 'axis'. Long arrows reflect a strong correlation with the ordination axes and thus environmental variables that are important in accounting for the pattern of community variation in the diagram (ter Braak, 1987). The latter is also indicated by the eigenvalues (scaled between 0 and 1) with the rather low values in Figure 17.4, while not atypical for ecological data (ter Braak, 1988), regarded as a consequence of the omission of bare ground sites and a focus upon floristic variation beyond the sacrifice zone (Perkins, 1991).

Figure 17.4 reveals a basic split in the data set between sites and species associated with low HUI on the lefthand side and those with high HUI on the right. The ungrazed control sites and those at low HUI (1500 m−5 km) thus occur in close proximity to each other in the biplot and are characterised by the valuable grazing grasses *Anthephora pubescens, Schmidtia papphophoroides* and *Eragrostis nindensis*, together with high shoot densities of *Pogonarthria squarrosa*, which may be thought of as the unpalatable grazing reserve (*sensu* Walker *et al.*, 1981). By contrast, at high HUI the vegetation is characterised by the unpalatable annuals *Perotis patens* and *Tragus berteronianus*, and the palatable and highly nutritious perennials *Digitaria eriantha* and *Brachiara nigropedata*. In summary, variation in the species composition of grasses on the ranches is characterised by a basic split between a low HUI *Anthephora* and a high HUI *Digitaria* community.

The variation in shrub canopy cover with distance away from the water trough is illustrated for all the piospheres studied in Figure 17.6. The control gradient provides a useful reference, against which the oldest ranch borehole (1), together with those of boreholes 3 and 5, can

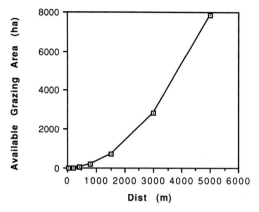

Figure 17.5   Variation in available grazing area with distance from the water trough

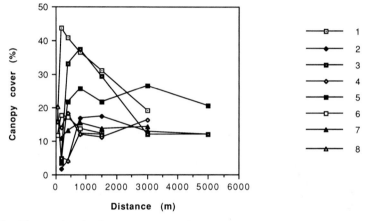

Figure 17.6   Change in shrub canopy cover (expressed as percentage ground shelter) with distance from boreholes

be seen to be characterised by markedly higher shrub cover, especially over 200–1500 m. However, the logical expectation that older boreholes are likely to be more encroached, while evident over the larger number of boreholes visited in the eastern Kalahari, is not fully supported by Figure 17.6. This is believed to be due to the fact that many shrubs had experienced pronounced dieback following the six year drought in the 1980s and while resprouting at the base, their canopy cover values (especially at borehole 2) were undoubtedly underestimated.

CCA of this shrub canopy cover data is similarly affected and believed to have weakened the significance of this age of borehole factor and produced the low eigenvalues in Figure 17.7. *Croton gratissimus, Boscia albitrunca, Rhigozum brevispinosum* and *Acacia ataxacantha*, the latter as a scrambling shrub, are most strongly associated with this age factor (Figure 17.6), while *Acacia erioloba, Maytenus heterophylla* and *Dichrostachys cinerea*, all characteristic encroachers, appear to reflect a soil pH gradient. This latter factor may result from relative nutrient enrichment in the zone beyond 0–50 m, though caution is required in attempting to explain all the observed variation in a purely HUI context. For example,

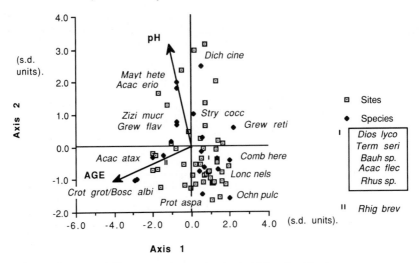

Figure 17.7 Canonical Correspondence Analysis biplot for shrub canopy cover data

the location of *Acacia fleckii*, a notable enchroacher, appears inexplicable in Figure 17.7 and relatively high shrub canopy cover values can occur in low HUI areas (Figure 17.6).

However, it should be stressed that an increased density of shrubs appears to be a characteristic feature of piospheres. At high HUI, shrub units were heavily browsed, often down to stumps, and appeared to act as focii for sand accumulation, which as stressed earlier was entirely absent beyond the sacrifice area. By contrast, at low HUI and out to a maximum of 2 km at borehole 1, the bush-encroached zone underwent a transition into the 'unpalatable grazing reserve'. This latter zone seems likely to come under heavy grazing pressure in drought years following its collapse of the highly productive grasses (the *Digitaria* group) in the 0–800 m zone, with cattle consequently being forced to graze at greater distances from the borehole. In this context an outward expansion of the bush-encroached zone seems likely to occur over time, with the question of the reversibility of this process becoming critical, unless a permanent decline in the per-hectare productivity of sandveld beef pastures is to occur.

DISCUSSION

The results of the study clearly indicate that sensitivity to disturbance impacted through the introduction of permanent cattle ranching in the semi-arid eastern Kalahari is not uniform or unindirectional among different environmental variables. Changes in soil and geomorphic variables (nutrient status, erosion, etc.) appear limited both in terms of degree and spatial extent of change. Changes in visible erosion activity are evident only at extreme HUI (i.e. around water points) in the sacrifice zones (Figure 17.8). Even this appears not to be a contributory factor to lowering nutrient status because the base values are so low to start with. It remains to be seen whether erosion increases in the future as utilisation of the cattle posts continues. Although the study followed two above-average wet seasons that came after a drought period, when the potential for water erosion would be expected to be at its greatest, sediment characteristics and topography mitigated against water erosion. The nature of the Kalahari may therefore reduce sensitivity to erosion and make findings from other semi-arid

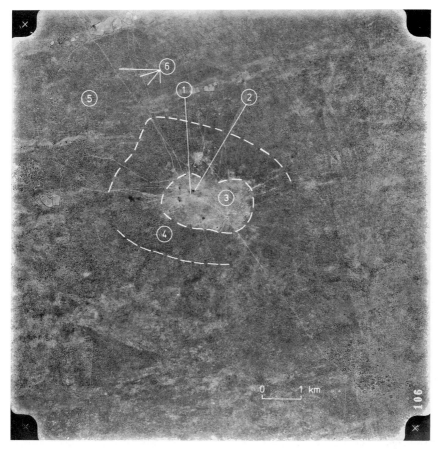

Figure 17.8 Aerial photograph of a Kalahari cattle post ranch. Dark patches near the borehole pump (1) are the dung-enriched floors of kraals (2), the pale area (3) is the 'sacrifice zone', devoid of vegetation. Bush encroachment (4) is evident from the presence of a large number of individual shrubs beyond the sacrifice zone in an intermitent area often devoid of grass. Less disturbed vegetation, showing the typical tonation of mixed grass and shrub savanna can also be identified (5). A large number of cattle tracks (6) radiate from the borehole

locations including those from the Botswana hardveld to the east (e.g. Biot, 1988), inapplicable to this situation.

The response of vegetation communities to disturbance is complex, and does not follow a simple pattern of reduced cover and reduced diversity. This may well be because, as Walker *et al.* (1981) note, such environments, being subjected to frequent natural disturbances such as fire and drought, are extremely resilient to permanent change (Holling, 1973). From a pastoral point of view, they may even benefit from it, because as noted above, with high HUI, the vegetation was dominated by the highly palatable and nutritious grass species, *Digitaria eriantha*. While it is important to stress that these results correspond to a good rainfall year, the notion of a unidirectional grazing intensity−palatability relationship which abounds in much of the range management literature (e.g. Stoddart, Smith & Box, 1975) is clearly overly simplistic.

As stressed earlier, the question of the stability of the bush-to-grass ratio and hence the permanence of bush encroachment is a critical one. The process has been modelled in terms of differential access to water, with grasses having competitive superiority in the upper layer and the woody component having almost exclusive use of that which infiltrates through (Walker et al., 1981). Irreversible alterations in infiltration rates, via reductions in soil depth by erosion losses, appear to be the critical factor that could effectively preclude the future recovery of the herbaceous layer.

The findings here suggest, however, that soil surface erosion is absent in all but the sacrifice zone, where small-scale ripples and dunes appear to result only from a local redistribution of material. Moreover, the increased cover associated with bush encroachment precludes the exposure of large tracts of bare ground, even in drought years, and so mitigates against sand movement over broad areas. In theory therefore, and in the absence of significant changes in soil structure, the process could be argued to be reversible, with bush-encroached savanna eventually reverting to tree savanna, and possibly following fire to open grassland.

Overall, the findings lend support to many of the issues recently raised by Abel & Blaikie (1989) regarding the sensitivity of semi-arid rangelands to disturbance and degradation. They argue that the term 'degradation' should be reversed for situations where irreversible changes, which reduce productivity, are imparted to the environment. Such changes relate to soil erosion, loss of nutrient status and possibly bush encroachment when the grass cover has been so removed that the rate of infiltration cannot support the recovery of the herbaceous layer (Walker & Noy-Meir, 1982).

There is, as yet, no evidence that such changes have occurred in the eastern Kalahari, though it is unclear how permanent the zones of bush encroachment are. The natural purturbations that affect the area, particularly drought (Perkins, 1990) and fire, may well make the eastern Kalahari environment resilient to degradation, as defined by Abel & Blaikie (1989). The disturbances caused by pastoralism must therefore be set against the background of natural fluctuations in the character and biomass of vegetation communities.

The linkages between different elements of the system are also less than straightforward in terms of their interpretation relative to degradation. Bush encroachment, while fairly unanimously viewed as degradation, may in fact impart a greater stability to geomorphic elements by increasing cover and limiting erosion. Interactions of this zone with drought, fire and grazing factors makes meaningful generalisation about the stability of bush-encroached land difficult, but has led to the assertion that its effective control demands an opportunistic system of management, more in keeping with the highly variable nature of savanna dynamics (Westoby et al., 1989).

Clearly, such relationships deserve closer investigation, but in this case they also bring into consideration the unique characteristics of the Kalahari. Very low relative relief and high infiltration capacities, even in the sacrifice zones around watering points where compaction might be expected, mitigate against overland flow contributing significantly, if at all, to erosion, making models of semi-arid vegetation—erosion interactions derived from areas of greater relief inapplicable (e.g. Thornes, 1990). Wind erosion may have a greater effect, especially given the aeolian origin of surface sediments, particularly if the size of sacrifice zones increase as the use of the water points continues into the future. Clearly, such issues require more examination as they significantly affect the sensitivity of geomorphic components to disturbance by pastoralism.

CONCLUSION

The work reported here indicates that the issue of environmental sensitivity to disturbance in semi-arid environments is a complex one. As well as different elements of the environment responding at different rates and in different ways, antecedent factors are of major importance. The topographical and soil characteristics of the Kalahari impart a greater resilience to degradation by erosion that in other areas with otherwise similar characteristics. As the system is naturally unstable, it displays a greater resilience to change than might commonly be expected. This also means that the time since pastoral activities commenced may not be a crucial factor influencing the degree of disturbance, because this simply represents a further layer of change in a system that is naturally dynamic.

REFERENCES

Abel, N. O. J. and Blaikie, P. M. (1989). Land degradation, stocking rates and conservation policies in the communal rangelands of Botswana and Zimbabwe. *Land Degradation and Rehabilitation,* **1,** 101–123.

Andrew, M. H. (1988). Grazing impact in relation to livestock watering points. *Trends in Ecology and Evolution,* **3,** 336–339.

Bhalotra, Y. P. R. (1985a). *Rainfall Maps of Botswana,* Department of Meteorological Services, Gaborone.

Bhalotra, Y. P. R. (1985b). *The Drought of 1981–5 in Botswana,* Department of Meteorological Serivces, Gaborone.

Biot, Y. (1988). Forecasting productivity losses caused by sheet and rill erosion. A case study from the Communal Areas of Botswana. Ph.D. Thesis, University of East Anglia.

Buckley, R., Gubb, A. and Wasson, R. (1987). Parallel dunefield ecosystems: predicted soil nitrogen gradient tested. *Journal of Arid Environments,* **12,** 105–110.

Buckley, R., Wasson, R. and Gubb, A. (1987). Phosphorus and potassium status of arid dunefield soils in central Australia and southern Africa, and biogeographic implications. *Journal of Arid Environments,* **13,** 211–216.

Carl Bro. (1982). *An Evaluation of Livestock Management and Production in Botswana with Special Reference to the Communal Areas,* Vols I–III, Evaluation Unit, Ministry of Agriculture and the Commission of the European Communities European Development Fund.

Cooke, H. J. (1985). The Kalahari today: a case of conflict over resource use. *Geographical Journal,* **51,** 75–85.

Debenham, F. (1952). The Kalahari today. *Geographical Journal,* **118,** 12–23.

Denbow, J. R. (1984). Prehistoric herders and foragers of the Kalahari: the evidence for 1500 years of interaction. In Schrire, C. (ed.), *Past and Present Hunter-gatherer Studies,* Academic Press, Orlando, FL. pp. 175–193.

Denbow, J. R. and Wilmsen, E. N. (1986). Advent and course of pastoralism in the Kalahari. *Science,* **234,** 1509–1515.

Georgiadis, N. J. (1987). Responses of savanna grasslands to extreme use by pastoralist livestock. Ph.D. thesis, Syracuse University, New York.

Holling, C. S. (1973). Resilience and stability of ecological systems. *Annual Review of Ecology and Systematics,* **4,** 1–23.

Hill, M. O. (1979). *DECORANA — A FORTRAN Program for Detrended Correspondence Analysis and Reciprocal Averaging,* Cornell University, Ithaca, NY.

Jerve, A. M. (1982). Cattle and inequality. A study in rural differentiation from southern Kgalagadi in Botswana. *DERAP Publications,* 143, Chr. Michelson Institute, Bergen.

Jongman, R. H. G., ter Braak, C. J. F. and van Tongeren, O. F. R. (1987). *Data Analysis in Community and Landscape Ecology,* Centre for Agricultural Publishing and Documentation (Pudoc), Wageningen, The Netherlands.

Leistner, O. A. (1967). The plant ecology of the southern Kalahari. *Memoir of the Botanical Survey of South Africa*, **38**.

Meigs, P. (1953). World distribution of homoclimates. *Reviews on Research on Arid Zone Hydrology*, UNESCO, Paris, pp. 203−209.

Mentis, M. T. (1984). Monitoring in South African Grasslands. *South African National Scientific Programmes Report* 91.

Perkins, J. S. (1990). Drought, cattle-keeping and range degradation in the Kalahari, Botswana. In Stone, G. J. (ed.), *Pastoralists' Response to Drought*, Aberdeen University Press, Aberdeen, in press.

Perkins, J. S. (1991). The impact of borehole dependent cattle grazing on the environment and society of the Kalahari sandveld, western Central District, Botswana. Ph.D. Thesis, University of Sheffield.

Skarpe, C. (1986). Plant community structure in relation to grazing and environmental changes along a north−south transect in the western Kalahari. *Vegetation*, **68**, 3−18.

Stoddart, L. A., Smith, A. D. and Box, T. W. (1975). *Range Management*, McGraw-Hill, New York.

ter Braak, C. J. F. (1987). The analysis of vegetation−environment relationships by canonical correlation analysis. *Vegetatio*, **69**, 69−77.

ter Braak, C. J. F. (1988). *CANOCO − A FORTRAN Program for Canonical Community Ordination by Partial Detrended Canonical Correspondence Analysis*, version 2.1. Technical Report LWA-88-02, Ministry of Agriculture and Fisheries. Wageningen, The Netherlands.

Thomas, D. S. G. (1988). Environmental management and environmental change: the impact of animal exploitation on marginal lands in central southern Africa. In Stone, G. J. (ed.), *The Exploitation of Animals in Africa*, Aberdeen University Press, Aberdeen, pp. 5−22.

Thomas, D. S. G. and Shaw, P. A. (1990). The deposition and development of the Kalahari Group sediments, central southern Africa. *Journal of African Earth Sciences*, **10**, 187−197.

Thomas, D. S. G. and Shaw, P. A. (1991). *The Kalahari Environment*, Cambridge University Press, Cambridge.

Thornes, J. B. (1990). The interaction of erosional and vegetational dynamics in land degradation: spatial outcomes. In Thornes, J. B. (ed.), *Vegetation and Erosion*, Wiley, Chichester, pp. 41−53.

Tolsma, D. J., Ernst, W. H. and Verwey, R. A. (1987). Nutrients in soil and vegetation around two artificial waterpoints in eastern Botswana. *Journal of Applied Ecology*, **24**, 991−1000.

Tyson, P. D. (1979). Southern African rainfall: past, present and future. In Hinchley, M. T. (ed.), *Proceedings of the Symposium on Drought in Botswana*, Botswana Society, Gaborone, pp. 45−52.

Tyson, P. D. (1986) *Climatic variability in Southern Africa*. Oxford University Press, Cape Town.

Tyson, P. D., Dyer, T. G. J. and Mametse, M. N. (1975). Secular changes in South African rainfall: 1880 to 1972. *Quarterly Journal of the Royal Meteorological Society*, **101**, 819−833.

United Nations (1977). *Draft Plan of Action to Control Desertification*, Conference on Desertification, Nairobi, 29 August−9 September 1977. A/CONF 74/L36, Nairobi, UNEP.

Walker, B. H. and Noy-Meir, I. (1982). Aspects of the stability and resilience of savanna ecosystems. In Huntley, B. J. and Walker, B. H. (eds), *Ecology of Tropical Savannas*, Springer-Verlag, Berlin, pp. 556−590.

Walker, B. H., Ludwig, D., Holling, C. S. and Peterman, R. M. (1981). Stability of semi-arid grazing systems. *Journal of Ecology*, **69**, 473−498.

Warren, A. and Agnew, C. (1988). An assessment of desertification and land degradation in arid and semi-arid areas. *Drylands Programme Issues Envelope* 2, International Institute for Environment and Development.

Weare, P. R. and Yalala, A. (1971). Provisional vegetation map. *Botswana Notes and Records*, **3**, 131−152.

Westoby, M., Walker, B. and Noy-Meir, I. (1989). Range management on the basis of a model which does not seek to establish equilibrium. *Journal of Arid Environments*, **17**, 235−239.

18 Successional Pattern and Landscape Sensitivity in the Mediterranean and Near East

MARK A. BLUMLER
Department of Geography, University of California, Berkeley, USA

ABSTRACT

Traditional views regarding successional relationships in Mediterranean/Near East summer-dry regions are not in accord with ecological theory or with empirical evidence, and interpretations of human impacts on the landscape are often simplistic. Phytosociological succession models are essentially Clementsian, and hold that maquis or forest is climax over much of the Basin, despite the complex variation in climate and soils. Supporting evidence for the model comes almost entirely from southern France, which is climatically marginal, or from rocky, untillable slopes. Grazing, woodcutting, and agriculture are assumed to cause massive soil erosion and vegetation retrogression in a quasilinear reversal of the successional sequence.

The course of succession after cessation of grazing was studied on a fertile dark rendzina in the Galilee of Israel. Perennial cover decreased slightly in the exclosure, especially in open soil areas, while tall annual grasses became dominant. Perennials were associated with rocks, but this was less true under grazing. That is, grazing apparently facilitates perennial establishment in open soil. Results suggest that vigorous, large-seeded annuals are an important component of 'climax' vegetation on deep fertile soils where winters are wet and summers dry. Dense shrubby vegetation is more likely where soil is poor or rocky.

These and other results suggest that humans have favoured some shrubs, herbaceous perennials, and annuals, and adversely affected others; and that in some places woody vegetation is more dense today than formerly, as a result of soil impoverishment and erosion (exposure of rocks). Soil and slope responses to agropastoralism may also have been more complicated than usually envisaged. Sensitivity of the Mediterranean landscape may be less than sometimes claimed.

INTRODUCTION

Humans have interacted with the environment in the great Mediterranean-type (summer-dry) region of the Old World, extending from the Atlantic Ocean to the Pamirs and Tien Shan, for many millennia (Naveh, 1984; Naveh & Dan, 1973). Pastoralism began in the Near East, and radiocarbon dates suggest that farming and plant domestication also were initiated in the Fertile Crescent and Levant earlier than in the other centres of agricultural origins such as China, south-east Asia, Mexico and Peru (Blumler & Byrne, 1991). Great civilisations arose and disappeared again long before the ascendance of non-Mediterranean Europe. Today, the Mediterranean and Near East (MedNE) have the appearance of a used-up land, with extensive areas of bare, limestone hills and scrubby vegetation; typically, scholars have

Landscape Sensitivity. Edited by D. S. G. Thomas and R. J. Allison
© 1993 John Wiley and Sons Ltd

concluded that agropastoralism has caused massive soil erosion, deforestation, vegetation degradation and desertification. Supporting empirical evidence is scanty, however, while terms such as 'degradation' and 'desertification' are often used in a loose and inconsistent manner (Rackham, 1982; Blumler, 1984, 1991a; Blaikie & Brookfield, 1987; Roberts, 1989). For instance, Dregne (1977) mapped not only the MedNE but also southern California as affected by desertification as severely as the Sahel, which is nonsense.

The disentanglement of natural and human causes of soil erosion is difficult (Vita-Finzi, 1969; Brice, 1978; Davidson, 1980; Wagstaff, 1981; Brookes, 1982; van Andel, Runnels & Pope, 1986; Blaikie & Brookfield, 1987; Roberts, 1989). Some of the MedNE's most spectacular erosional landforms appear to be natural, and have been surprisingly little affected by agriculture and grazing (Wise, Thornes & Gilman, 1982; Gilman & Thornes, 1985). Limestone naturally tends to give rise to rocky, shallow soils due to irregular solution weathering and the enormous quantities of rock that must be dissolved before soil can build up (but cf. Drew, 1982). Reconstruction of the hypothetical former soil cover, and its degree of development on rocky slopes is to some extent based on the assumptions that dense climax woodland formerly blanketed the landscape, and that natural disturbance of the climax was minimal. As discussed below, these assumptions are overdue for revision.

Scholars often claim that 'deforestation' has occurred if there is evidence of woodcutting or browsing, without considering the regenerative abilities of the vegetation (e.g. Zohary, 1973; Thirgood, 1981; additional references in Blumler, 1984); but Rackham (1982) pointed out that maquis shrubs can be grazed very close yet still grow up into woodland if left alone. It also must be considered that the semi-arid climate of the MedNE would naturally support less lush vegetation than is found in the humid temperate zone (Meher-Homji, 1965). Pollen diagrams often fail to support the prevailing view that the land is less worked today than formerly: for instance, well-known diagrams from Morocco (Lamb, Eicher & Switsur, 1989), the Zagros (van Zeist, 1967) and the Argolid (Pope & van Andel, 1984) do not show any significant decrease in the ratio of arboreal to non-arboreal taxa during the Holocene. Changes in arboreal species-composition are frequent in pollen diagrams, but it is often difficult to relate these changes to human activities or to establish that they represent degradation. For example, shifts from oak to pine are sometimes regarded as evidence for degradation (e.g. Zohary, 1973; Wojterski, 1990), yet pine forests probably incorporate at least as much biomass per unit area as oak woodland.

It is commonly assumed, too, that grazing (by goats, especially) has eliminated the most palatable species, and cause their replacement by 'anti-pastoral' (spiny or poisonous) vegetation (Zohary, 1962, 1973; Rackham, 1982). However, in a recent study spininess did not correlate with grazing pressure (Noy-Meir, Gutman & Kaplan, 1989). Alternatively, it may be that spines serve a cooling function (Warming, 1895), or that they are associated with soil calcium (Blumler, 1991a). Other lines of evidence against the hypothesis that grazing has eliminated many species are the lack of plausible candidate taxa (those suggested by Rackham, for instance, are poorly adapted to summer drought), the absence of supporting evidence in the pollen record, and the MedNE's spectacular species-diversity, on both local and regional scales. The belief that domestic grazing must have caused wholesale extinctions also presupposes that disturbance is unnatural, no longer a popular viewpoint among English-speaking ecologists. Semi-arid regions, including the MEdNE, are generally thought to be very sensitive to human impacts as well as climatic change, but Naveh (1967, 1973, 1982; Naveh & Whittaker, 1979) has pointed out that resilience of MedNE vegetation is often very high; in part, this is due to the presence of protected and diverse microsites provided by rock outcrops, which may have increased in density as a result of soil erosion (Blumler, 1991a).

Succession

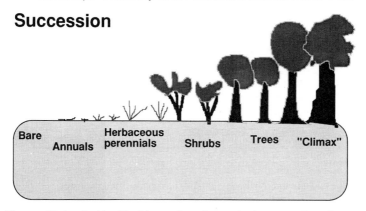

Bare Annuals Herbaceous perennials Shrubs Trees "Climax"

Figure 18.1   A simplified but otherwise typical succession diagram

Because agropastoralism has had a pervasive influence, it is difficult to obtain high-quality empirical evidence on potential natural vegetation in the MedNE. Hence, interpretations of human impacts on vegetation are based upon the often uncritical application of succession models originally developed in Europe and the eastern United States — regions that do not experience summer drought. If present-day vegetation differs from the predicted natural condition, it is often argued that human influence must be the cause. Vegetation that deviates from model expectation is also sometimes cited as supporting evidence for anthropogenic soil erosion. However, the circumstances that favour trees, shrubs and herbs in competition with one another or in response to grazing and woodcutting change as climate and soil conditions change, so reliance upon Euro-American models can be misleading. Furthermore, succession theory is undergoing radical revision at present, though the implications are not yet appreciated fully by students of dry, especially seasonally dry, environments. In this chapter, theoretical and empirical evidence is presented suggesting that traditional models of successional sequences in summer-dry regions are topsy-turvy with respect to reality. In particular, annual plants can be important under all disturbance regimes and can even comprise 'climax' vegetation if seasonal drought is severe. Hence, human impacts on vegetation, and perhaps also on soils, have been more complicated and less negative than generally assumed.

SUCCESSION THEORY

Traditional notions of succession are intuitive and predate Clements (1916; Glacken, 1967; Worster, 1977). Every ecology text includes a diagram similar to that of Figure 18.1, which corresponds to the Clementsian model. Succession begins when disturbance creates an unvegetated surface; in the traditional model, the ground is usually colonised first by relatively short-lived, short, herbaceous plants, and gradually replaced (in the absence of disturbance) by taller, long-lived, and woody plants until the biggest, 'baddest' individuals that the climate will support eventually come to dominate. The linear and progressive nature of the succession model is in accord with early and mid-20th-century notions of progress — notions that have now largely disappeared. Also incorporated is the assumption that natural disturbance is rare, hence that 'climax' is the norm — except when humans muck up the landscape. This assumption is an outgrowth of the Judeo-Christian human—nature split, and Romantic belief in the 'forest primeval'. It is now known, however, that succession does not always lead to dominance by the tallest woody plants in an area; the probable replacement of redwoods and other conifers by pygmy forest on marine terraces along the California coast is a classic counter-example (Jenny, Arkley & Schultz, 1969; Westman, 1975). In addition, there is a growing awareness that disturbance is pervasive in nature (Grubb, 1977; Connell, 1978); in fact, it is questionable whether *any* species can be adapted to entirely undisturbed conditions (Blumler, 1991a; Blumler, Olsvig-Whittaker & Naveh, n.d.). Accordingly, succession theory has changed. The seminal paper, by Connell & Slatyer (1977), proposed three succession models:

1. *Facilitation*: early successional plants alter environment in a manner that is favourable to later successional taxa (this, especially, is the Clementsian model).
2. *Tolerance*: early successional plants alter environment in a manner that has no effect on later successional taxa, which slowly replace the pioneer species.
3. *Inhibition*: early successional plants compete successfully with seedlings of later successional taxa, which are able to establish only after disturbance.

In all of these models early successional plants create conditions that are unfavourable for the establishment of their own offspring; Connell & Slatyer presented evidence suggesting that all three models may apply at times.

In the 'inhibition' model, however, succession comes very close to tautology. Disturbance by definition initiates succession, but the successional sequence cannot proceed unless there is further disturbance; how, then, can one distinguish disturbances that 'initiate' succession from those that allow it to continue, except by the characteristics of the plants that are favoured? Connell & Slatyer (1977, p. 1123) assumed that succession-initiating disturbances are grander in scale than the often local disruptions that allow establishment of climax trees, arguing in effect that local disturbances necessarily favour long-lived species. However, sometimes succession is tautologically defined as a change from herbaceous to woody vegetation without regard to severity and timing of disturbance (Miles, 1978, p. 63; Blumler, 1991a). Grazing, for instance, typically creates only patchy openings in the vegetative cover, yet removal of grazing is almost universally seen as initiating succession. Although modern succession theory is more subtle than the traditional model, it continues to incorporate linearity and progression as the prevailing paradigm; the difficulty we have in moving away from this conception is perhaps rooted in the very structure of our language (Blumler, 1991a; cf. Whorf, 1939, 1941, 1956).

In the MedNE, succession theory has been dominated by phytosociology, which is inherently Clementsian (though a more modern concept is beginning to take form [e.g. Barbero *et al.*, 1990], and there has long been a recognition that edaphics play a larger role than Clements realised). Most scholars assume that succession proceeds as follows:

Rock or bare ground $\rightarrow$ Annuals $\rightarrow$ Perennial herbs (steppe) $\rightarrow$ Low shrubs (batha) $\rightarrow$ Tall shrubs (maquis) $\rightarrow$ Trees (forest)

and that human impacts cause regression in a quasilinear reversal of the successional sequence. Annuals are ruderals only, while tall woody plants are climax. Empirical support for the model comes primarily from studies carried out in the northern Mediterranean (especially southern France), which is climatically marginal, or on rocky, untillable slopes (Braun-Blanquet, 1932; Braun-Blanquet, Roussine & Nègre, 1951; Horvat, Glavac & Ellenberg, 1974; Tomaselli, 1977; Houssard, Escarre & Romane, 1980; Jackson, 1985), with the results being extrapolated to the entire Basin (e.g. Quézel & Barbero, 1982; Barbero *et al.*, 1990; Wojterski, 1990). Deviations of vegetation pattern from Clementsian expectation are usually ascribed to human impacts (Zohary, 1962, 1973, 1983; Pignatti, 1978; Chalabi, 1982), often without any evidence.

Litav (1965; Litav, Kupernik & Orshan, 1963) in Israel and Rackham (1982) in southern Greece offered alternatives to the prevailing succession model. Litav (1965, 1967; Litav & Orshan, 1971; Litav, Kupernik & Orshan, 1963) provided experimental as well as distributional evidence suggesting that tall annual grasses can outcompete subshrubs on good soils in the absence of disturbance. He accepted that maquis is climax over much (but not all) of the mediterranean-climate territory of Israel, but argued that tall annual grasses and perennial geophytes such as *Hordeum bulbosum* would have constituted the subclimax after removal of maquis, with subshrubs becoming important only after massive soil erosion and overgrazing. Although this conclusion is still controversial, Israeli ecologists now generally accept that succession often remains 'arrested' at herbaceous or dwarf shrub stages despite no disturbance (Naveh, 1982; Zohary, 1982; Rabinovitch-Vin, 1983). Rackham similarly reported that while maquis grows up into woodland when grazing and cutting cease, garrigue is converted into annual-dominated steppe. He suggested that garrigue is eliminated by the increased moisture use of tall maquis, leaving only enough water for patches of steppe. However, Litav's experimental results suggest the alternative hypothesis that direct competition for water between shrubs and annuals is responsible for the vegetation change.

CLIMATE, SOILS AND VEGETATION DYNAMICS

In humid regions, forests tend to dominate fertile soils in the absence of disturbance, but in semi-arid regions competition between woody plants and herbs is more equal, and is strongly influenced by soil conditions (Walker *et al.*, 1981). Grasses are favoured if water is concentrated in upper soil layers (as on heavy soils) and woody plants if water penetrates deeper (Walter, 1971; Walker *et al.*, 1981). Paradoxically, moisture availability is often higher in coarse-textured soils and around rocks than in deep, fine-textured materials; this is especially true where seasonal drought is severe, because heavy soils can dry out completely. This is known as the 'inverse texture effect' (Noy-Meir, 1973). In addition, as Blumler (1984, 1991a) pointed out, fertility favours herbs because they have high relative growth rates due to low investment in lignin (see also Specht, 1969), and their rapid use of soil moisture also can preclude deep percolation; for annual plants in particular this is a mechanism by which severe drought stress can be induced in neighbouring perennials during a long dry season. On the other hand, infertile substrates tend to support a high proportion of woody plants, though these will be low-growing. Rock outcrops also favour woody species, and perennials generally, because they reduce seedling competition and serve as a natural mulch against water loss (in addition to offering protection against grubbing and grazing).

Although annual plants are usually thought of as ruderals, one would expect on theoretical grounds that increasing seasonal drought would expand their ecological amplitude (Westoby, 1979/80; Blumler, 1984). For instance, Raunkiaer (1934; cf. Blumler, 1984, 1991a) pointed out that annuals increase in importance as summer drought becomes more severe, and argued that annuals are best adapted to withstand drought because the seed is the most drought-resistant plant organ. Annuals also compete strongly during the rainy season (Blumler, 1984): they

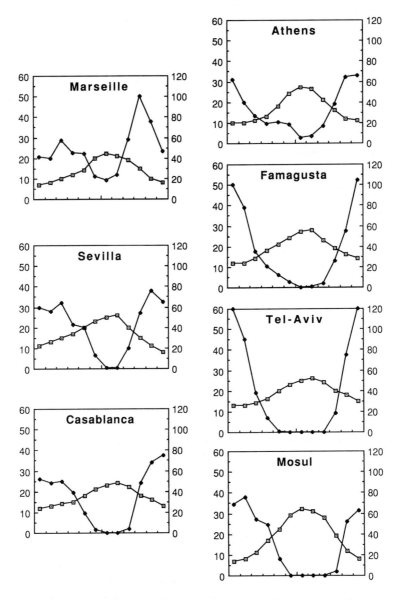

Figure 18.2 Climate diagrams for selected MedNE stations, indicating the tendency for seasonality of precipitation and summer drought to increase towards the south and east. Mean monthly temperature in °C (left) and precipitation in mm (right) on the vertical axis, and month on the horizontal axis (redrawn from Walter, Harnickell & Mueller-Dombois, 1975)

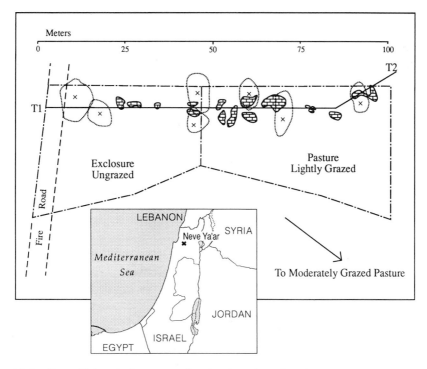

Figure 18.3 Neve Ya'ar study areas, showing location of rocks and trees along the permanent transect (T1–T2)

have relative growth rates even higher than those of perennial herbs (Muller & Garnier, 1990), especially those xerophytic species adapted to withstanding prolonged seasonal drought, and so should win in seedling competition. As dry-season rainfall increases, however, advantage shifts to perennials that do not need to restrict their germination and growth to the wettest season.

The Mediterranean region is climatically and edaphically heterogeneous. Summer drought generally increases to the south and east (Figure 18.2), with the most extreme contrast between wet winter and dry summer being found in the eastern Mediterranean and also in the Fertile Crescent (Perrin de Brichambaud & Wallén, 1963; Blumler, 1984, 1991a); not surprisingly, annual plants are particularly dominant there. Southern France receives considerable summer rain, and is a poor analog to these 'true' Mediterranean-type regions that experience five months or so of hot, rainless weather. Based on these theoretical considerations and the empirical evidence for annual competitiveness provided by Litav and also many researchers in California and other summer-dry regions, Blumler (1984) proposed that annuals can actually be competitive dominants if seasonal drought is sufficiently severe.

SUCCESSION AND THE IMPACT OF GRAZING: AN EMPIRICAL TEST

The hypothesis that under severe seasonal drought and on productive soil without disturbance tall, large-seeded, annual grasses can outcompete all perennial species was tested at an experimental range in the Lower Galilee of Israel. The results are reported in more detail elsewhere (Blumler, 1991a; Blumler, Olsvig-Whittaker & Naveh, n.d.), and are merely summarised here. The study site, the Neve Ya'ar Agricultural Experimental Station, receives a mean annual precipitation of 578 mm, but none falls between May and September. Soil is dark rendzina — a fertile, slightly basic, heavy clay loam formed on calcrete ('nari') overlying Eocene chalk (Dan *et al.*, 1975). Soil depth averages about 25 cm, and the nari outcrops as scattered, loaf-shaped rocks over about 20% of the surface. Vegetation is a deciduous oak (*Quercus ithaburensis*) woodland (about 25% canopy cover) with a species-rich (about 150 species/0.1 ha) understory dominated by annuals; woody plants are almost completely restricted to nari outcrops. Evergreen sclerophylls are rare on this soil type, except on steep slopes where the underlying chalk is exposed, and it is generally accepted that succession does not proceed to maquis or evergreen woodland (Eig, 1933; Oppenheimer, 1950; Zohary, 1962, 1982; Naveh & Whittaker, 1979; Blumler, 1991a, 1991b). At Neve Ya'ar, grazing pressure since 1969 has been less than would be likely in a natural ecosystem, yet the pastures are dominated by annuals.

Vegetation of an exclosure laid out in 1981 was compared with an adjacent, lightly grazed pasture, and another, more moderately grazed area (Figure 18.3). All study areas are on level ground. Figure 18.4 shows that tall annual cereal grasses (wild barley, *Hordeum spontaneum*, and wild oats, *Avena sterilis*) became progressively more dominant in the exclosure, until they occupied 87% of the cover; in open soil in full sun, they achieved 96% cover (Figure 18.5). The grazed pasture had a lower relative cover of annuals, and a much lower wild cereal cover. Examination of a permanent transect through the exclosure demonstrated that between 1982 and 1988 frequency of perennials in small quadrats decreased in open soil and increased around rocks (Figure 18.6); yet most of the perennial grasses believed capable of replacing annuals in the absence of disturbance (Naveh, 1957; Litav,

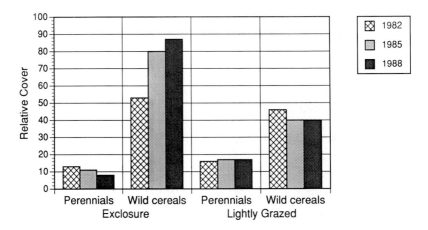

Figure 18.4 Changes in understory species-composition between 1982 and 1988 in the exclosure and adjacent lightly grazed pasture at Neve Ya'ar (from Blumler, Olsvig-Whittaker & Naveh, n.d.)

1967; Litav & Seligman, 1969, 1970; Noy-Meir, Gutman & Kaplan, 1989) were on-site. Perennial species were highly significantly associated with rocks in the exclosure (Table 18.1 and Figure 18.7). A few of these species were less significantly associated with rocks, or found preferentially in open soil, in grazed pastures. That is, grazing may facilitate establishment of perennials away from rocks, presumably by removing competing annuals, i.e. in a manner analogous to the facilitation of shrub invasion of grasslands in many semi-arid regions (e.g. Walker *et al.*, 1981). Certainly cattle prefer annuals to many adult perennials at Neve Ya'ar, though it is not known whether perennial seedlings are also unpalatable. Also,

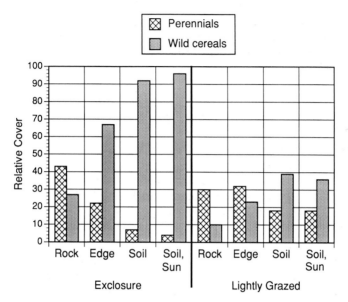

Figure 18.5 Relationship between microhabitat, grazing, and species-composition at Neve Ya'ar (1988 data). Rock edge defined as a 15 cm border (from Blumer, Olsvig-Whittaker & Naveh, n.d.)

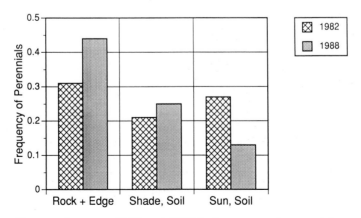

Figure 18.6 Changes between 1982 and 1988 in frequency of perennial species in 20 × 20 cm quadrats along a permanent transect in the exclosure, in relation to microhabitat (from Blumler, Olsvig-Whittaker & Naveh, n.d.)

Table 18.1 Association of perennial species with microhabitats, in relation to grazing

Species	U	LG	U + LG	MG	All
Quercus ithaburensis	0.0001	0.01	0.0001	0.0001	0.0001
Capparis spinosa	0.0001	0.0001	0.0001	—	0.0001
Micromeria myrtifolia	0.001	0.0001	0.0001	0.0001	0.0001
Rhamnus palaestinus	0.0001	0.0001	0.0001	0.0001	0.0001
Alcea setosa	0.0001	N	0.0001	NS	NS
Anemone coronaria	0.0001	0.0001	0.0001	0.0001	0.0001
Carlina curetum	N	N	N	0.01	
Cyclamen persicum	0.0001	0.0001	0.0001	0.0001	0.0001
Echinops adenocaulos	N	N	N	NS	
Eryngium creticum	N	0.0005(O)	0.005(O)	—	0.005(O)
Gynandriris sisyrinchium		0.05			
Hordeum bulbosum	NS	0.25(O)	0.01(O)		
Hyparrhenia hirta	0.0001	—	0.0001	0.0001	0.0001
Ornithogalum narbonense		0.05			
Pallenis spinosa	N	0.01(O)	0.025(O)	—	0.025(O)
Piptatherum holciforme	N	—	N	0.005	0.001
				0.05(SH)	0.05(SH)
Thrincia tuberosa	0.001	NS	0.05	NS	0.05
Verbascum sinuatum	N	N	N	N	0.05

U, ungrazed; LG, lightly grazed; MG, moderately grazed pastures.
Association is with rocks unless otherwise indicated (O, open soil, SH, shade): $P <$ values given in table (chi-squared test); NS, not significant, N, sample size too small for significance test.

the rotational systems that are so common in the Near East (and at Neve Ya'ar) might favour perennials if their season of germination differs from that of the annuals. Finally, grazing may be able to reduce annual seed output sufficiently in any given spring that seedling density the following fall will be low, allowing perennials to establish in open spots. However, the complexity of the outcropping pattern at Neve Ya'ar, which is not entirely consistent from pasture to pasture, and the limited number of pastures studied precludes conclusiveness.

Litav (1967) noted that maquis shrubs had spread only on the rock walls of abandoned agricultural terraces in the Judean Mountains, while annuals remained dominant in deep soil pockets; he also found a statistically significant association of perennials with rock outcrops. More recent Israeli studies on fertile soils have also found that annuals remain dominant indefinitely after reduction of elimination of human impacts (Zohary, 1969; Noy-Meir & Seligman, 1979; Noy-Meir, Gutman & Kaplan, 1989; Shmida & Ellner, 1984; Kutiel, 1985). In studies that do not attempt to separate out the influence of rock outcrops, perennial cover sometimes increases, though the increase is usually slight. For instance, if a pair of sites with 35% rock cover is removed from the data set of Noy-Meir, Gutman & Kaplan (1989), then there is little tendency for increase in perennial cover with relaxation of grazing pressure (Table 18.2). The conclusion of Noy-Meir, Gutman & Kaplan that reductions in grazing favoured *Hordeum bulbosum* more than the wild cereals is contradicted by their own data (Blumler, 1991a). Furthermore, annuals dominated most of their exclosures.

Theoretical models of vegetation dynamics assume that though annuals may be able to outcompete perennial seedlings, they cannot replace mature perennials, so occasional chance establishment of perennials should lead to their eventual dominance in the absence of major disturbance (Connell & Slatyer, 1977; Westoby, 1979/80; Shmida & Ellner, 1984). This

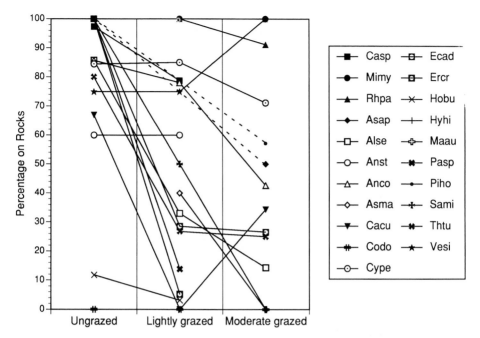

Figure 18.7 The relationship between grazing pressure and tendency of perennial species to occur around rocks at Neve Ya'ar. Legend gives first two letters of genus and species for each taxon, e.g. 'Casp' = *Capparis spinosa*, 'Rhpa' = *Rhamnus palaestinus* (cf. Table 18.1). Rock cover is slightly higher in the exclosure, but the dramatic increase in percentage of individuals in open soil under grazing cannot be explained solely on this basis (from Blumler, 1991a)

is a key tenet of the inhibition model, for instance. The transect data reported here (Figure 18.5) suggest that annuals may be able to outcompete even tall adult perennials if seasonal drought is sufficiently severe (presumably by complete utilisation of soil moisture during the rainy season). In short, these data suggest that certain annual plants can be 'climax' at Neve Ya'ar and in similar environments (which are widespread in Israel and throughout the Fertile Crescent [Harlan, 1967; D. Zohary, 1969; M. Zohary, 1973]); at the very least, the wild annual cereals seem able to hold their ground against subshrubs and many, if not all, herbaceous perennials, as Litav first pointed out.

Table 18.2 Lifeform and grazing intensity at paired sites along fence rows in the Upper Jordan Valley (calculated from Noy-Meir, Gutman & Kaplan, 1989)[a]

Lifeform	No grazing	Relative cover light−moderate grazing	Heavy grazing
N	8	13	7
Wild cereals	43.5	19.8	2.9
Other annuals	30.2	56.9	71.4
Total annuals	73.8	76.8	74.3
Perennials	26.6	23.5	25.4

[a] Excluding the most rocky ('RP', with 35% rock cover) of their 15 pairs of sites.

A short-term study such as this one cannot shed a great deal of light on oak regeneration. We observed no successful seedling establishment, and other studies have demonstrated that oak establishment in dense annual grassland in seasonally dry environments is problematic (Kaplan, 1984; Blumler, 1991a, 1991b). Since even the oaks are closely associated with nari outcrops at Neve Ya'ar, it is uncertain whether they would be present at all if soil were deeper. If oak regeneration does occur eventually, increased shade might eliminate the annuals: it is a general rule in the MedNE that annual plants increase in importance as the overstory becomes sparse or absent (Braun-Blanquet, Roussine & Nègre, 1951; Zohary, 1962), probably because slow-growing stress-tolerators are favoured over fast-growing annuals under low light conditions (Grime, 1979) and because seasonal soil moisture regime is moderated in shade. An eventual thickening up of the oak stand would be in accord with the inhibition model, except that succession would proceed directly from annuals to forest without any intermediate stage, while shorter perennials would spread only *after* the trees. On both theoretical and distributional grounds, however, one would expect woodlands to become more open as seasonal drought becomes more severe (Oppenheimer, 1950; Blumler, 1984, 1991b).

DISCUSSION

There are, then, both theoretical and empirical reasons for believing that 'succession' in seasonally dry regions, to the extent that it can be said to occur at all, is a very different process from that usually envisioned by MedNE scholars, and that seasonality of precipitation regime (as mediated by edaphics) strongly influences the competitive abilities of species of different lifeforms. Where seasonal drought is extreme, a closed, wooded climax on deep, fertile soils is much less likely than on rocky or poor soils or where there is subirrigation. Shrubs, in particular, require disturbance to invade grassland on good soils, where annuals can be expected to dominate under most circumstances. Human impacts may cause changes in species dominance, but not in the simple linear manner of the traditional succession/regression model. Grazing, for instance, favours some members of all lifeforms, and only slightly favours annuals as a group. As in other semi-arid regions (Walker *et al.*, 1981), grazing can lead to shrub invasion of grassland (Anonymous, 1951; Litav, Kupernik & Orshan, 1963; Le Houérou, 1981). Traditional agriculture has similar effects: for instance, Eig (1926) pointed out that the Arab plough removes annuals but leaves deep-rooted perennials; this has allowed several woody weeds to spread (Zohary, 1950). If grazing or agriculture lead to soil impoverishment or massive erosion, annuals probably would be hurt, not favoured. Abundant herbs, especially annuals, are sometimes taken as evidence for degradation or desertification, but the opposite may often be the case. For instance, because agricultural soils are comparatively fertile and deep, they should be dominated naturally by herbs, especially annuals, in summer-dry regions (excepting areas with a high water table). In the MedNE it is often pointed out that woody or herbaceous perennial plants are present primarily around rocks, and argued that a formerly dense perennial cover has retreated to the rocks because they offer protection against grazing, root-grubbing and the plough (Wright, McAndrews & van Zeist, 1967; Nemati, 1977; Noy-Meir & Seligman, 1979; Benabid, 1982; Kaplan, 1984; additional references in Blumler, 1984). But rocks also benefit perennials by reducing competition from annuals and mitigating summer drought stress, so it may be that on some soil types perennials were *never* important away from rocks. If it is true that agropastoralism has caused massive soil erosion or impoverishment (and it is probably judicious to conclude that there has been at least some negative impact [Davidson, 1980]), then one would expect that vegetation would have responded with a shift to denser but shorter woody growth, e.g. maquis instead of deciduous oak parkland.

These conclusions suggest that some revision may be in order of prevailing interpretations of the climatic implications of Holocene vegetation changes. There is considerable evidence that early Holocene vegetation in the MedNE was different than it is today (Pons, 1981, 1985; van Zeist & Bottema, 1982; Huntley & Birks, 1983; Roberts, 1989). In some areas, woody vegetation remained unimportant, while elsewhere deciduous oak woodland was in place. Migrational lag effects due to the restriction of evergreen sclerophylls to isolated refuges during the late Pleistocene were undoubtedly important, but probably not the entire story. It is generally believed that winter-deciduous trees require either cold winters or humid conditions including summer rain (Box, 1982; Roberts, 1989), but this ignores the importance of deciduous parkland in the Fertile Crescent, characterised by semi-arid conditions and the most Mediterranean (summer-dry) climate in the world. On the other hand, present-day sub-Mediterranean regions with significant summer rain, such as monsoon Afghanistan, are covered by evergreen maquis-type vegetation, and the most mesic sectors of Mediterranean California are also dominated by evergreen species (Freitag, 1971; Minnich, 1985; Blumler, 1991a, 1991b). Quaternary extinctions of conifers in Europe may have allowed deciduous forest to become 'climax' under mesic Mediterranean conditions, but it is also possible that deciduous oak woodland prevailed during the early Holocene because of either colder winters or hotter, drier summers, combined with comparatively deep soils.

Calculation of early Holocene solar radiation receipt based on Milankovitch cycles (COHMAP Members, 1988) confirms the likelihood of greater temperature extremes; it is also plausible that summer drought would have intensified, because of the effects of such a regime on the strength and migration of the Azores semipermanent high-pressure cell (cf. Byrne, 1987). If so, open deciduous parkland (not dense forest) with an annual-dominated understory, such as is found in the Fertile Crescent today, might have characterised much of the Mediterranean Basin. Maquis may have replaced deciduous oak park later because of anthropogenic soil erosion, as sometimes suggested (Aymonin, 1976; Pons, 1981; Roberts, 1989; Barbero et al., 1990), or because of higher winter temperatures and summer rain. It is unlikely that tropical moisture could have spread as far as the northern Mediterranean in the early Holocene, as Roberts (1989, pp. 89–90) hypothesised, and in any case this only would have favoured evergreen species; however, there is some suggestive evidence for a mid-Holocene increase in summer rainfall (Bintliff, 1982). Arboreal vegetation remained sparse until the mid-Holocene throughout much of the Fertile Crescent, despite winter precipitation that must have been equal to that experienced today (van Zeist, 1967, 1969; van Zeist & Bottema, 1982). Perhaps soil erosion or a shift to a regime with reduced summer temperatures and some summer rainfall allowed a thickening up of woodland.

What are the implications in terms of natural soil properties and sensitivity of landforms to human impacts? One hesitates to add to the already heavy load of 'ill-informed speculations' (Thornes, 1990b, p. xiii) about influences of vegetation on geomorphological processes; however, a few comments may be in order with the caveat that empirical data are clearly needed. The dense woody climax cover normally envisioned by Clementsian phytosociologists would have minimized soil erosion and favoured soil development. Alternatively, if parkland vegetation with an annual understory was widespread in the early Holocene, one would expect somewhat greater natural rates of erosion, higher soil pH and perhaps less pronounced horizonation. Of course, erosion under ungrazed annual vegetation is not nearly as rapid as in agricultural systems in which the soil is bare for part of the rainy season; but on the other hand, natural disturbances such as wild ungulate grazing and fire would have exposed the surface episodically. With removal of the protective cover of grass litter, soils become vulnerable to the intense storms that typically open the rainy season in many parts of the MedNE, prior to re-establishment of the annual sward (in a few weeks to months, depending on rainfall and temperature). On hard limestone and dolomite, however, exposure of rocks reduces sensitivity due to natural terracing and improved infiltration as a result of formation of cracks at the rock−soil interface. At the same time, aeolian input of base-rich fine materials from hyper-arid regions is of major importance in semi-arid border regions such as Israel (Yaalon & Ganor, 1973, 1975; Danin, Gerson & Garty, 1983), and even in the northern Mediterranean (Macleod, 1980; Mizota, Kusakabe & Noto, 1988). Presumably there was less aeolian transport during the Saharan early and mid-Holocene wet phase (cf. COHMAP Members, 1988), while irrigation agriculture and overgrazing of the deserts may perhaps exacerbate modern rates. To some extent aeolian deposition should counteract both profile development and erosion (Dan, Yaalon & Koyumdjisky 1972), while the hyper-arid source regions suffer little, being water- rather than nutrient-limited. In short, soil conditions at the dawn of the Neolithic may have been more similar to the present that usually assumed, while the millennia since have not witnessed an entirely unilinear, catastrophic process of deterioration. This argument ignores the possibly widespread former existence of deep, well-developed, relict soils; these may have disappeared as a result of climatic and vegetational adjustments relatively early in the Holocene, eroded away over a longer period on poorly vegetated slopes, or in many cases were probably truncated by human impacts.

Although it is recognised that there are complicating effects of growth form and disturbance adaptation (Thomas, 1988; Brown, 1990), models incorporate the simplifying assumption that cover and biomass are inversely correlated with erosion; also, cover, productivity and biomass are assumed to be correlated (e.g. Thornes, 1990a). If vegetation becomes dominated by annuals, however, the relationship between productivity and biomass is decoupled: even very productive Mediterranean grasslands with high cover have total biomass similar to that reported by Francis & Thornes (1990) for 'degraded' matorral with cover so low that soil loss is rapid. On the one hand, then, low biomass values do not in themselves indicate degradation, while on the other hand the most erosion-resistant vegetation may not be 'climax' especially on fertile substrates. In the particular case of south-eastern Spain studied by Francis & Thornes (1990), Thornes's (1990a) model may have greater applicability because the bimodal rainfall regime favours perennials over annuals, compared to my study site in Israel with its pronounced unimodal precipitation pattern and protracted summer drought (Blumler, 1984).

There is no desire here to pretend that degradation is not a problem in the MedNE. At the same time, much of the vegetation evidence for degradation is problematic at best, in part because MedNE vegetation is more resilient than even Naveh (1967, 1973, 1982) has recognised. One wonders, in fact, whether humid temperate realms are not in some respects more sensitive to human impacts, contrary to the widespread belief that semi-arid regions are especially fragile. Regardless, it seems that in the past negative feedbacks operated to constrain human ability to destroy the MedNE environment, while today the dramatic increase in inputs of energy and capital from elsewhere create a potentially more dangerous situation.

ACKNOWLEDGEMENTS

Fieldwork in Israel was funded in part by a University of California Regents Traveling Fellowship, and a fellowship from The Technion. Z. Naveh and L. Olsvig-Whittaker constructed the exclosure at Neve Ya'ar and graciously provided baseline data for the study summarised here. They also assisted in the research, along with A. Mann, P. Kutiel and A. Liston, to whom I am most grateful.

REFERENCES

Anonymous (1951). *Pasture and Fodder Development in Mediterranean Countries*, OEEC Technical Assistance Mission, Paris, no. 56.

Aymonin, G. G. (1976). Deux singulières chênaies en 'zone bioclimatique méditerranéenne'. *Bulletin de la Société Botanique de France*, **123**, 79−82.

Barbero, M., Bonin, G., Loisel, R. and Quézel, P. (1990). Changes and disturbances of forest ecosystems caused by human activities in the western part of the mediterranean basin. *Vegetatio*, **87**, 151−173.

Benabid, A. (1982). Bref aperçu sur la zonation altitudinale de la végétation climacique du Maroc. *Ecologia Mediterranea*, **8**, 301−315.

Bintliff, J. L. (1982). Palaeoclimatic modelling of environmental changes in the East Mediterranean region since the last glaciation. In Bintliff, J. L. and van Zeist, W. (eds), *Palaeoclimates, Palaeoenvironments and Human Communities in the Eastern Mediterranean Region in Later Prehistory*, BAR, Oxford, pp. 485−527.

Blaikie, P. and Brookfield, H. (1987). *Land Degradation and Society*, Methuen, London, p. 296.

Blumler, M. A. (1984). Climate and the annual habit. Unpublished Ma thesis, University of California.

Blumler, M. A. (1991a). Seed size and environment in Mediterranean-type grasslands in California and Israel. Unpublished Ph.D. thesis, University of California.

Blumler, M. A. (1991b). Winter-deciduous *versus* evergreen habit in mediterranean regions: a model. *Proceedings of the Symposium on California's Oak Woodlands and Hardwood Rangeland Management*, Davis, CA, USDA, Pacific Southwest Forest and Range Experiment Station, General Technical Report, PSW-126, Berkeley, CA, pp. 194−197.

Blumler, M. A. and Byrne, R. (1991). The ecological genetics of domestication and the origins of agriculture. *Current Anthropology*, **32**, 23−54.

Blumler, M. A., Olsvig-Whittaker, L. and Naveh, Z. (n.d.). Between a rock and a hard place: succession, vegetation resilience, and microsite heterogeneity on calcareous substrate in Israel. In preparation.

Box, E. O. (1982). Life-form composition of mediterranean terrestrial vegetation in relation to climatic factors. *Ecologia Mediterranea*, **8**, 173−181.

Braun-Blanquet, J. (1932). *Plant Sociology*, McGraw-Hill, New York.

Braun-Blanquet, J., Roussine, N. and Nègre, R. (1951). *Les groupements Végétaux de la France Méditerranéenne*, CNRS, Montpellier.

Brice, W. L. (1978). *The Environmental History of the Near and Middle East since the Last Ice Age*, Academic Press, London.

Brookes, I. A. (1982). Geomorphological evidence for climatic change in Iran during the last 20 000 years. In Bintliff, J. L. and van Zeist, W. (eds), *Palaeoclimates, Palaeoenvironments and Human Communities in the Eastern Mediterranean Region in Later Prehistory*, BAR, Oxford, pp. 191−228.

Brown, A. G. (1990). Soil erosion and fire in areas of Mediterranean type vegetation: results from chaparral in southern California, USA, and matorral in Andalucia, southern Spain. In Thornes, J. B. (ed.), *Vegetation and Erosion*, Wiley, Chichester, pp. 269−287.

Byrne, R. (1987). Climatic change and the origins of agriculture. In Manzanilla, L. (ed.), *Studies in the Neolithic and Urban Revolutions. The V. Gordon Childe Colloquium, Mexico, 1986*, BAR, Oxford, pp. 21−34.

Chalabi, M. N. (1982). Sols et climax forestiers en Syrie littorale. *Ecologia Mediterranea*, **8**, 137−141.

Clements, F. E. (1916). Plant succession: an analysis of the development of vegetation. *Carnegie Institute of Washington Publications*, **242**, 1–512.

COHMAP Members (1988). Climatic changes of the last 18 000 years: observations and model simulations. *Science*, **241**, 1043–1052.

Connell, J. H. (1978). Diversity in tropical rain forests and coral reefs. *Science*, **199**, 1302–1310.

Connell, J. H. and Slatyer, R. O. (1977). Mechanisms of succession in natural communities and their role in community stability and organization. *American Naturalist*, **111**, 1119–1144.

Dan, J., Yaalon, D. H. and Koyumdjisky, H. (1972). Catenary soil relationships in Israel 2. The Bet Guvrin catena on chalk and nari limestone crust in the Shefela. *Israel Journal of Earth-Sciences*, **21**, 99–114.

Dan, J., Yaalon, D. H., Koyumdjisky, H. and Raz, Z. (1975). *The Soils Map of Israel (1:500 000)*, Ministry of Agriculture, Bet Dagan.

Danin, A., Gerson, R. and Garty, J. (1983). Weathering patterns on hard limestone and dolomite by endolithic lichens and cyanobacteria: supporting evidence for eolian contribution to terra rossa soil. *Soil Science*, **136**, 213–217.

Davidson, D. A. (1980). Erosion in Greece during the first and second millenia BC. In Cullingford, R. A., Davidson, D. A. and Lewin, J. (eds), *Timescales in Geomorphology*, Wiley, New York, pp. 143–158.

Dregne, H. E. (1977). Desertification of arid lands. *Economic Geography*, **53**, 322–331.

Drew, D. P. (1982). Environmental archaeology and karstic terrains: the example of the Burren, Co. Clare, Ireland. In Bell, M. and Limbrey, S. (eds), *Archaeological Aspects of Woodland Ecology*, BAR, Oxford, pp. 115–127.

Eig, A. (1926). A contribution to the knowledge of the flora of Palestine. *Zionist Organization and Hebrew University Institute of Agriculture and National History Agricultural Experiment Station Bulletin*, **4**.

Eig, A. (1933). A historical-phytosociological essay on Palestinian forests of *Quercus aegilops* L. ssp. *ithaburensis* (Desc.) in past and present. *Beihefte zum Botanischen Zentralblatt*, **51B**, 225–272.

Francis, C. F. and Thornes, J. B. (1990). Runoff hydrographs from three mediterranean vegetation cover types. In Thornes, J. B. (ed.), *Vegetation and Erosion*, Wiley, Chichester, pp. 363–384.

Freitag, H. (1971). Studies in the natural vegetation of Afghanistan. In Davis, P. H., Harper, P. C. and Hedge, I. C. (eds), *Plant Life of South West Asia*, Royal Botanical Garden, Edinburgh, pp. 89–106.

Gilman, A. and Thornes, J. B. (1985). *Land-use and Prehistory in South-east Spain*, George Allen & Unwin, London.

Glacken, C. J. (1967). *Traces on the Rhodian Shore*, University of California, Berkeley.

Grime, J. P. (1979) *Plant strategies and vegetation processes*. Wiley, Chichester.

Grubb, P. J. (1977). The maintenance of species richness in plant communities: the importance of the regeneration niche. *Biological Reviews*, **52**, 107–145.

Harlan, J. R. (1967). A wild wheat harvest in Turkey. *Archaeology*, **20**, 197–201.

Horvat, I., Glavac, V. and Ellenberg, H. (1974). *Vegetation Südosteuropas*, Gustav Fischer, Jena, p. 768.

Houssard, C., Escarre, J. and Romane, F. (1980). Development of species diversity in some Mediterranean plant communities. *Vegetatio*, **43**, 59–72.

Huntley, B. and Birks, H. J. B. (1983). *An Atlas of Past and Present Pollen Maps for Europe: 0–13 000 Years ago*, Cambridge University Press, Cambridge.

Jackson, L. E. (1985). Ecological origins of California's Mediterranean grasses. *Journal of Biogeography*, **12**, 349–361.

Jenny, H., Arkely, R. and Schultz, A. M. (1969). The pygmy forest podsol ecosystem and its dune associates of the Mendocino coast. *Madroño*, **20**, 60–74.

Kaplan, Y. (1984). The ecosystem of the Yahudia Nature Reserve with emphasis on dynamics of germination and development of Quercus ithaburensis Decne. Unpublished Ph.D. thesis, University of Wageningen, Wageningen.

Kutiel, P. (1985). Competition between annual plants along a soil depth gradient. Unpublished Ph.D. thesis, Hebrew University, Jerusalem.

Lamb, H. F., Eicher, U. and Switsur, V. R. (1989). An 18 000-year record of vegetation, lake-level and climatic change from Tigalmamine, Middle Atlas, Morocco. *Journal of Biogeography*, **16**, 65–74.

Le Houérou, H. N. (1981). Long-term dynamics in arid-land vegetation and ecosystems of North Africa.

In Goodall, D. W., Perry, R. A. and Howes, K. M. W. (eds), *Arid-land Ecosystems: Structure, Functioning, and Management*, Cambridge University Press, New York, pp. 357−384.

Litav, M. (1965). Effects of soil type and competition on the occurrence of *Avena sterilis* L. in the Judean Hills (Israel). *Israel Journal of Botany*, **14**, 74−89.

Litav, M. (1967). Micro-environmental factors and species interrelationships in three batha associations in the foothill region of the Judean hills. *Israel Journal of Botany*, **16**, 79−99.

Litav, M. and Orshan, G. (1971). Biological flora of Israel. I. *Sarcopoterium spinosum* (L.) Sp. *Israel Journal of Botany*, **20**, 48−64.

Litav, M. and Seligman, N. (1969). Competition between *Oryzopsis holciformis* (M.B.) Hack and annual plants. I. Performance of *Oryzopsis* as affected by presence of annual plants, nutrient supply and differential sowing time. *Israel Journal of Botany*, **18**, 149−169.

Litav, M. and Seligman, N. (1970). Competition between *Oryzopsis holciformis* (M.B.) Hack and annual plants. II. Performance of *O. holciformis* and two annuals when grown in mixtures of different ratios. *Israel Journal of Botany*, **19**, 517−528.

Litav, M., Kupernik, G. and Orshan, G. (1963). The role of competition as a factor determining the distribution of dwarf shrub communities in the Mediterranean territory of Israel. *Journal of Ecology*, **51**, 467−480.

Macleod, D. A. (1980). The origin of the red Mediterranean soils in Epirus, Greece. *Journal of Soil Science*, **31**, 125−136.

Meher-Homji, V. M. (1965). Aridity and semi-aridity: a phyto-climatic consideration with reference to India. *Annals of the Arid Zone*, **4**, 152−157.

Miles, J. (1978). *Vegetation Dynamics*, Chapman & Hall, London.

Minnich, R. A. (1985). Evolutionary convergence or phenotypic plasticity? Responses to summer rain by California chaparral. *Physical Geography*, **6**, 272−287.

Mizota, C., Kusakabe, M. and Noto, M. (1988). Eolian contribution to soil development on Cretaceous limestones in Greece as evidenced by oxygen isotope composition of quartz. *Geochemical Journal*, **22**, 41−46.

Muller, B. and Garnier, E. (1990). Components of relative growth rate and sensitivity to nitrogen availability in annual and perennial species of *Bromus*. *Oecologia*, **84**, 513−518.

Naveh, Z. (1957). Report on range research. FAO working party on Mediterranean pasture and fodder development, fifth meeting, Tel-Aviv, progress report.

Naveh, Z. (1967). Mediterranean ecosystems and vegetation types in California and Israel. *Ecology*, **48**, 445−459.

Naveh, Z. (1973). The ecology of fire in Israel. *Proceedings of the Annual Tall Timbers Fire Ecology Conference*, Vol. 13, pp. 131−170.

Naveh, Z. (1982). Mediterranean landscape evolution and degradation as multivariate biofunctions: theoretical and practical implications. *Landscape Planning*, **9**, 125−146.

Naveh, Z. (1984). The vegetation of the Carmel and Nahal Sefunim and the evolution of the cultural landscape. In Ronen, A. (ed.), *The Sefunim Cave on Mt. Carmel and its Archaeological Findings*, BAR, Oxford, pp. 23−63.

Naveh, Z. and Dan, J. (1973). The human degradation of Mediterranean landscapes in Israel. In di Castri, F. and Mooney, H. A. (eds), *Mediterranean-type Ecosystems: Origin and Structure*, Springer-Verlag, Berlin, pp. 373−390.

Naveh, Z. and Whittaker, R. H. (1979). Structural and floristic diversity of shrublands and woodlands in northern Israel and other mediterranean countries. *Vegetatio*, **41**, 171−190.

Nemati, N. (1977). Range rehabilitation problems of the steppic zone of Iran. *Journal of Range Management*, **30**, 339−342.

Noy-Meir, I. (1973). Desert ecosystems: environment and producers. *Annual Review of Ecology and Systematics*, **4**, 25−52.

Noy-Meir, I. and Seligman, N. (1979). Management of semi-arid ecosystems in Israel. In Walker, B. H. (ed.), *Management of Semi-arid Ecosystems*, Elsevier, Amsterdam, pp. 113−160.

Noy-Meir, I., Gutman, M. and Kaplan, Y. (1989). Responses of Mediterranean grassland plants to grazing and protection. *Journal of Ecology*, **77**, 290−310.

Oppenheimer, H. R. (1950). The water turn-over of the Valonea oak. *Palestine Journal of Botany (Rehovot)*, **7**, 177−179.

Perrin de Brichambaud, G. and Wallén, C. C. (1963). A study of agroclimatology in semi-arid and arid zones of the Near East. *World Meteorological Organization Technical Note*, 56.

Pignatti, S. (1978). Evolutionary trends in Mediterranean flora and vegetation. *Vegetatio*, **37**, 175–185.

Pons, A. (1981). The history of the Mediterranean shrublands. In di Castri, F, Goodall, D. W. and Specht, R. L. (eds), *Mediterranean-type Shrublands*, Elsevier, New York, pp. 131–138.

Pons, A. (1985). La paléoécologie face aux variations spatiales du bioclimat méditerranéen. *Bulletin de la Société Botanique de France*, **131**, 77–83.

Pope, K. O. and van Andel, T. H. (1984). Late Quaternary alluviation and soil formation in the southern Argolid: its history, causes and archaeological implications. *Journal of Archaeological Science*, **11**, 281–306.

Quézel, P. and Barbero, M. (1982). Definition and characterization of mediterranean-type ecosystems. *Ecologia Mediterranea*, **8**, 15–29.

Rabinovitch-Vin, A. (1983). Influence of nutrients on the composition and distribution of plant communities in mediterranean-type ecosystems of Israel. In Kruger, F. J., Mitchell, D. T. and Jarvis, J. U. M. (eds), *Mediterranean-type Ecosystems: The Role of Nutrients*, Springer-Verlag, Berlin, pp. 74–85.

Rackham, O. (1982). Land-use and the native vegetation of Greece. In Bell, M. and Limbrey, S. (eds), *Archaeological Aspects of Woodland Ecology*, BAR, Oxford, pp. 177–198.

Raunkiaer, C. (1934). *The Life Forms of Plants and Statistical Plant Geography*, Clarendon Press, Oxford.

Roberts, N. (1989). *The Holocene*, Basil Blackwell, Oxford.

Shmida, A. and Ellner, S. (1984). Coexistence of plant species with similar niches. *Vegetatio*, **58**, 29–55.

Specht, R. L. (1969). A comparison of the sclerophyllous vegetation characteristic of mediterranean type climates in France, California, and southern Australia. I. Structure, morphology, and succession. *Australian Journal of Botany*, **17**, 277–292.

Thirgood, J. V. (1981). *Man and the Mediterranean Forest*. Academic Press, London.

Thomas, D. S. G. (1988). The biogeomorphology of arid and semi-arid environments. In Viles, H. A. (ed.), *Biogeomorphology*, Basil Blackwell, Oxford, pp. 193–221.

Thornes, J. B. (1990a). The interaction of erosional and vegetational dynamics in land degradation: spatial outcomes. In Thornes, J. B. (ed.), *Vegetation and Erosion*, Wiley, Chichester, pp. 41–53.

Thornes, J. B. (1990b). *Vegetation and Erosion*, Wiley, Chichester.

Tomaselli, R. (1977). Degradation of the Mediterranean maquis. UNESCO, *MAB Technical Notes*, **2**, 33–72.

van Andel, T. H., Runnels, C. N. and Pope, K. O. (1986). Five thousand years of land use and abuse in the southern Argolid, Greece. *Hesperia*, **55**, 103–128.

van Zeist, W. (1967). Late Quaternary vegetation history of western Iran. *Review of Palaeobotany and Palynology*, **2**, 301–311.

van Zeist, W. (1969). Reflections on prehistoric environments in the Near East. In Ucko, P. J. and Dimbleby, G. W. (eds), *The Domestication and Exploitation of Plants and Animals*, Duckworth, London pp. 35–46.

van Zeist, W. and Bottema, S. (1982). Vegetational history of the eastern Mediterranean and the Near East during the last 20 000 years. In Bintliff, J. L. and van Zeist, W. (eds), *Palaeoclimates, Palaeoenvironments and Human Communities in the Eastern Mediterranean Region in later Prehistory*, BAR, Oxford, pp. 277–321.

Vita-Finzi, C. (1969). *The Mediterranean Valleys*, Cambridge University Press, Cambridge.

Wagstaff, J. M. (1981). Buried assumptions: some problems in the interpretation of the 'Younger Fill' raised by recent data from Greece. *Journal of Archaeological Science*, **8**, 247–264.

Walker, B. H., Ludwig, D., Holling, C. S. and Peterman, R. M. (1981). Stability of semi-arid savanna grazing systems. *Journal of Ecology*, **69**, 473–498.

Walter, H. (1971). *Ecology of Tropical and Subtropical Vegetation*. Oliver & Boyd, Edinburgh.

Walter, H., Harnickell, E. and Mueller-Dombois, D. (1975). *Climate-diagram Maps of the Individual Continents and the Ecological Climatic Regions of the Earth*, Springer-Verlag, Berlin.

Warming, E. (1895). *Plantesamfund: Grundträk af den Ökologiska Plantegeografi*, Philipsen, Copenhagen.

Westman, W. E. (1975). Edaphic climax pattern of the pygmy forest region of California. *Ecological Monographs*, **45**, 109–135.

Westoby, M. (1979/80). Elements of a theory of vegetation dynamics in arid rangelands. *Israel Journal of Botany*, **28**, 169–194.

Whorf, B. L. (1939). The relation of habitual thought and behaviour to language. In Spier, L. (ed.), *Language, Culture, and Personality*, Sapir Memorial Publication Fund, Menasha, WI, pp. 75–93.

Whorf, B. L. (1941). Language, mind and reality. *The Theosophist (Madras)*, **63**(1;2), 281–91; 25–37.

Whorf, B. L. (1956). In Carroll, J. B. (ed.), *Language, Thought and Reality*, MIT and Wiley, New York.

Wise, S. M., Thornes, J. B. and Gilman, A. (1982). How old are the badlands? In Bryan, R. and Yair, A. (eds), *Piping and Badland Erosion*, Geobooks, Norwich, pp. 259–277.

Wojterski, T. W. (1990). Degradation stages of the oak forests in the area of Algiers. *Vegetatio*, **87**, 135–143.

Worster, D. (1977). *Nature's Economy: The Roots of Ecology*, Sierra Club, San Francisco.

Wright, H. E., McAndrews, J. H. and van Zeist, W. (1967). Modern pollen rain in western Iran, and its relation to plant geography and Quaternary vegetational history. *Journal of Ecology*, **55**, 415–443.

Yaalon, D. H. and Ganor, E. (1973). The influence of dust on soils during the Quaternary. *Soil Science*, **116**, 146–155.

Yaalon, D. H. and Ganor, E. (1975). Rates of aeolian dust accretion in the Mediterranean and desert fringe environments of Israel. *Congress of Sedimentology*, **2**, 169–174.

Zohary, D. (1969). The progenitors of wheat and barley in relation to domestication and agricultural dispersal in the Old World. In Ucko, P. J. and Dimbleby, G. W. (eds), *The Domestication and Exploitation of Plants and Animals*, Duckworth, London, pp. 47–66.

Zohary, M. (1950). The segetal plant communities of Palestine. *Vegetatio*, **2**, 387–411.

Zohary, M. (1962). *Plant Life of Palestine*. Ronald Press, New York.

Zohary, M. (1973). *Geobotanical Foundations of the Middle East*, Gustav Fischer, Stuttgart, 2 vols.

Zohary, M. (1982). *Vegetation of Israel and Adjacent Areas*, Dr L. Verlag, Wiesbaden.

Zohary, M. (1983). Man and vegetation in the Middle East. In Holzner, W., Werger, M. J. A. and Kusima, I. (eds), *Man's Impact on Vegetation*, Junk, The Hague, pp. 287–296.

Part 3

SENSITIVITY AND BUILT
ENVIRONMENTS

19 The Environmental Sensitivity of Blistering of Limestone Walls in Oxford, England: A Preliminary Study

HEATHER A. VILES

Jesus College, University of Oxford, UK

ABSTRACT

Blistering is an important weathering process affecting gypsum-encrusted urban walls that results in unsightly, and often costly, damage. Despite many studies, the detailed mechanisms involved in blister development remain imperfectly understood. In this chapter, a simple three-stage model of blister development is presented, based on a review of available information, and is then used as a basis for speculations on the environmental sensitivity of the blistering process. The bulk of the chapter reports on an archival case study of the blistering of limestone walls in Oxford, England, using photographs, coal consumption records and meteorological data. Analysis of three series of photographs from two walls in the city centre spanning a maximum of 75 years (1863–1937) shows that blistering activity is patchy over space and time. There are no clear relations between blistering and any of the environmental variables. Many blisters are extremely stable over the period of record, despite relatively high, and variable, levels of atmospheric pollution. Microclimatic conditions experienced by individual walls are thought to exert an important control on blistering. It is concluded that the later stages of blister development observed on the photographs are relatively insensitive to any single environmental change. A number of processes are involved in blister expansion, and these react differently to the various environmental fluctuations recorded.

INTRODUCTION

Limestone walls, like natural rock outcrops, are subjected to a range of weathering processes. The consequences of building stone weathering (also known as building stone decay) can be very severe, resulting in damage to important historical monuments and in expensive repairs. Much research effort, involving geomorphologists among others, has therefore gone into investigating the exact causes and natures of the processes responsible. One important consideration for those people involved with building management is the environmental sensitivity of building stone weathering. We need to know more about the response of weathering systems to environmental changes, such as declining air pollution levels, in order to make the most effective management decisions.

Winkler (1973) suggests that urban building stone weathering has many similarities to weathering of natural outcrops under desert conditions. Urban limestone buildings in the temperate zone are certainly affected by a wide range of chemical, physical and biological

Landscape Sensitivity. Edited by D. S. G. Thomas and R. J. Allison
© 1993 John Wiley and Sons Ltd

weathering processes in addition to limestone dissolution in carbonated water, which is usually regarded as the major weathering process on natural outcrops in the same zone. Weathering of urban limestone buildings involves the interaction of a range of weathering agents with a variety of stone types, under different microenvironmental conditions. Rainfall, polluted air, temperature changes, salts, groundwater and stone-dwelling organisms are all important weathering agents. All three components of the weathering system (agents, stone and microenvironmental conditions) are vital in determining the rate and nature of the resultant weathering. Livingston (1986) indicates the importance of internal characteristics in predisposing some stone types to high rate of weathering. He calls such predisposing characteristics 'inherent vice'. Winkler (1973) describes the importance of different types of water movements within porous stone to the progress of weathering. Water moves upwards and downwards, towards and away from the interior, under diffusion and capillarity, bringing reactive compounds into contact with the stone.

Several workers have attempted to list and categorise weathering processes affecting limestone buildings. Jaynes & Cooke (1987, Figure 1), for example, provide a neat division of weathering processes caused by polluted air. They contrast sulphation (the reaction of calcium carbonate with sulphuric acid to form gypsum, or calcium sulphate) in sheltered and unwashed areas with a suite of solution processes in areas exposed to rainfall. A wider range of processes is identified by the *Building Research Station (BRS) Digest* (1965), including solution, skin formation, soluble salt weathering, frost weathering, thermal expansion, metal corrosion, lichen and bacterial action. Some of these processes lead to what might be called 'secondary weathering processes', such as the blistering and exfoliation of gypsum 'skins', which is the main focus of the rest of this chapter.

GYPSUM CRUST FORMATION AND BLISTERING

The development, and subsequent erosion of, blackened crusts containing high concentrations of gypsum is one of the most important processes affecting sheltered parts of buildings in polluted atmospheres. Blistered stonework from Oxford, England, is shown in Figure 19.1. The *BRS Digest* (1965, p. 3) provides the following description:

> Calcium sulphate tends to form a hard, glossy skin on limestone wherever the stone is not freely washed by rain. . . . This sulphate skin tends to be harmful rather than protective. Some varieties of stone offer a good resistance; with others, skin formation leads to a spontaneous blistering and scaling of the surfaces.

Arkell (1947a, p. 152) states, with particular reference to Headington freestone walls in Oxford, that:

> exposure to the weather forms a hard, impermeable, black crust, which later blisters; the blisters then burst and the skin curls up and peels off.

In order to provide an explanation of how such blistering works, we need to consider two stages, i.e. the formation of a gypsum crust, and the subsequent breaching of this crust.

Gypsum crusts are widely found on urban limestone buildings and most studies investigating their formation have used detailed examination of real crustal material, rather than laboratory experimental studies. Two important considerations are, first, how the gypsum is formed,

Figure 19.1   Detailed view of a blistered Headington freestone wall, Oxford, England. Lens cap is approximately 5.5 cm in diameter

and secondly, how the crust and associated underlying zone, develop. Fassina (1988) shows that sulphuric acid reacts with calcium carbonate to produce gypsum (calcium sulphate), with the sulphuric acid coming either from oxidation of sulphur dioxide in atmospheric droplets (i.e. in acid rain) or from oxidation of sulphur dioxide on the limestone surface (e.g. from dry deposition) in the presence of catalysts (metal oxides and carbonaceous particles) and moisture. Gypsum crusts develop most favourably in areas with high general humidity (e.g. frequent fogs), high atmospheric sulphur dioxide concentrations and easily available metal oxides and carbonaceous particles. It is also important that rainwashing is limited, as gypsum is more soluble than calcite. Spectacular gypsum crusts are reported from Venice, Italy (del Monte & Vittori, 1985), Oviedo, Spain (Heath, 1986) and many other towns.

Charola & Lewin (1979) carried out detailed scanning electron microscopy investigations of gypsum crusts developed on marble from Venice and showed that groups of platy gypsum crystals formed framboidal (raspberry-shaped) accretions. They found, as did Camuffo, del Monte & Sabbioni (1983), that gypsum crystal growth tends to occur mainly at the substrate—crust boundary. Saiz-Jimenez & Bernier (1981), from studies in Seville, Spain, suggest that gypsum crystals grow preferentially at sites of structural defects in the stone surface. Camuffo, del Monte & Sabbioni (1983) and Domaslowski (1982) all point to the importance of stone porosity and surface roughness in encouraging the growth of gypsum crystals. These authors also point out that the resultant gypsum crust forms a higher surface than the original stone surface, despite any dissolution of the original stone material.

Several investigators point out the importance of weathering within and beneath the crust, when water is able to circulate, producing internal disintegration. Where other compounds, such as the salt sodium chloride, are present they may encourage dissolution of gypsum within the crust (Evans, 1970). Del Monte & Vittori (1985) stress the importance of the below-crust microenvironment, and show how, in Venice, solution at the stone boundary produces a layer of highly reactive microcrystalline calcite. As they say (p. 117):

> The crust then, is not simply the direct effect of the attack: it both facilitates the damaging
> chemical reactions and prolongs them since, in practice, it reduces water evaporation.

Manning (1989) also suggests that the below-crust layer is weakened by moisture and salt action.

Several hypotheses, which are not necessarily mutually exclusive, have been put forward to explain why, on certain stone types, gypsum crusts tend to blister and exfoliate. Arkell (1947a, p. 153), for example, says that:

> The blistering of the skin, with formation of an empty cavity behind it, seems to be due
> to the fact that the calcium sulphate skin has different physical properties from the stone
> ... behind it and so reacts independently to changes of temperature and moisture. ...
> In particular the skin expands more than the stone when heated by the sun, and the blistering
> is a natural response to the conflict of forces so set up.

Sengupta & de Gast (1972) discuss three major mechanisms that may be involved. First, blistering may result from the growth of crystals immediately behind the skin, but they say that experimental results (which they do not reference or describe) do not back this up. Secondly, an increase of volume occurs on the conversion of calcium carbonate to calcium sulphate. Livingston (1986) regards this as a likely mechanism of spalling and quotes the molar volumes involved as 37.7 cm^3 mol^{-1} for calcium carbonate and 60.0 cm^3 mol^{-1} for hydrated calcium sulphate. Finally, as Arkell (1947a) suggested, the linear coefficient of expansion of gypsum is, according to Sengupta & de Gast (1972) approximately five times that of calcium carbonate.

To date there is no convincing evidence to disprove any of these hypotheses and the exact mechanisms involved in blistering on buildings are still unclear. It is not certain why some stone types are particularly prone to blistering. Schaffer (1932) observed blistering on Portland Stone in London, which is usually regarded as a very durable stone. Butlin et al. (1988) report severe blistering on the West Front of Lincoln Cathedral, built of a local Jurassic limestone that is no longer quarried but is probably of medium durability. Arkell (1947a) shows how prone the extremely non-durable Headington freestone is to blistering. It is reasonable to conclude that blistering is most likely to occur where the contrast between crust and stone characteristics is most pronounced — either because the stone is particularly soft, or because the crust is particularly hard and well-developed.

Blistering also occurs on other building stones, and a range of similar processes that occur on building stones and natural outcrops — e.g. salt burst, contour scaling and salt fretting and spalling — affect sandstones, limestones, granite and marble. Winkler (1973) discusses the importance of subflorescences of salts in producing such phenomena, where salts crystallise immediately below a weathered stone surface forming a thin, indurated skin, which then, as with gypsum crusts, can peel off. Such subflorescences are common in urban and desert areas.

A MODEL OF BLISTERING

The fragmentary information available on gypsum crust development and blistering, some of which has been reviewed in the previous section, can be used to develop a preliminary

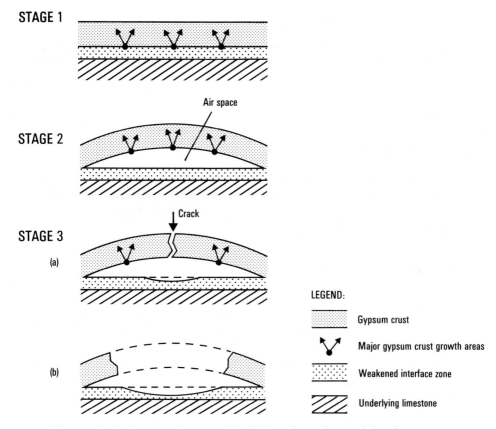

Figure 19.2   A three-stage model of blister formation and development

model of blistering. A three-stage model is presented in Figure 13.2 and the stages described in more detail below.

First, a gypsum crust is established on a suitably sheltered limestone surface from the reaction of sulphuric acid with calcium carbonate. This crust contains a range of other pollutants (e.g. carbonaceous particles and salts) as well as fragments of the original stone. The crust grows primarily from the stone–crust interface zone, and as it grows the underlying stone becomes weakened by dissolution and salt attack. In the second stage, the crust bulges to form a blister as it begins to part company from the underlying stone. The mechanisms involved are still debated, but the results are clear with a bowed crust overlying an air space and weakened stone surface.

The third stage of blistering involves the 'bursting' and subsequent exfoliation of the blister. Here, an initial crack develops in the blister surface as some sort of stability threshold is breached. At this stage, water can penetrate through the crack in the crust, further weathering the stone beneath and aiding removal of disintegrated material. Support for the crust from below thus decreases, and water and mechanical stresses weather the crust, leading to exfoliation of the blistered surface. The strength of the crust thus declines dramatically as soon as it is breached. The model therefore proposes that the first two stages of the blistering process are primarily caused by the continuing transformation of calcite to gypsum, with a wider range of processes being implicated in the third stage.

We can use this preliminary model to examine the potential environmental sensitivity of the blistering process. Manning (1989) has proposed that past high levels of atmospheric sulphur dioxide have produced gypsum crusts on many buildings that are only now beginning to blister and exfoliate, despite vast improvement in air quality. This idea amplifies the 'memory effect' concept proposed by R. U. Cooke (in BERG, 1989) which basically suggests that pollutants stored within porous building materials can continue weathering, or be reactivated, many years after emplacement. Therefore, there is no necessary, immediate correlation between declining air pollution levels and declining stone decay rates.

If the three-stage model of blister development is realistic, we can investigate the reality of these claims for a 'memory effect'. A decrease in atmospheric sulphur dioxide, though producing a decline in gypsum formation, may not be matched by a decline in blistering activity if stage three has been reached and the system has become unstable and dominated by a wider range of processes. If, however, stage three has not been reached a decline in atmospheric sulphur dioxide may slow down the rate of gypsum production, and therefore decrease the chance of the gypsum crust becoming unstable and blistering. However, where carbonaceous particles or other sources of sulphur are important, gypsum production may be unaffected.

In reality the situation is likely to be more complicated than presented in the preceding discussion, as other features of the environment are highly likely to fluctuate as well as atmospheric sulphur dioxide concentrations, and also affect the blistering process. High humidity has been shown to encourage the production of gypsum, and therefore a decline in sulphur dioxide levels, coupled with an increase in humidity, may not lead to the expected decline in gypsum production. An increase in the frequency and magnitude of temperature fluctuations may encourage blister development by increasing the stresses between crust and underlying stone. An increase in rainfall washing over the surface will accelerate dissolution of the gypsum crust, therefore decreasing the likelihood of blister formation. If the stage three threshold has been breached, an increase in dissolved salt concentrations, rainfall or temperature changes will all favour rapid exfoliation of the blistered area.

Arkell (1947a) suggested that the blistering process was a cyclical one, with exfoliation leading to a new surface upon which a new gypsum crust would develop and subsequently blister. According to our three-stage model, this would only occur if levels of atmospheric sulphur dioxide remained relatively high — otherwise a new gypsum crust would not be established. If atmospheric sulphur dioxide levels decrease during the three stages, and remain low, we would not expect blistering to begin again.

Data from experimental and field studies of the blistering process are urgently needed to test the validity of this model and to investigate further the sensitivity of the blistering process. The archival investigation of the blistering of two Oxford walls reported below goes some way toward testing the model.

ARCHIVAL STUDIES OF THE ENVIRONMENTAL SENSITIVITY OF BLISTERING

Archival studies have a special place in building stone decay research, given that many buildings have well-documented histories. Photographic records can be particularly useful. Winkler (1973) uses sequential photographs to illustrate the progress of decay on several buildings. Fassina (1988, p. 188) uses a series of four photographs taken between 1900 and 1970 of a sculpture on the façade of St Mark's basilica, Venice, to show how decay has accelerated since industrialisation in the city began in the 1930s. In both these examples

Figure 19.3 Photograph of the Trinity College rowing eight, summer 1882, as sampled in T1

photographs are used in a purely qualitative way, but there is great scope for quantiative analyses. In a different context, Pentecost (1991) has shown the value of such an approach, using repeated photographs to investigate the rate of rock weathering by granular disintegration on sandstone outcrops. Blistering is an especially suitable process to monitor using photographic methods, as it has highly visible effects — at least during stage three of the blistering process.

In Oxford several walls have been the background for annual photographs of student groups (e.g. college members, rowing crews) since the 19th century, as illustrated by Figure 19.3. Change in the decay of these walls can be compared from year to year, with changes in the dimensions of breached blisters particularly easy to identify. Thus it is possible to track the progress of a sample of blisters from initial cracking through to complete exfoliation.

Clearly, a photographic record of blistering such as that discussed above has many imperfections. It is not possible to view the earlier stages of crust and blister development and the quality of photographs (especially in the early days of photography) may not be good enough to permit clear observations. Photographs that were not taken primarily to record blistering will often have blisters out of focus, or obscured by the major subjects of the photo. However, despite these inadequacies such a photographic record does extend our knowledge of blister activity back into the 19th century and provides a source of evidence.

It is obviously of great interest to compare the photographic record of blistering with changing environmental characteristics, such as rainfall, humidity, temperature fluctuations and atmospheric sulphur dioxide concentrations. Two main types of record are available over the *ca.* 100 year timescale, i.e. direct meteorological records and indirect data. In Oxford, most meteorological variables have been observed since at least the 1880s, though some have much shorter records available. Fog frequency, for example, has only been recorded since the 1920s, and smoke and sulphur dioxide concentrations were not recorded until the 1950s. Before this time, various proxy data can be used to provide estimates of sulphur dioxide levels.

Brimblecombe (1982) has investigated the use of various indirect sources of evidence of atmospheric sulphur dioxide and smoke concentrations, including deaths from various respiratory diseases and coal consumption figures. In London, for example, he uses figures on the total amount of coal imported into the city to estimate sulphure dioxide concentrations, using an equation that requires information on the diameter of the city and the type of coal (i.e. its sulphur content). In Oxford, data on the total amount of coal imported to the city are difficult to obtain as by the mid-19th century there were many different sources, and modes of transport, of coal.

Several of the Oxford colleges kept detailed records of coal consumption on an annual basis. Such coal records can be used in two main ways. First, they can be used to estimate smoke and sulphur dioxide concentrations in the vicinity of the college itself. This only works if the well-documented assumption that colleges were self-fouling (i.e. the smoke and sulphur dioxide produced from their chimneys was the major contributor to stone decay on college walls) is accepted. Beckinsale & Combey (1958, p. 182) claim that

> the quadrangular shape of colleges ensured that at least one side of the quadrangle was liable to be fouled by the smoke-barrage from the opposing façade. This self-fouling . . . was most operative on windy and wet days with a turbulent air-movement.

If this is true, the coal consumption data can be put into a modified version of Brimblecombe's equation. Alternatively, coal consumption records from individual colleges can be used as estimators of city-wide climatic conditions (and other influencing factors such as population size) and, by implication, smoke and sulphur dioxide levels. This only works if one assumes that colleges burnt more coal when it was colder and less when it was warmer. The truth of this assumption could be tested by comparing coal records from two or more colleges over the same timespan and seeing how well they correlate with each other, and with climatic fluctuations. Unfortunately, the coal records so far unearthed in this study are highly fragmentary and overlapping series are rare.

In either of the two cases presented above it is highly likely that coal records give an imperfect record of smoke and sulphur dioxide concentrations at the detailed scale required to compare with blistering of individual walls. Much more work needs to be done to trace, and interpret, more coal records for Oxford.

THE OXFORD CONTEXT

A wide range of building stones has been used in Oxford and may be seen in the city fabric today (Jenkins, 1988). For many centuries, however, locally quarried stones were most popular, including two types of stone from the nearby Headington quarries. Headington hardstone is a highly durable stone, suitable for plinths, and Headington freestone is a much more variable stone used for ashlar. Arkell (1947a) states that Headington stone is first recorded in Oxford in the 14th century in New College bell tower, but its use declined after the mid-18th century when its inferior qualities were recognised. Arkell (1947a, p. 50) quotes Robert Plot's observations of Headington stone (presumably freestone) in the mid-17th century as: 'the stone cuts very soft and easy . . . very porous, and fit to imbibe lime and sand, but hardening continually as it lies in the weather'.

The quarries are now abandoned and much of the decayed Headington freestone on Oxford buildings has been replaced with more durable stone. However, blackened and blistered Headington freestone walls are still visible in many places (see Figure 19.1).

Headington hardstone and freestone are both Jurassic limestones from the Corallian beds (North, 1930) of the Middle Oolite series (Watson, 1911). They were deposited in coral-free channels around the reefs that formed the local Coral Rag deposits. The major contents are finely ground shell and coral debris, mixed with a small and variable quantity of ooliths. According to North (1930) the resultant Headington freestone is a soft 'somewhat cellular, earthy limestone'.

A STUDY OF TWO OXFORD WALLS

Methods

Analysis of Photographs

So far in this ongoing project two Headington freestone walls have been studied in some detail. One is a north-facing 18th century wall in Fellows Building, Corpus Christi College, and the other is an east-facing corner of Garden Quadrangle, Trinity College, built in the late 17th century (Royal Commission on Historical Monuments England, 1939). Both colleges are in the city centre area: Corpus Christi is situated towards the south and Trinity towards the north. Photographs are available for the Corpus Christi College wall over the period 1886–1937 (known hereafter as CC1). For the Trinity College wall, two series of photographs are available covering different parts of the same wall for the periods 1863–1937 (ground-floor ashlar, referred to subsequently as T1) and 1899–1902 (first floor ashlar, referred to subsequently as T2). Figure 19.3 shows an example of one of the photographs sampled. Both walls have since been refaced in more resistant stone. The Corpus Christi wall was refaced sometime after 1957 in Clipsham stone; and the Trinity College corner was refaced between 1958 and 1966 in Bath (Box) stone, with a Portland plinth and some Clipsham patches (Oakeshott, 1975) and now looks as shown in Figure 19.4.

A sample of photographs from each wall was analysed using the following methods, which had to be developed to deal with fragile photographs that could not be traced over or removed from the college archives. A grid on a transparent plastic sheet was placed over each photograph and measurements made of (a) several invariant features found in the whole series of photos (e.g. window and door frames) in both horizontal and vertical dimensions; and (b) the vertical and horizontal dimensions of a sample of blisters.

Sampling was constrained in two major ways. Suitable photographs were not always available for each year (often colleges had two or more walls used for team photographs; some years creeper obscured large areas of wall). CC1 was sampled at approximately three-year intervals, T1 at approximately five-year intervals and T2 at two-year intervals. Also, many blisters were not visible in all photographs (because of people standing in the way or growth of creepers) and so the size and nature of the blister sample was constrained in each photograph series. For CC1 10 blisters were sampled over a 42 year period, for T1 16 blisters were sampled over 75 years and for T2 44 blisters were sampled over 14 years. In fact, for T1 and T2, several blisters became permanently obscured during the period of record, decreasing sample sizes to 14 and 25 respectively.

Figure 19.4   Trinity College corner sampled in T1 (as shown in Figure 19.3) and T2, after refacing

Careful study of sequential photographs revealed clearly when blisters expanded, so the problems of accuracy caused by each photograph having been taken from slightly different positions are not crucial. Measurements have been standardised (using the measurements of invariant features) but have not been scaled up to actual size.

Coal Consumption Data

Coal consumption records are patchily available for both colleges over the period of the photographic record. Those from Corpus Christi College, however, give cost only — and so estimates of consumption in tonnes have had to be made using coal prices obtained from other college records (who might have used different suppliers and therefore have been charged slightly different amounts). In several cases data are only available for part of the year and so the annual total will be underestimated.

Records of coal consumption for Trinity College are available over the period 1863−1868 and 1912−1937. Sadly, the account book for the crucial middle period of record is missing from the college archives. Coal records from Corpus Christi College are available from 1867−1895. After this period, the coal accounting methods in this college changed and the records are more difficult to handle. The two coal consumption series only overlap, therefore, for 1867 and 1868 and unfortunately the 1868 figure for Trinity is an underestimate based on partial data only for that year, so comparison of the consistency of the two time series is impossible.

Data Comparison and Analysis

Because of the somewhat fragmentary nature of much of the data collected, only simple analyses have been attempted. Figure 19.5 shows the type of raw data produced in the form

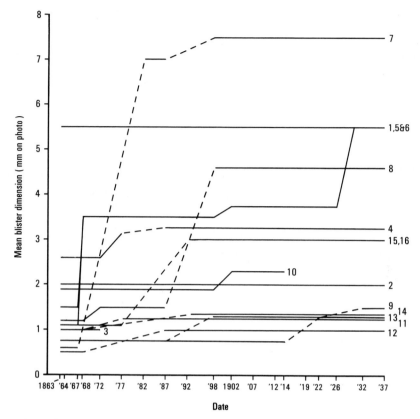

Figure 19.5 Graph of blister dimension changes (mean of horizontal and vertical dimensions), 1863–1937, Trinity College (T1). Numbers at the righthand end of lines on graph are blister sample numbers

of a graph of blister dimension changes over time. The blister dimension data as shown in Figure 19.5 were reduced to nominal form, i.e. presence or absence of a change in dimension at a particular date. From this, numbers of blisters changing dimension at each sampled date could be calculated for each photo series, as shown in Table 19.1.

Figure 19.6 shows graphically the change in blister activity at each sampled date for the longest photograph series, T1. Also plotted for comparison are figures on various environmental variables and coal consumption collected over the winter (October–March, following college terms) preceding the date of the photograph. These data were graphed in order to see whether there were any clear relationships between blister activity and environmental characteristics over the preceding few months.

In order to compare all the blister data more easily with coal consumption and meteorological data, the records from the three photograph series were combined. As different sampling intervals had been used the data were grouped into five-year periods. Expressing the number of blisters changing in any five-year period as a percentage of the observed blister sample size then gives a measure of blister activity, as shown in Table 19.2. These figures, and five-year mean data on air temperatures, rainfall, ground frosts and coal consumption are presented graphically in Figure 19.7.

Table 19.1 Summary of blister dimension changes for CC1, T1 and T2

Location		CC1	T1	T2
Sample size[a]		10	14–16	20–44
No. of dimension	0	5	3	16
changes per	1	1	7	20
blister	2	1	4	6
	3	2	2	2
	4	0	0	0
	5	0	0	0
	6	1	0	0
Total no. of blister changes		15	21	37
Timespan of record		1896–1937	1868–1937	1889–1902
Length of record (years)		42	75	14
Approximate sampling interval (years)		3	5	2

CC1		T1		T2	
Sampled dates	Blister changes	Sampled dates	Blister changes	Sampled dates	Blister changes
1896	—	1863	—	1889	—
1898	2	1864	0	1890	3
1899	1	1867	2	1892	3
1905	1	1868		1894	6
1909	1	1872	1 ⎫ 1	1896	7
1911	2	1877	2 ⎬ 1 ⎫ 1	1898	5
1913	1	1882	1 ⎭ ⎬	1900	5
1920	1	1887	2 ⎭	1902	2
1922	3	1892	3		
1924	1	1898		Total	37
1926	0	1902	3		
1928	1	1907	0		
1930	0	1912	0		
1932	0	1914	0		
1934	1	1919			
1937	0	1922			
		1926			
Total	15	1932	2		
		1937	0		
		Total	21		

[a] T1 and T2 sample sizes decrease over time as blisters disappear from view and do not reappear.

Results

Figure 19.5 shows graphically the changing dimensions of blister samples from T1. Dotted lines indicate where the exact date of a dimensional change is unknown. Changes in both horizontal and vertical dimensions were recorded from the photographs, but in these figures the data are presented as mean width/height dimensions (in standardised photograph mm).

Table 19.2   Five-year means of blister activity data from CC1, T1 and T2

Period	Sample size[a]/photo series used	Blister activity[b] (%)
1863–7	48 (T1)	4
1868–92	32 (T1)	6
1873–7	15 (T1)	20
1878–82	15 (T1)	7
1883–7	15 (T1)	7
1888–92	144 (T1, T2)	7
1893–7	92 (CC1, T2)	17
1898–1902	125 (CC1, T1, T2)	18
1903–7	25 (CC1, T1)	4
1908–12	35 (CC1, T1)	9
1913–17	25 (CC1, T1)	4
1918–22	48 (CC1, T1)	10
1923–7	34 (CC1, T1)	3
1928–32	44 (CC1, T1)	9
1933–7	34 (CC1, T1)	3

[a] Sample size here refers to total sample size over five-year period, and varies according to number of sample points for each photo series included.

[b] Percentage blister activity calculated as number of blister dimension changes observed in five-year period divided by total number of blister observations in that period (i.e. sample size as described above) × 100.

A few blisters changed dramatically in size during the period of record, some coalesced with others, but most showed relatively modest changes in dimensions. From these figures we can summarise the nature and timing of blister changes as shown in Table 19.1. In Table 19.1 blister changes are presented for each photo series. The data show that for CC1 out of a sample of 10 blisters, five showed no dimensional change over the 42 year period, one changed once, two three times and one six times. For T1, out of 16 blisters, only three showed no change over 75 years, with the majority changing once. For T2, out of 44 blisters sampled, 16 showed no change, and the majority changed only once.

The changes in blister dimensions are clustered in time, showing spasms of activity, but these seem to vary from wall to wall. CC1, T1 and T2 all show evidence of a peak of blister activity around the turn of the century, but differences in record lengths and sampling dates make more detailed comparisons difficult.

Table 19.2 shows a combined blister activity data set for all three photo series. Despite differences in sampling frequency, and ages of the two walls, it seems reasonable to combine the three data sets to give a total sample of up to 70 blisters (bearing in mind the changes in sample size of T1 and T2 over time as some blisters became permanently obscured and therefore were removed from the sample), with 73 individual blister dimension change observations. Five-year means of blister activity have been calculated from the three data sets. Column 2 in Table 19.2 shows the total sample size and which photo series have been used for each five-year period. The early period (1863–1887) is dominated by CC1, 1888–1902 contains information from all three series (or various combinations of these), and the later period (1903–1937) is dominated by CC1 and T1. Any conclusions drawn from this combined data set on blister activity must be made bearing this in mind.

Figure 19.6 shows the variability of blister activity as recorded by the T1 photograph series and Figure 19.7 shows the smoothed series for all blister data over the same period.

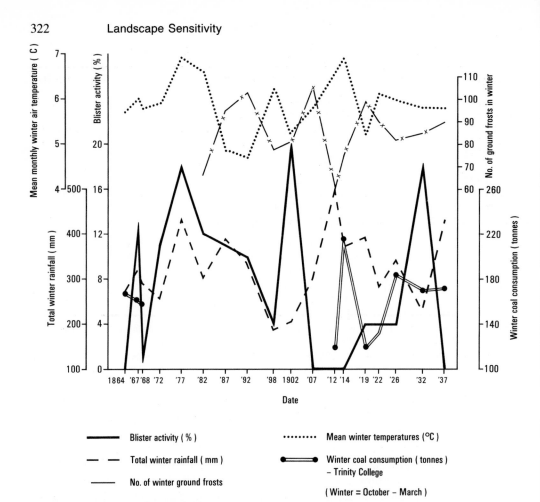

Figure 19.6 Graph of changing blister activity, total winter rainfall, mean winter air temperature, number of winter ground frosts and winter coal consumption for T1. Winter data are for the period October–March preceding the date of the photograph

Discussion

The data set collected so far has many shortcomings, which must be elucidated before any conclusions are drawn. The analysis of the photographic records of blistering by simply measuring the changing dimensions of a sample of blisters may not be the best approach. It would also be useful to collect data on the density of blistering on different walls and to see how this changes over time. Larger samples, collected from different parts of the wall (e.g. ashlar, window frames, string courses), would also provide more useful information. Sampling intervals should be standardised, as far as possible, between all the different data collected.

The use of coal consumption records as proxy data for atmospheric sulphur dioxide concentrations has already been criticised in general terms, but some further points are relevant here. There is a large problem with missing data, and the many different methods of recording coal puchases make interpretation difficult even when records are available. Meteorological

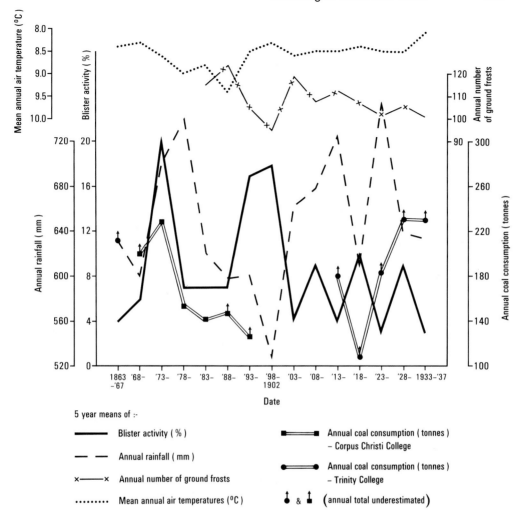

Figure 19.7 Graph of five-year means of annual rainfall, mean annual air temperature, number of ground frosts, coal consumption and blister activity changes (using data from CC1, T1 and T2)

data are much more easily obtainable for the relevant period and sampling intervals, but it has proved very difficult to decide on which meteorological variables, at which level of resolution, are the vital ones. This, of course, is not surprising, given our lack of knowledge on the blistering process. Finally, there are many alternative methods of analysing the data sets collected and it is difficult to know which method is the most appropriate. Considering the record of blistering from one wall (using one photo series as shown in Figure 19.6) with allied coal records from the same college results in a small sample. Combining the blister records from the two, different, walls and using five-year mean data, produces a larger, more 'smoothed' data set (as shown in Figure 19.7), but may not lead to any more profound insights.

The problems of the data set discussed in the previous paragraph are common to any archival data set — especially in the early days of a study — and should not be seen as insurmountable.

Looking first at the blister data sets for the two walls separately, it is possible to identify some important trends. The small sample (10 blisters) from Corpus Christi College shows half of the blisters remaining unchanged over 41 years, and with the rest fairly evenly spread over one, two, three and six changes. For the Trinity College wall, however, both photo sets reveal a different pattern, with under half remaining unchanged, nearly half changing once and around a quarter to one-third changing two or three times. These differences between the two walls may be the result of three things, i.e. sample size differences, microclimatic differences (the Corpus Christi wall is north facing, and the Trinity wall is east facing) and/or age differences (the Trinity wall is older than the Corpus Christi one). The more active pattern of blistering on the Trinity College wall might be explained by the surface being younger, more unstable and blistering not having progressed so far. Conversely, temperature changes may be more dramatic on east-facing walls and therefore encourage blistering on the Trinity wall.

The clustering over time in blistering activity recorded on both walls is of interest. For the Corpus Christi wall, the 15 recorded blister dimension changes are fairly evenly spread over the period of record (1896−1937) with a minor peak in 1922. For the Trinity College wall, photo series T1 shows a peak of activity between 1867 and 1902, with a secondary peak between 1919 and 1932. Photo series T2 backs up this spasm of activity in the late 19th century, with a particular peak in the 1890s. These differences again suggest that microclimatic, age and sample size differences may be important in conditioning the responses of the two walls.

Using the most detailed blister data set (T1) and comparing it with coal consumption data and meteorological data (Figure 19.6) indicates some possible relationships between blister activity and climate and air pollution in the preceding winter months. Three types of relationship might be expected, i.e. direct (positive or negative), lagged or none, and these relationships may be meaningful or spurious. Up until 1902 when there is a major peak of blistering activity, the graph of blister activity is positively and directly correlated with total rainfall during the preceding winter and mean winter air temperatures. After 1902 these relationships disappear and, in fact, a negative, direct relationship between blister activity and rainfall can now be discerned. There is no relationship visible between tonnes of coal burnt by the college in the preceding winter and blister activity, nor between number of ground frosts the previous winter and blistering activity.

If the pre-1902 relationships between blistering and rainfall and temperature are significant, they suggest that blister exfoliation is encouraged in warm, wet winter conditions — which is contrary to expectations. The change in situation around 1902 may be explained by a 'blister activity spasm', resulting in very different conditions on the wall afterwards. The reason for this blister activity spasm is unclear.

Combining the three blister data sets and examining the relationship of five-year mean data with meteorological and coal consumption data (Figure 19.7) shows that there are no clear-cut relationships. As Figure 19.7 demonstrates, overall blistering activity peaks in 1873−7 and 1893−1902. Rainfall reaches a low point in 1898−1902 and tends to show a generally inverse relationship with blistering activity (especially from the 1890s onwards). Coal consumption shows a generally direct, unlagged relationship with blistering activity until the 1890s. When the coal record begins again in 1913 the relationship becomes less clear, though a generally positive link, lagged by five years, may be seen. Temperature and frost data show that 1888−92 was a particularly cold spell, but blistering activity over this period shows no increase. During the next 10 years blistering does peak, so there may be a lagged relationship.

There are a few, tentative, conclusions that we can draw from these data sets. First, microclimatic factors, history of blistering and age of walls seem to be important to explaining blistering activity on an individual wall. Secondly, blistering activity seems also be to influenced more generally by rainfall, air pollution and temperature, but the influences are not particularly strong and seem to change in importance over time.

Finally, it is worth seeing what the findings of this archival study imply for the model of blister development and its environmental sensitivity proposed in Figure 19.2. The archival research only tackles the third stage of blistering, and suggests that, once a crack in the gypsum crust appears, exfoliation proceeds really quite slowly (under the environmental conditions experienced in Oxford between 1863 and 1937). Evidence from this study shows that many blisters are extremely stable, even given the relatively polluted atmospheric conditions undoubtedly experienced in Oxford before the Clean Air Acts of the 1950s.

This study also shows that exfoliation of blisters occurs in measurable, discrete 'spasms' rather than as a continuous process. The unclear relationship of blister exfoliation with any environmental variables found in the Oxford case backs up the suggestion that the third stage of blistering is accomplished by a range of weathering processes, not necessarily related to sulphuric acid-induced formation of gypsum. At a general level, therefore, blistering of gypsum-encrusted limestone walls appears not to be highly sensitive to environmental change. However, at a more detailed level microenvironmental changes, and historical factors, are likely to affect the suite of weathering processes responsible for blistering, but such influences will be much harder to recognise in the archival record.

ACKNOWLEDGEMENTS

I would like to thank the Leverhulme Trust for a generous grant which facilitated this research and the President and Fellows of Trinity College, for permission to reproduce Figure 19.3. Clare Hopkins, Trinity College archives and Christine Butler, Corpus Christi College archives, gave invaluable assistance with the photographic studies and collection of coal consumption data. Lorraine Carter and Andrew Goudie looked after Amy while I did the work for this chapter. Professor R. U. Cooke provided highly pertinent comments on an earlier draft of this paper.

REFERENCES

Arkell, W. J. (1947a). *Oxford Stone*, Faber & Faber, London.
Arkell, W. J. (1947b). *The Geology of Oxford*, Clarendon Press, Oxford.
Beckinsale, R. P. and Combey, W. (1958). Smoke over Oxford. *Smokeless Air*, **105**, 178–186.
BERG (Building Effects Review Group) (1989). *The Effects of Acid Deposition on Buildings and Building Materials in the United Kingdom*, HMSO, London.
Brimblecombe, P. (1982). Early urban climate and atmosphere. In Hall, A. R. and Kenward, H. K. (eds), *Environmental Archaeology in the Urban Context*, CBA Research Report no. 43, pp. 10–25.
Building Research Station (1965). The weathering, preservation and maintenance of natural stone masonry. *Building Research Station Digest* no. 20.
Butlin, R. N., Coote, A. T., Yates, T. J. S., Cooke, R. U. and Viles, H. A. (1988). A study of the degradation of building materials at Lincoln Cathedral, Lincoln, England. *Proceedings, VIth International Congress on the Deterioration and Conservation of Stone, Torun, Poland, 1988*, pp. 246–255.
Camuffo, D., del Monte, M. and Sabbioni, C. (1983). Origin and growth mechanisms of the sulfated crusts on urban limestone. *Water, Soil and Air Pollution*, **19**, 351–359.
Charola, A. E. and Lewin, S. Z. (1979). Efflorescences on building stones — SEM in the characterisation and elucidation of the mechanisms of formation. *Scanning Electron Microscopy*, **1**, 379–386.

del Monte, M. and Vittori, O. (1985). Air pollution and stone decay: the case of Venice. *Endeavour*, **9**, 117–122.

Domaslowski, W. (1982). *La conservation preventative de la pierre*, UNESCO, Paris.

Evans, I. S. (1970). Salt crystallisation and rock weathering: a review. *Revue de Geomorphologie Dynamique*, **19**, 153–177.

Fassina, V. (1988). Stone decay of Venetian monuments in relation to air pollution. In Marinos, P. G. and Koukis, G. O. (eds), *Engineering Geology of Ancient Works, Monuments and Historical Sites*, Balkema, Rotterdam, pp. 787–796.

Heath, M. (1986). Polluted rain falls in Spain. *New Scientist*, 18 September, 60–62.

Jaynes, S. M. and Cooke, R. U. (1987). Stone weathering in southeast England. *Atmospheric Environment*, **21**, 1601–1622.

Jenkins, P. (1988). Geology and the Buildings of Oxford. *Oxford Region Thematic Trail no. 2*, Thematic Trails, Oxford Polytechnic.

Livingston, R. A. (1986). Architectural conservation and applied mineralogy. *Canadian Mineralogist*, **24**, 307–322.

Manning, M. I. (1989). Effects on structural materials. In Longhurst, J. W. S. (ed.), *Acid Deposition: Sources, Effects and Controls*, British Library Technical Communications, London.

North, F. J. (1930). *Limestones: Their Origins, Distribution, and Uses*, Thomas Murby, London.

Oakeshott, W. F. (ed.) (1975). *Oxford Stone Restored*. Oxford University Press for the Trustees. Oxford Historic Buildings Fund, Oxford.

Pentecost, A. (1991). The weathering rates of some sandstone cliffs, Central Weald, England. *Earth Surface Processes and Landforms*, **16**, 83–91.

Royal Commission on Historical Monuments England (1939). *An Inventory of the Historical Monuments in the City of Oxford*, HMSO, London.

Saiz-Jimenez, C. and Bernier, F. (1981). Gypsum crusts on building stones: A scanning electron microscope study. *Sixth Triennial Meeting, ICOM Committee for Conservation, Ottawa*, 81/10/5-1–5-9.

Schaffer, R. J. (1932). *The Weathering of Natural Building Stone*. Building Research Special Reports no. 18, DSIR, HMSO, London.

Sengupta, M. and de Gast, A. A. (1972). Environmental deterioration and evaluation for dimension stone. *Canadian Mining and Metallurgical Bulletin*, January, 54–58.

Watson, J. (1911). *British and Foreign Building Stones*, Cambridge University Press, Cambridge.

Winkler, E. M. (1973). *Stone: Properties, Durability in Man's Environment*, Springer-Verlag, Vienna.

20 Sensitivity of South Australian Environments to Marina Construction

NICK HARVEY

Mawson Graduate Centre for Environmental Studies, The University of Adelaide, Australia

INTRODUCTION

In the last 10 years there has been a proliferation of private marina development proposals for the South Australia coast, together with a smaller number of freshwater marina proposals. The large number of proposals has focused attention on planning and approval processes, including potential environmental impact, but there has been no detailed investigation of the impacts of existing marinas or the sensitivity of the various environments in which they have been constructed.

Recent South Australian studies have examined the environmental and strategic planning issues associated with coastal marina proposals for recreational craft (Harvey & Swift, 1990a), the legislative and approval processes for these proposals (Harvey & Swift, 1990b), and also the engineering aspects of these proposals (Swift & Harvey, 1990). No similar studies have been conducted for freshwater marinas in South Australia.

The majority of the recent South Australian marina proposals have been subject to environmental impact assessment (EIA) procedures and, in particular, have had legislative requirements to produce environmental impact statements (EISs), which provide baseline environmental data and an assessment of potential impacts. Earlier marina proposals were not subject to the same legislative requirements because they pre-dated EIA legislation and generally had less-detailed environmental studies. Notwithstanding the environmental impact data that have been produced for EISs and various environmental and feasibility studies, the monitoring of impacts for marinas that have already been constructed has either been *ad hoc* or non-existent until environmental problems have arisen. This lack of monitoring and environmental auditing is not restricted to marina projects or to South Australia, as noted by Buckley (1989).

The South Australian coast (Figure 20.1) is approximately 3700 km long and contains a variety of coastal landforms, including cliffed coasts, rocky outcrops, mangroves, mudflats, extensive sandy beaches, coastal dunes, two major gulfs (Spencer Gulf and Gulf St Vincent) and a number of offshore reefs and islands (the largest being Kangaroo Island). Consequently there is a great variety in coastal environments around the South Australian coast. Davies (1977) notes the predominantly high-energy deep water wave levels off the South Australian coastline and the lack of any major rivers (apart from the River Murray) discharging to the sea.

The implication of this for marina development is that, unlike the east coast of Australia, there are very few sheltered estuaries in South Australia so that recreational craft activity

Landscape Sensitivity. Edited by D. S. G. Thomas and R. J. Allison
© 1993 John Wiley and Sons Ltd

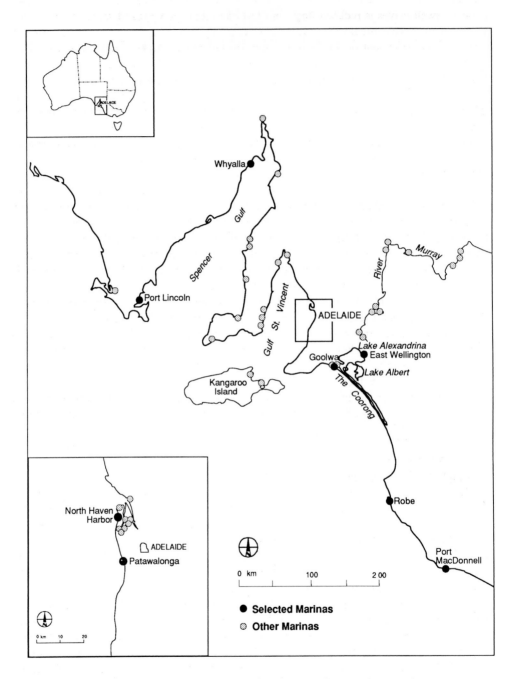

Figure 20.1   Location map, showing major South Australian marinas and permanent moorings and the eight marinas selected for discussion in this paper

is predominantly focused within the two gulf regions. Even within the gulfs, where the impact of ocean swell waves is reduced, there are few natural anchorages and wave conditions are such that expensive breakwater protection is usually required either for the marina and residential development or for the protection of a channel entrance.

The alternative boating environment is along South Australia's only major river, the River Murray (Figure 20.1). Although the River Murray and its tributaries drain an area of 105.6 $\times$ 10^6 ha and is South Australia's most important water resource, its catchment is mainly in the eastern states of Australia and there are no major tributaries downstream from the South Australian border (Harvey, 1988). In South Australia the River Murray has a low-gradient 630 km course from the border to the sea and enters lakes Alexandrina and Albert, and the Coorong Lagoon, before flowing out to the sea near Goolwa.

The River Murray is important for recreational boating, though, unlike the coastal environment, the lower-energy river environment permits the use of different boat types such as houseboats, water ski-boats, canoes and riverboats. However, the River Murray is essentially discrete from the coastal boating environment because the river mouth is generally unsafe for navigation (Harvey, 1988). In addition a number of barrages across the lakes and lower reaches of the river separate the fresh river water from seawater incursions and maintain the freshwater lakes at a constant level. Some of these barrages do have navigable locks, but these are mainly used to gain access to the mouth of the river and the Coorong Lagoon. The River Murray provides a sheltered boating environment compared to the coast and river flow is regulated by a series of locks and weirs along the length of the river. Towards the coast, the river flows into the more exposed boating environment of lakes Alexandrina and Albert, but the lower reaches of the river and the Coorong Lagoon provide a more sheltered boating environment.

This chapter assesses the sensitivity of various coastal and fluvial environments to marina development in South Australia. Case studies are used to illustrate the contrasts between the different types and scales of marina developments and their impacts on varied South Australian environments.

MARINAS IN SOUTH AUSTRALIA

The number of South Australian registered recreational boats (excluding commercial craft) is low (47 000) compared to some European countries of smaller land area such as France (500 000 pleasure craft) or Spain with a pleasure craft fleet estimated to be just over 100 000 (Qanir, 1989), though Qanir does not give a definition of pleasure craft. However, on a per capita basis the South Australian boat ownership figure of 32.8 per 1000 population is higher than both France (9.3 per 1000) or Spain (2.7 per 1000).

The small boating population in South Australia is catered for by a wide range in marina sizes so that the larger marinas are comparable in size with some of the larger European facilities. For example, North Haven Harbor in South Australia has a water area of 48 ha, which is larger than the Brighton Marina (38.4 ha) in England. However, because of the different layout of the two marinas the potential berthing capacity of North Haven Harbor (containing a public and a private marina) is about two-thirds the capacity of the Brighton marina (containing an inner and outer harbour), which has been described as the largest single harbour for private craft in Europe (Brighton Marina Visitors' Guide).

In South Australia, the bulk of recreational boats (94.4%) are stored on land. The remaining 5.6% occupy an estimated 2600 wet berths. About half of these (1340) occur in the Adelaide metropolitan region where there are a number of medium to large coastal marinas (Harvey & Swift, 1990a). Outside the metropolitan area there are few fully serviced private marina facilities, though some private berths are incorporated within commercial marina facilities. Elsewhere, boat moorings are scattered and generally of the swing or fore and aft type.

Eight of the larger marina facilities (six coastal and two freshwater), involving significant construction works, have been selected (Figure 20.1) to illustrate variable impacts on different environments. These marinas were selected on the basis that they all involved significant construction work, and are representative of the South Australian gulf environments (including east and west coast gulf environments, and estuarine and open coast), the south-east coast (both open coast and natural basin) and the river environment (both floodplain and lower reaches). All these marinas differ in their berthing capacities (Table 20.1) and have been constructed in environments ranging from river floodplains, coastal wetlands, and sandy coasts to exposed coasts.

The berth numbers in Table 20.1 refer to the existing berths plus the potential berthing capacity of the basin given its existing bathymetry. For example, North Haven Harbor consists of two separately run marinas having a collective total of 578 occupied berths, but the dredged basin is large enough to accommodate construction of at least another 600 berths. In addition it should be noted that four of the marinas are associated with waterfront residential development (Port Lincoln, North Haven, Goolwa and East Wellington), so that the water area of some of these marinas is partly dedicated to canals and waterways instead of berthing facilities.

ENVIRONMENTAL SENSITIVITY TO MARINA CONSTRUCTION

The impact of marinas on the environment can be divided into two stages. First, the construction stage will have immediate construction impacts such as dredging, blasting, construction of protective walling and breakwaters, reclamation and dewatering. Construction of large marinas will often be staged over a number of years but the impacts for each stage are usually readily identifiable.

Construction impacts are most noticeable in terms of the amount of dredged material, the size of the basin created, land reclaimed and also the amount of protective walling or breakwaters required (Table 20.2). The 2140×10^3 m^3 of material dredged for the North Haven Harbor basin or the 1.55 km rock breakwater at Port MacDonnell represent major construction impacts on the environment. However, each of the marinas has had different impacts on the environment because of differences in marina type, size and location.

Secondly, there will be a number of sometimes less-identifiable impacts, which for the purposes of this chapter can be called long-term impacts ($10^1 - 10^2$ years), even though their timeframe may be short in geomorphological terms. These are usually the ongoing impacts such as sediment accretion, erosion, subsidence, changes in flora and fauna (both marine and terrestrial), changes to water flow and currents (whether fluvial or marine), water quality and changes to groundwater flows.

The long-term impacts ($10^1 - 10^2$ years) vary between marinas, though the relatively recent construction (see Table 20.1) of most of the marinas limits the availability of long-term impact data. For example, the older coastal marinas such as the metropolitan coast marinas and the

Table 20.1   Marinas used in this study

Marina location (See Figure 1)	Existing berths Recreational (potential in parenthesis)	Commercial	Associated residential development	Date completed
Spencer Gulf				
Port Lincoln	50 (100)	58 (58)	yes	1987
Whyalla	35 (140)	—	no	1988
Metropolian Coast				
North Haven	578 (1200)	15	yes	1975
Patawalonga	143 (300)	—	no	1964
Southeast Coast				
Robe	10 (30)	40 (150)	no	1964
Port MacDonnell	22 (50)	70 (120)	no	1977
River Murray				
Goolwa	340 (340)[a]	—	yes	1989
East Wellington	62 (280)	—	yes	1990

[a] This relates to the current basin, but there is approval to expand the basin to allow for a potential berthing capacity of 1150 berths.

Port MacDonnell marina have revealed long-term sediment accretion, whereas the newer Spencer Gulf marinas have had less than five years for identification of such trends.

The eight marinas selected are exposed to a variation in physical processes. The six coastal marinas experience tidal differences ranging from 0.9 m for the south-east ports of Robe and Port MacDonnell, up to 2.4 m for the metropolitan marinas in Gulf St Vincent. The tides generally increase in range toward the head of the gulfs so that Port Lincoln at the entrance to Spencer Gulf has a tidal range of 1.4 m compared to Whyalla with a range of 2.3 m. In contrast, the two river marinas have constant water levels throughout most of the year, though the East Wellington marina may occasionally experience raised river levels of over 2 m during flood conditions.

Table 20.2   Construction impacts of marinas

	Basin size (ha)	Basin dredged sediment (10^3 m^3)	Breakwater protection (km)	Basin edge protection (km)	Reclaimed area (ha)
Spencer Gulf					
Port Lincoln	15.2	760	1.2	4.1	31.6
Whyalla	4.6	150	0.8	0.3	2.63
Metropolitan Coast					
Patawalonga	5.6	—	0.32	0.1	—
North Haven	48	2140	1.35	2.8	48
Southeast Coast					
Robe	3.94	—	0.2	0.43	—
Port MacDonnell	71.9	—	1.55	—	—
River Murray					
Goolwa	4.7	200	0.15	0.85	17.8
East Wellington	62.2	180	—	1.4	12.2

There are also differences in wave energy between the eight sites. The sites in the south-east (Robe and Port MacDonnell) have low-energy wave environments (<2 m) because they are sheltered from the moderate- to high-energy waves ($2-4$ m) generated by the predominant south-west winds. Robe is protected by a headland and Port MacDonnell by the presence of offshore submerged calcarenite reefs. However, the metropolitan marinas (North Haven and the Patawalonga) are on a sandy coast that is exposed to moderate wave energy from the south-west and consequently are both in a zone of northerly littoral drift. The two Spencer Gulf marinas (Port Lincoln and Whyalla) are both sheltered from the predominant south-westerlies. Port Lincoln marina is in a low-energy bay with a direct wave approach whereas Whyalla marina is on an exposed gulf coast where the weaker southerlies (than the south-westerlies) produce a north-easterly littoral drift. The two river marinas (Goolwa and East Wellington) are both in sheltered environments with wave action restricted to the fetch across the width of the river ($0.5-1.0$ km).

The implication of the above variation in processes is that each marina will require different forms of basin construction and levels of protective works. These in turn will affect factors such as the physical impact of construction (in terms of environmental modification), the littoral drift and consequent sediment erosion/accretion, tidal exchange and integrity of adjacent coastal ecosystems. The sensitivity of these environments to change is illustrated by examining the impacts from the marinas.

Spencer Gulf

The largest of the Spencer Gulf marinas, at Port Lincoln, was constructed in a sheltered coastal wetland containing mud flats, samphire swamps and coastal heath (Figure 20.2). The site contained predominantly unconsolidated Holocene estuarine sediments and late Pleistocene aeolian dune sediments, which are consolidated in parts (as calcrete) but at this location were relatively easy to excavate. The immediate construction impact was the dredging of 760×10^3 m^3 of sediment in the dry, after dewatering, and the construction of 5.3 km of protective walling, including 3.8 km of rock and concrete edge treatments, 0.3 km of steel sheet piled wharf and 1.2 km of rock sea-wall (Bagnall, 1986). The effect of all this was the loss of 27 ha of intertidal mudflats and 20 ha of coastal lowland. However, prior to marina construction there was reduced tidal flushing of the wetland because of road construction and the vegetation had suffered from human impact (such as the use of off-road vehicles and the dumping of rubbish). The area not dredged for the marina basin and waterways was covered with dredge material as landfill.

The long-term impacts of the marina are difficult to determine since it was only completed in 1987. One positive impact appears to be the increased tidal flushing to the wetlands behind the marina, which has resulted in an increased density of samphire and general improvement in the habitat value of the wetland. This indicates the sensitivity of the intertidal environment to variations in tidal flushing. No data are yet available on any sediment accretion/erosion patterns.

In the northern Spencer Gulf, the Whyalla marina was constructed on a more exposed gulf coast in contrast to the sheltered coastal wetland at Port Lincoln. The marina site is adjacent to a cliffed headland of Middle Proterozoic age. Immediately south of the headland is a wide Holocene coastal plain with mangroves and intertidal sand and mud deposits. The construction impacts of the marina were the construction of two breakwaters (a 0.35 km western breakwater — part of which already existed — and a new 0.45 km eastern breakwater) using rock blasted

Figure 20.2 Oblique view of Port Lincoln marina, showing flooded samphire swamp in the background (photograph: David Kelsey)

from the adjacent headland, reclamation of 2.63 ha of land using 150×10^3m of dredged material, the loss of 3.5 ha of seagrass and the creation of a new beach on the northern side of the marina with approximately 60×10^3m^3 of dredged sand. The dredging was conducted using a standard cutter suction dredge.

As with the Port Lincoln marina, the recent completion (1988) makes it difficult to examine long-term impacts. However, the presence of a north-easterly migrating sand bar, with an approximate volume of 280×10^3m^3, is likely to impact on the marina by the turn of the century and require a major dredging operation. This sand bar is associated with a wide intertidal plain to the south of the marina site, and the interruption of this sand is a long-term event that should not be confused with the sensitivity of the average annual littoral drift rates at the actual marina site. These annual rates are relatively low (2×10^3 m^3 year^{-1}) and the interruption by the breakwaters of this littoral drift is unlikely to cause scour to the beach on the northern side.

Metropolitan Coast

On the metropolitan coast the most noticeable impact of marina construction has been the build up of sediment due to the interruption of the dominant northerly littoral drift (Figures 20.3 and 20.4). Both North Haven Harbor and the Patawalonga marina have been constructed in predominantly Holocene coastal sediments along a shoreline that has well-documented

Figure 20.3 Vertical aerial photograph of North Haven Harbor showing dominant northerly littoral drift (north at top of photograph). The tree-lined road to the east of the marina indicates the former coastline (source: Department of Lands, Adelaide, South Australia; reproduced by permission)

sediment accretion rates since the end of the post-glacial Marine Transgression (Harvey & Bowman, 1987). These relatively high rates indicate a sensitivity of this environment to any artificial interruptions to the natural drift pattern. The coastal dunes, which provided a natural reservoir of sand, have essentially been removed from the system because of urban encroachment, so that without a significant input of sand to the system there are now sand management problems for the coast, which essentially has a finite sand supply.

The 48 ha basin for North Haven Harbor was excavated in the dry by bunding and dewatering of intertidal and subtidal sediments at the northern end of the metropolitan coastline. Dredging operations encountered problems removing an estimated 2140×10^3 m^3 of sediment because of the presence of organic matter (largely detrital seagrass) together with the sand from the harbour basin. Although this material has not been dated, it is likely that much of the sediment was recently deposited due to the trapping effect of reclaimed land and a major breakwater to the north of the new harbour (Bourman & Harvey, 1986). The basin edges of the new harbour have been armoured by 2.8 km of rock protection and the harbour entrance is protected by two rock breakwaters, a 0.6 km southern breakwater (with a 0.2 km extension within the entrance) and a 0.55 km northern breakwater.

Figure 20.4   Oblique view of the Patawalonga marina looking south, showing the locked basin, sand accumulation on southern side of the breakwater, erosion on the northern side, and the offshore bar resulting from northerly littoral drift (photograph: Nick Harvey)

Apart from the immediate construction impact of basin excavation and rock protection, there are now emerging long-term impacts of sediment accretion. As noted in Harvey & Bowman (1987), the net littoral drift of sand in the area has various estimates ranging from 30 to 50 $\times$ 10^3 m^3 $year^{-1}$, though recent calibrated estimates based on wave-energy data suggest figures as high as 68 $\times$ 10^3 m^3 $year^{-1}$ (Reidel & Byrne and Kinhill Stearns, 1985). Many of the maintenance problems could have been avoided if the entrance had been located to the north in the initial construction phase. Consequently there has been sediment accretion against the southern breakwater and harbour entrance, which now requires an ongoing sand management programme, to the extent that the relocation of the harbour entrance was considered a possible strategy to cope with this long-term impact (Reidel & Byrne and Kinhill Stearns, 1985). Between 1981 and 1985 a total of almost 200 $\times$ 10^3 m^3 had to be dredged to maintain a navigable entrance to the marina. Over the last 10 years it has been necessary to dredge an average of 37 $\times$ 10^3 m^3 $year^{-1}$. In addition there is a major problem with detrital seagrass accumulation on the windward side of the southern breakwater.

Further south, but also on the metropolitan coast, the Patawalonga marina has experienced similar problems with the predominantly northerly littoral drift of sand. The marina, unlike the North Haven Harbor is constructed in the Patawalonga Creek, which was originally an ephemeral stream but has now become a major stormwater outlet for urban Adelaide. The marina berths are within the Patawalonga Lake created by a lock structure at the mouth of the creek. The major construction impact associated with this marina (apart from the creation of an artifical lake) has been the construction of a 0.2 km southern rock breakwater and a 0.12 km northern concrete jetty in 1964 to protect the entrance channel (Figure 20.4).

The long-term impact of the breakwater construction has been the accumulation of sand ($\sim$200$-$300 $\times$ 10^3 m^3) on the southern side of the breakwater, erosion on the beaches to the north of the Patawalonga and the creation of an offshore bar at the mouth of the Patawalonga. Although the build-up of sand on the southern side of the breakwater and the start of erosion on the northern beaches occurred within five years, the presence of the offshore bar (as a navigation hazard) and the continuing erosion problems north of the Patawalonga represent long-term impacts that require coastal management strategies. It is estimated that $10-20 \times 10^3$ m^3 year^{-1} of sand now by-passes the mouth of the creek. In response the state government has used sand from south of the breakwater in its sand replenishment programmes (average of 31×10^3 m^3 year^{-1} 1974–89) and continues to deposit sand (average 44×10^3 m^3 year^{-1} 1974–90) on the beaches north of the Patawalonga. It is also possible that elevated groundwater levels from the locked Patawalonga and irrigation of a golf course behind dunes to the north may be contributing to the sensitivity of beach erosion at this location.

South-east Coast

In the south-east of the state, the Port MacDonnell marina has the largest single rock breakwater (1.55 km) of all the South Australian marinas. Although no basin was dredged for the marina, the breakwater itself represents a significant visual impact on the coastal environment because of the use of a dark basaltic rock in contrast to the lighter-coloured local Pleistocene aeolianite. The construction of the breakwater on this open coast was possible because of the presence of offshore aeolianite reefs, which break the high-energy southern ocean waves. Consequently the wave energy is low to very low at the shoreline (Short & Hesp, 1980).

Since 1977, when the breakwater was completed, there has been a progressive build-up of an incipient tombolo in the lee of the breakwater. Short & Hesp (1980) recognised the potential for sand accumulation inside the harbour but assumed an accretion rate of 80×10^3 m^3 year^{-1} based on a nine-month sediment accumulation survey up to August 1978. Using this assumed rate they estimated that the beach would built out to the breakwater with a total shoreline accretion of 765×10^3 m^3 by about 10 years after the sediment survey. About 10 years later Maunsell & Partners (1989) estimated that the total volume of sediment accumulation between 1975 and 1987 was only 104×10^3 m^3, some 600×10^3 m^3 less than Short & Hesp's (1980) estimates. This illustrates the danger of speculating on long-term impacts based on short-term data.

The reason for this discrepancy appears to relate to the source of sediment, which was also discussed by Short & Hesp (1980). They concluded that the breakwater position prevented sediment arriving from the west and that sediment from the east was unlikely given the downdrift position of the beach from the dominant wave approach. The most likely sediment source according to Short & Hesp is offshore from the beach, i.e. in the wave shadow of the breakwater and also seaward of it. Given this conclusion it would appear that Short & Hesp may have been correct in identifying the source of sediment but overestimated its potential supply. The sensitivity of the system was such that after the breakwater construction there was a rapid readjustment in sediment distribution by the formation of an incipient tombolo but subsequent accretion has slowed down.

On the western side of the breakwater there has been significant beach progradation as sand has accumulated against the breakwater. Beach profiles have been conducted across this sand accumulation but the variable length of the profiles makes it difficult to quantify. In addition to the progradation of the beach, there has been siltation within the harbour, in the lee of the breakwater, of 22.6×10^3 m^3 between 1972 and 1987 (Maunsell & Partners, 1989).

Another south-east coast marina, at Robe, has had a less dramatic impact than the Port

MacDonnell marina. The Robe marina, unlike Port MacDonnell has been constructed in a natural lagoonal depression. The surrounding area is predominantly consolidated Pleistocene aeolianite, which also forms offshore reefs in places, similar to that occurring offshore from Port MacDonnell. The construction impacts of the Robe marina have been the cutting of a channel (removing approximately 10.2×10^3 m^3 of material), in 1964, through a low-lying section of Pleistocene dunerock to the coastal lagoon behind, and the building of a 0.2 km breakwater to protect an entrance channel. Sections of the lagoon basin have also been deepened. In addition, quarrying of the granite rock used for the breakwater destroyed a nearby outcrop of geological significance. Under current environmental scrutiny this may not have been permitted. The granite is also darker than the local aeolianite and creates a visual contrast in the environment.

The long-term impact of the Robe marina has largely been the accumulation of sand against the channel entrance breakwater by the interruption of the natural littoral drift. Short & Hesp (1980) estimated that 10×10^3 m^3 of sand had accumulated on the western side of the breakwater and was causing beach erosion downdrift. Fotheringham (1983) indicates that by 1983 approximately 30×10^3 m^3 of sediment had been dredged from the breakwater since its construction and that on average it was necessary to remove approximately $5-6 \times 10^3$ m^3 of sediment every three years. Most recently Fotheringham (personal communication) has calculated that the accumulation of sediment against the breakwater has been occurring at a rate of approximately 5×10^3 m^3 year^{-1} and has destroyed a local reef of environmental significance. The environmental sensitivity is not restricted to the local marine life as this sediment build up has also resulted in starving the nearby town beach, which in 1987 required the construction of a 0.2 km groyne and the dumping of 15×10^3 m^3 of sand for beach replenishment. In addition, there has been shoaling within the marina and formation of a small beach.

River Murray

The River Murray marinas are quite different to the coastal marinas in their construction, since the wave climate is considerably less and water levels are relatively constant due to river regulation. For the same reason there are fewer effects of river erosion and deposition, and low river flows are common because of the multiplicity of upstream users. As noted by Harvey (1988) low river flows have resulted in at least 20 periods where no water flowed out to sea for 100 or more days consecutively. The potential for low to medium return period flooding on the river has also been substantially reduced by the construction of increased upstream dam storages. However, large floods such as the 1956 flood (with an estimated 160 year return period) would not be contained by these storages.

At the lower end of the River Murray a marina has been constructed on a large island in the final bend of the river at Goolwa. The island is formed from consolidated Pleistocene dune barrier sediments with intervening swales. Adjacent to the marina is one of the barrages separating the fresh river water from the saltwater of the estuary. Freshwater is maintained at a relatively constant level behind the barrage (though flooding can elevate river levels by a metre or so at this location) but under natural conditions without the artificial barrage the marina would be in a saline estuarine environment. At this location wave energy generated by the south and south-westerly winds is restricted by coastal dune barrier protection and a fetch across the river of approximately 1.0 km.

The construction impact of the marina has been the excavation of the 4.7 ha basin (Figure

Figure 20.5 Oblique view of the Goolwa marina showing the marina basin with minimal rock protection and fixed wooden piles, pontoons and boardwalk structures (photograph: Nick Harvey)

20.5) within the swale depressions behind a consolidated dune barrier, and the use of the 200×10^3 m^3 of sediment to build up surrounding areas for marina facilities and residential development (Figure 20.7). A small (0.1 km) entrance breakwater has been constructed and 0.85 km of rock walling protect the basin edges. The rock protection needed is considerably less than the coastal marinas because of the lower wave energy and the lack of tidal fluctuation. Consequently the rock protection is of small-sized stones from the local consolidated dune rock (calcrete).

Long-term impacts are difficult to determine because of the relatively recent completion of the marina basin (1989). There is unlikely to be any significant sediment accumulation either against the entrance breakwater or within the basin. The only significant physical impact is likely to be a deterioration in water quality because of inadequate flushing, though this has not yet been a problem. It should be noted that the natural environmental sensitivity at this site has already been modified by the artificially raised freshwater levels (maintained for a period of 50 years) and the alteration of former estuarine conditions.

Further up the River Murray, a marina has been constructed at East Wellington on the northern side of Lake Alexandrina. The marina basin of 62.2 ha has been excavated from low-lying alluvial sediments of the flood plain. In addition, the river levee has been breached in two places for access to the basin. Wave energy is restricted by a 0.5 km fetch across the river.

As with the Goolwa marina, the relatively recent construction of the marina makes it difficult to speculate on long-term impacts. However, there have already been problems of erosion along some of the basin edges due to the long fetch across the basin and the resultant wave attack on the exposed basin edges. Although the river levees protect the marina basin from river-generated waves, the artificial basin constructed behind the levees has at least four times the fetch across the river and consequently the basin edges are more sensitive to wave erosion. This floodplain environment would have been more sensitive to flood erosion under a natural river system, particularly with the breached levees and large marina basin at this site. However, the regulated river flow has now reduced the sensitivity of this environment to such erosion from all but the large floods.

DISCUSSION

It is apparent from the impact data that the South Australian marinas have affected different environments in different ways. For example, the Port MacDonnell breakwater represents a major visual intrusion on the south-east coast, particularly because of the size of the breakwater and the use of dark basaltic rock, which contrasts with the local dune calcarenite. There was no significant alteration to the actual coastline in terms of construction impacts but the long-term impacts of sediment accretion behind the breakwater and the tombolo formation are gradually altering the shape of the coastline. The sensitivity of the coastal sediment budget in the area is demonstrated by the rapid sediment accretion after breakwater construction but reduced sedimentation rates as the environment adjusts to its new state.

In contrast the Port Lincoln marina does not have a major open-sea breakwater and sand accretion problem (though long-term sediment trends will not be available for some time), but the construction impacts of dredging and land reclamation have significantly altered a coastal wetland area with a resultant loss of intertidal seagrass mudflats and samphire swamps. The sensitivity of the coastal wetland to tidal flushing rates and the increased flushing capacity of the dredged marina basin has actually improved flushing of the remaining samphire swamp area, so that the vegetation is now more diverse and of greater habitat value. The success of this marina has encouraged proposals for further development that would engulf even more wetland and vegetation, but the recognition of the sensitive nature of this environment is likely to be incorporated in conditions for further development at the site.

The two metropolitan marinas have contrasting levels of construction impacts between the major rock protection works at North Haven Harbor of 1.35 km of breakwater and 2.8 km of edge treatments and the 48 ha of reclaimed land compared with the small 0.32 km channel breakwater at the Patawalonga. Similarly the 2140×10^3 m^3 of dredged basin material from North Haven Harbor represent a major impact compared with the use of the regulated Patawalonga lake. However, both marinas are located on an easterly sandy gulf coast that is sensitive to interruptions of the predominant northerly littoral drift. This coast has a sediment supply, albeit a relatively finite one, which is exposed to a moderate wave energy produced by the predominant south-westerlies. Consequently both marinas have experienced similar problems with longshore drift and have required dredging of sand accumulation against their southerly breakwaters.

The Whyalla marina, like North Haven Harbor, has used much of the dredged material as adjacent landfill. At Whyalla 2.63 ha of reclaimed land for marina related facilities compares with the 48 ha of reclaimed land at North Haven Harbor where the land has been turned into prime waterfront real estate. Although the Whyalla marina site has lower annual littoral drift rates to the metropolitan coast marinas, there is a long-term environmental sensitivity related to the Holocene coastal environment to the south. It is predicted that there will be major accumulation of sand within 30 years, due to a mobile sand bank that is moving in a north-easterly direction from the Holocene intertidal plain towards the cliffed headland.

The Robe marina, like the Port Lincoln marina, has been constructed inland as opposed to the open-coast marinas elsewhere. It has not destroyed seagrass flats and samphire swamps as has occurred at Port Lincoln but it has occupied a natural coastal lagoon and created an artifical opening to the sea. The breakwater constructed to protect the opening has resulted in littoral drift problems of sand accretion and downdrift beach erosion similar (but reduced in scale) to the metropolitan coast marinas.

Unlike the coastal marinas the two river marinas have not experienced long-term impacts of sediment accretion, and though they have only recently been completed they are unlikely to experience such problems in the future. These marinas do not have the need for entrance breakwaters, and rock protection of basin edges is at a reduced scale compared with the coastal marinas, not only because of lower energy conditions but also because of relatively constant water levels in the river. However, their construction impacts have produced large 4−60 ha basins that have removed floodplain wetlands or samphire swamps in natural depressions, and used dredged material for reclamation of surrounding land. The long-term impacts of the river marinas tend to be related more to problems with water quality and the need for adequate flushing rather than the sediment accretion and the need for dredging that occurs in the coastal marinas. However, the sensitivity of the fluvial environment has already been modified by river flow regulation so that problems of erosion, flooding or low river levels have all been reduced.

The eight sites selected indicate that, apart from the immediate impact of environmental modification by marina construction, each site has a different environmental sensitivity. The most sensitive geomorphic process affected by the coastal marinas in this study appears to be the interruption to the littoral drift of sediment. The sensitivity of this depends on the location of the marina relative to predominant winds, the local bathymetry, the consequent wave refraction and the potential supply of sediment. The two estuarine sites provide a contrast in their level of impact and indicate the sensitivity of coastal wetlands to variations in tidal flushing. At Port Lincoln the resultant increase in tidal flushing has had a positive effect on the adjacent samphire vegetation, whereas the locked Patawalonga has reduced tidal flushing, which together with input of urban stormwater has resulted in poor water quality, accretion of contaminated silt and loss of the original vegetation.

CONCLUSION

It is apparent that marinas have both construction impacts and long-term impacts on the environment. These impacts vary with both the type of marina and the environment itself. The level of impact also depends to a great extent on previous environmental studies such as EIS documents, the ability of these to recognise 'sensitivity' in the environment, and the extent to which decision-makers impose conditions to take note of this sensitivity. It is likely that the inclusion of EIA in South Australian planning legislation in 1982 and the heightened environmental awareness of the 1980s has been associated with a greater awareness of environmental sensitivity.

This awareness was clearly not evident from some of the earlier marina developments at North Haven, where littoral drift of sand and seagrass is a major problem, at the Patawalonga and Robe marinas, where littoral drift is not only a problem but is also causing downdrift beach erosion problems, and at the Port MacDonnell marina, which has a tombolo developing inside the breakwater.

More recent coastal marina developments at Port Lincoln and Whyalla have had EISs required and consequently are associated with a greater awareness of the environmental sensitivity. Although both marinas have only recently been completed, the environmental processes and the potential long-term impacts are better researched and understood. The construction impacts on the environment may still be significant but the more recent marina proposals are less likely to be approved if they are located in sensitive environments than was the case for the earlier marina proposals.

On the River Murray the marinas selected are also more recent and have been subject to environmental scrutiny. Neither of the selected river marinas has hydrological problems with marina excavation or with saline groundwater flows which could occur further upstream. The essentially offstream marina basins are not affected by fluvial processes to the same extent that the coastal marinas are affected by coastal processes. The regulated river flow has reduced the sensitivity of the environment to flood erosion so that water quality problems and flushing rates are more important for the marinas than any fluvial erosion or deposition.

In conclusion, there is a clear need to incorporate an understanding of environmental sensitivity in the approval process for marina developments. In particular, there is a need for prediction of potential impacts of geomorphic processes after the construction stage and also a need for detailed monitoring programmes geared to amelioration measures. Although many marina developments have a design life of 100 years, which is short in a geomorphological timeframe, it is evident that significant geomorphic changes occurring in this timeframe can result in costly management problems, which could be avoided with a greater awareness of the sensitive nature of various environments.

ACKNOWLEDGEMENTS

I would like to thank Peter Swift and Doug Fotheringham for their assistance in the compilation of some of the data and for helpful discussions on the manuscript. I would also like to thank M. Barnhurst, A. Reid and R. Taylor for their assistance in data collection and two anonymous reviewers for their useful comments on the manuscript.

REFERENCES

Bagnall, M. (1986). Lincoln Cove marina development — Porter Bay, Lincoln, South Australia, *Proceedings 1st Australasian Port, Harbour and Offshore Engineering Conference*, Institution of Engineers, Australia, pp. 287–292.

Bourman, R. P. and Harvey, N. (1986). Landforms. In Nance, C and Speight, D. L. (eds), *A Land Transformed: Environmental Change in South Australia*, Longman Cheshire, Melbourne, Chap. 4.

Buckley, R. (1989). Precision in environmental impact prediction: first national environmental audit, Australia, *Resource and Environmental Studies no. 2*, Centre for Resource and Environmental Studies, Australian National University.

Davies, J. L. (1977). The coast. In Jeans, D. N. (ed.) , *Australia: A Geography*, Sydney University Press, Chap. 7, pp. 134–152.

Fotheringham, D. (1983). Foreshore erosion at Town Beach, Robe. Coastal Management Branch Technical Report, 83/5, South Australian Department of Environment and Planning.

Harvey, N. (1988). Coastal management issues for the mouth of the River Murray, South Australia. *Coastal Management*, **16**(2), 139–149.

Harvey, N. and Bowman, G. M. (1987). Coastal management implications of a Holocene sediment budget: Lefevre Peninsula, South Australia. *Journal of Shoreline Management*, **3**, 77–93.

Harvey, N. and Swift, P. (1990a). Coastal marinas in South Australia: environmental issues and strategic planning. *Australian Geographer,* **21**(2), 141−150.

Harvey, N. and Swift, P. (1990b). Contrasts in processing coastal marina applications in South Australia. *Australian Planner,* **28**(3), 44−47.

Maunsell and Partners (1989). Port MacDonnell fishing boat slipping and storage investigation. Report prepared for the Department of Marine and Harbors, South Australia.

Qanir, A. (1989). Planning of marina developments along the Moroccan coastline. *Ocean and Shoreline Management,* **12**, 561−570.

Reidel & Byrne and Kinhill Stearns (1985). North Haven, South Australia, Marina entrance shoaling study. Report prepared for North Haven Trust, Australia.

Short, A. D. and Hesp, P. A. (1980). Coastal engineering and morphodynamic assessment of the coast within the South East Coast Protection District, South Australia, Report prepared for the Coast Protection Board, South Australia.

Swift, P. M. and Harvey, N. (1990). Developing coastal marinas in South Australia. *Proceedings of 3rd Australasian Ports and Harbours Conference*, Institution of Engineers, Australia, pp. 97−101.

Index

Index compiled by Phil Porter